Titanium

A Technical Guide

Written or compiled by
Consulting Editor

Matthew J. Donachie, Jr.

The Hartford Graduate Center

ASM INTERNATIONAL™
Metals Park, OH 44073

First printing, March 1988
Second printing, August 1989

Library of Congress Catalog Card Number: 88-070043

ISBN: 0-87170-309-2

Printed in the United States of America

ASM INTERNATIONAL™

Metals Park, OH 44073
Tel: (216) 338-5151

Cover: SR-71 Blackbird advanced reconaissance aircraft.
(Courtesy of Lockheed California Company)

Dedication

I would like to dedicate this book to my mother, Viola Donachie. She was married to a long-time ASM INTERNATIONAL member for almost 50 years before his death; she was also a mother of two Ph.D. metallurgists. My mother married into a family of wire drawers and tinkerers and came to appreciate the field of metallurgy long before many of you who read this were born. Her patience with her husband, Steve, known to many ASMers as Scotty, her dedication to his career, and her understanding as her two sons puttered around the laboratory and followed in Steve's footsteps endeared her to all of us. I am delighted to be able to recognize her contributions to my career.

M. J. D., Jr.

Preface

Titanium and its alloys are nearing the end of their fourth decade of commercial-industrial service. Very few books on titanium have appeared in the literature, although over the past two decades there have been a number of symposium volumes which have covered advances in the industry. ASM INTERNATIONAL has decided to provide some more convenient and practical reference volumes for a wide-ranging group of industrially significant subjects. Titanium has been chosen as the first topic in this important new series.

This technical guide to titanium and its alloys is meant to provide the most complete titanium summary possible within its 13 chapters and 10 appendices. The book is working guide to the field of titanium metallurgy. The appendices, in particular, should provide a substantial fund of knowledge or source information to amplify topics not fully covered in the text. No book can be all-inclusive, and the present volume is no exception. Every effort, however, has been made to condense and review the significant features of the metallurgy and application of titanium and its alloys. Considerable reduction of material was necessary in order to accomplish the task and to develop a manageable volume. User comment about the book is welcome so that future editions may be improved. The publisher also will find such information important in order to make other books of this series as valuable to the user as possible.

I would recommend Chapter 1, Executive Summary, to the first-time user of titanium and its alloys. In presentational form it is a concept adapted from the corporate world to suit the needs of those who want rapid information summaries and may not (at the moment) have the time or resources to devote to a more exhaustive perusal of the subject.

No effort of this magnitude is the work of one individual; many behind-the-scenes associates are responsible for the final book. However, I particularly would like to acknowledge two persons for their contributions. The first is my wife Martha. As she has so many times in the past, Martha again has patiently put up with my evenings, nights, and weekends of effort to bring this volume to fruition. The second is Jack R. Braverman, Senior Editor, ASM INTERNATIONAL. Jack has done an outstanding job in assembling the appendices, editing the volume, and otherwise processing the vast amount of material which goes into a book such as this. Its successful publication is a tribute to his effort, and I wish to thank him for it.

Manchester, CT   *M. J. D., Jr.*
1987

Acknowledgements

The editor wishes to thank the dozens of individuals who contributed their time and attention to the gathering of material for this technical guide. In particular there are the following.

Bernard de Gelas, Deputy Commercial Director, CEZUS, France. **Sang J. Kim**, Professor, Department of Metallurgy, Seoul National University, Seoul, Korea. **E. L. Thellmann**, Manager, Applied Materials and **James J. Snyder**, Senior Metallurgical Engineer, Gould, Inc., Cleveland, Ohio. **C. Nunez,** Professor, University of Barcelona, Spain. **Ritesh P. Shah,** Department of Metallurgy, M. S. University, Valsad, India. **Len Connor** and **Harry Reid**, American Welding Society, Miami, FL. **George J. Esseff, Jr.**, Supra Alloys, Inc., Camarillo, CA. **Anthony C. Barber**, Technical Director, IMI Titanium, Ltd., Birmingham, England. **Stanley Abkowitz**, President, Dynamet Technology, Inc., Burlington, MA. Cathy Bomber, Publications Editor, Teledyne Wah Chang, Albany, OR. **Hiroshi Tanikawa**, Manager, Showa Denko America, Inc., New York, NY. Crystal L. Revak, Manager, Marketing Communications, RMI Company, Niles, OH. **Linda Gibb**, Manager, Advertising, The Metallurgical Society, Inc., Warrendale, PA. **P. B. Belk**, President, Universal Wire Works, Inc., Houston, TX. **Richard E. Leopold**, President, Vulcanium Corp., a div. of Industrial Titanium Corp., Northbrook, IL. **V. Loechelt** and **A. Rendigs**, Messerschmitt-Bolkow-Blohm GmbH, Bremen, Federal Republic of Germany. **Dr. Antonio Forn**, Escola Universitaria Politechnica, Vilanova, Spain. **Dr. T. R. Anantharaman**, Professor, Department of Metallurgical Engineering, Banaras Hindu University, India. **Robert Davies**, General Sales Manager, Titanium Industries, Fairfield, NJ. **Koichi Ashihara**, Advertising Manager, Kobe Steel, Ltd., Tokyo, Japan. **O. Edward Nelson**, Marketing Manager, Oregon Metallurgical Company, Albany, OR. **Ilse A. Minkenberg**, Manager, Public Relations, Howmet Turbine Components Corp., Greenwich, CT. **Georgi M. Hockaday,** Executive Associate, Titanium Development Association, Dayton, Oh. **Young-Sub Kim**, Dept. of Materials Science and Engineering, Case-Western Reserve University, Cleveland, OH.

A special note of acknowledgment is made to ASM INTERNATIONAL's *Metals Handbook* series, Ninth Edition, from which parts of this guide were drawn. In particular, appreciation is expressed to the ASM Handbook Committee, the contributors to the series, and the staff who produce the volumes.

Contents

4 Wrought Alloy Processing 37

5 Heat Treating 57

6 Machining 75

9 Powder Metallurgy 113

10 Joining 131

13   Recent and Future Advances

A   Selected References for Additional Reading

B   Glossary

C   Machining Data

H **Listing of Manufacturers, Suppliers, Services** **343**

I **Standards and Specifications** **349**

J Designations, Applications, Properties 443

Index 453

Chapter 1

Executive Summary

HOW AND WHEN TO USE THIS CHAPTER

One of the central problems associated with an extended treatment offered by any reference book is the well-known difficulty a reader experiences in trying to find a specific topic. This is especially true when only a quick, superficial understanding is required. A purchasing agent looks for instant, decision-supporting information. The engineer needs a quick memory refresher about matching alloy types with proven designs. The fabricator needs a brief summary of the latest techniques in a given field. There never seems to be adequate time available to research a subject thoroughly.

This executive summary chapter supports such types of demanding needs by providing a shorthand overview of the major topics discussed throughout the book. It introduces the reader, in a simple, direct manner, to an exceptionally wide variety of topics. It also maximizes a reader's ability to use the volume in the most efficient way, since it suggests this technical guide's overall scope.

Whether the user is relatively unfamiliar with titanium alloys or an expert, frequent referencing of the topics contained within this guide is probable, because the book offers an extensive, single-volume coverage of titanium-related data, both theoretical and practical.

Be sure to refer to the Table of Contents and the Index since they give valuable insights to the location of specific topics.

FACTS AND/OR BENEFITS

Titanium has long had appeal to metal designers; there exists, after all, nearly 40 years of modern industrial practice to support its use. The following is a brief listing of some significant facts and/or important benefits offered by titanium alloys.

- Ti's density is about 60% of that of steel.

- Ti's cost is about 1.3 times that of stainless steel.

- Ti's modulus is 55% that of steel.

- Ti is exceptionally corrosion resistant; this feature often exceeds that of stainless steel in most environments, including the human body.

- Ti may be forged by means of standard techniques.

- Ti is castable, although the investment casting method is preferred. (Investment casting has a lower cost than conventional forged/wrought fabricated structures. The method may be very helpful when intricate shapes are involved.)

- Ti may be processed by means of powder metal technology. (The powder may cost more, yet P/M offers property and processing improvements—plus an overall cost-savings potential.)

- Ti is joinable. (This may be achieved by means of fusion welding, brazing, adhesives, diffusion bonding, and fasteners.)

- Ti is formable and readily machinable, assuming reasonable care is taken.

- Ti is available in a wide variety of types and forms, and so no real development is required in most cases.

Photo courtesy of Lockheed California Co.

Figure 1.1 - SR-71 aircraft; the first use of beta Ti alloys in aerospace structural applications

BASIC TITANIUM METALLURGY

Structures

Titanium has two elemental crystal structures: one is body-centered cubic, the other close-packed hexagonal. (See Figure 3.1.) The cubic structure is found only at high temperatures, unless the titanium is alloyed to maintain the cubic structure at lower temperatures. (Although the melting point of titanium is in excess of 1660°C (3000°F), titanium's alloys are used at temperatures only to about 538°C (1000°F).

Titanium's two crystal structures are commonly known as alpha (for hexagonal) and beta (for cubic). The alpha and beta "structures"—sometimes called systems or types—are the basis for the generally accepted three classes of titanium alloys. These are alpha, alpha-beta, and beta.

Some relationships among the various classes or subclasses of titanium alloys are schematically shown in Figure 1.2. The figure also indicates the effects that structures have on alloy characteristics. Representative alloy compositions are also indicated.

Commercially pure (CP) titanium is, of course, alpha in structure. Additions of alloying elements produce the range of possible microstructures in titanium alloys.

A variation of alpha-beta alloys recognizes the wide range of alloy chemistry and structure possible. It is called near-alpha.

In similar fashion there are near-beta alloys, along with beta. These all are alloys which retain an essentially beta structure to room temperature.

Beta structures are generally referred to as metastable beta.

Ti and Ti Alloy Characteristics

The commercially pure (CP) titanium, along with the alpha and near-alpha titanium alloys, generally demonstrate the best corrosion-resistance qualities. They are the most weldable of the titanium/titanium alloy family.

Pure titanium usually has some amount of oxygen alloyed with it. The strength of CP is affected by the interstitial (oxygen and nitrogen) element content.

Alpha alloys usually have high amounts of aluminum which contribute to oxidation resistance at high temperatures. (Alpha-beta alloys also contain, as the principal element, high amounts of aluminum, but the primary reason is to stabilize the alpha phase.)

Alpha alloys cannot be heat treated to develop higher mechanical properties, because they are single-phase alloys. The addition of certain alloying elements to titanium permits the alloys to be heat treated or processed in the temperature range where the alloy is two-phase (alpha and beta). The two-phase condition permits the structure to be refined and, by permitting some beta to be retained temporarily at lower temperature, enables optimum control of the microstructure during subsequent transformation when the alloys are "aged" after cooling from the forging or solution heat-treatment temperature.

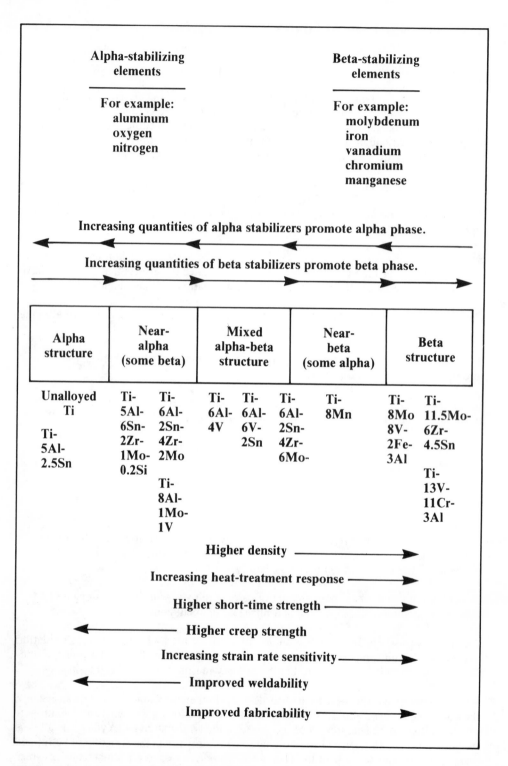

Figure 1.2 - Schematic relationships: titanium alloying effects on structure; selected alloy characteristics

The alpha-beta alloys, when properly treated, have an excellent combination of strength and ductility. They are stronger than the alpha or the beta alloys.

The beta alloys are metastable; that is, they tend to transform to an equilibrium, or balance of structures. The beta alloys generate their strength from the intrinsic strength of the beta structure and the precipitation of alpha and other phases from the alloy through heat treatment after processing.

The most significant benefit provided by a beta structure is the increased formability of such alloys relative to the hexagonal crystal structure types (alpha and alpha-beta).

GETTING THE MOST OUT OF Ti ALLOYS

In order to realize the greatest potential that either titanium or titanium alloys can bring to a specific application, it is of great value to have a few, simple rules of thumb in mind initially before a design is actually begun. Some of the more important guidelines are indicated here.

Wrought titanium alloys are the most readily available of all the types. They also have the greatest experience factor. Consider using castings, however, for savings in weight and cost. Cast-plus-HIP (hot isostatic pressed) material can attain the same strength levels as wrought alloys.

Powder alloys are coming into their own as an accepted technique, but it may be that they will never be inexpensive. Given the initial expense of powder, there is also the built-in, and possibly cost-offsetting, near-net shape (NNS) capability that powder offers. This implies at very least a potential for overall lower costs when amortized over the entire project.

If a powder titanium alloy is even a possible candidate material, begin planning for its use during the initial design stage; do not wait and try to fit it into later developmental stages.

Powder allows more exotic titanium alloys to be mixed. Nevertheless, it is wise when making a titanium alloy selection to use the more common alloys, regardless of the material type, unless uncommon properties are absolutely needed. (Ti-6Al-4V must have widespread advantages, or else it would not be so commonly used.)

Handbooks, reference material, etc. all are valuable, but there is no substitute for personal contact with a supplier or fabricator. (A partial list of titanium suppliers and primary metal fabricators appears in Appendix H.)

Do not depend on properties that assume unusual forming conditions and/or unrealistic casting yields. If your design admits of no flexibility with respect to property level realization, the design may be irreversibly compromised later.

Do not buy an aerospace specification unless there is an absolute need for the properties and performance it provides. When using titanium in noncritical applications, choose less stringent specifications, when possible, to save money and time.

6

Executive Summary

RECENT AND FUTURE DEVELOPMENTS

The following is a listing of a few topics which briefly indicate titanium's recent, but lesser-used, developments. It also suggests the probable direction titanium alloy progress will take.

- Textured alloys. These are technically feasible, although there is no driving factor behind the concept.

- Superplastic forming. This technique is being used, but not widely.

- Advanced P/M. There have been various laboratory demonstrations, but a correct property base does not yet exist.

- Rapid solidification rate (RSR) processing. This process is still experimental.

- Higher-temperature capability. Although beta processed alloys exist, aluminides are still being developed and tested.

Refer to Chapter 13, "Recent and Future Advances," for more details.

INTERESTING BACKGROUND FACTS

Titanium is the ninth most abundant element on the planet; it makes up approximately 0.6% of the earth's crust. It occurs in nature only in chemical combination, usually with oxygen and iron. The present mineral sources of titanium are rutile, ilmenite, and leucoxene, an alteration product of ilmenite. Rutile is 93-96% titanium dioxide; ilmenite contains 44-70% titanium dioxide. Leucoxene concentrates contain up to 90% titanium dioxide. The formula for rutile — that is, titanium dioxide — is TiO_2. For ilmenite it is $(Fe,Mg,Mn)TiO_3$.

Principal world producers of ilmenite and titanium slag made from ilmenite are Australia, Canada, Norway, the Republic of South Africa, the United States, and the U.S.S.R. Main producers of rutile are Australia, Sierra Leone, and the Republic of South Africa. Titanium is produced mainly by the U.S.S.R., United States, Japan, the United Kingdom, and China.

The United States mines about one-third of its titanium raw material requirements. The balance is imported mainly from Australia in the form of rutile, synthetic rutile, and ilmenite. It is also imported from Canada in the form of titanium slag. For economic reasons, U.S. producers of titanium metal have relied almost entirely on imported natural rutile, mainly from Australia.

There are estimations that in the year 2000, United States titanium metal demand will be about 45 000 short tons, equivalent to an average annual growth rate of 5.5% from 1983. The probable domestic demand for nonmetal forms of titanium in 2000 is estimated to be 700 000 short tons, representing an average annual growth rate of 1.8% from 1983. The probable demand for titanium metal in 2000 in the rest of the world is 160 000 short tons; the probable nonmetal demand is estimated at 2.2 million short tons.

The primary producers of titanium sponge and ingot in the United States in 1985 were:

- Titanium Metals Corp., Timet Div.

- RMI Company

- Oremet Titanium Corp.

- International Titanium Inc.

- Oregon Metallurgical Corp.

In 1983, 13% of the world titanium supply came from rutile and 87% from ilmenite. Of the total world supply, 4.7% was converted to titanium metal. About 37% of this metal was produced from rutile, consuming about 14% of world rutile production. The metal produced in the U.S.S.R. and China was 63% of the world total and is believed to have been manufactured from high TiO_2 slag made from ilmenite.

World mine production of titanium was a little more than 1.6 million tons in 1983, following a rather constant level of about 1.8 to 2.1 million tons per year in the 1978-82 period.

In 1983, Australia produced an estimated 24% of the titanium content of world ilmenite production. Some was used in Australian pigment plants, but most was exported to the United Kingdom, the United States, Spain, France, and other countries. Canada accounted for about 22% of the world's production of titanium contained in ilmenite. Virtually all of the Canadian ilmenite was converted to titanium slag and shipped in that form to the United States, the Federal Republic of Germany, and other European countries and to titanium pigment plants in Canada. The Republic of South Africa provided 15% of the titanium content of the world's ilmenite production in the form of slag containing 85% TiO_2.

The United States produced about 5% of the world titanium raw material supply in 1982 and consumed nearly all of its own production through 1983. Other major producing nations were Norway, the U.S.S.R., Finland, and India. Norway exported much of its production to countries in Northern Europe. Probably all of the ilmenite produced in the U.S.S.R. was used in that country or by other centrally planned economy countries in Eastern Europe.

Domestic mines produced almost one-third of total United States' requirements for ilmenite and slag in 1982. Imports of major quantities of slag from Canada continued through 1983, and rutile requirements were met predominantly by imports from Australia, including ilmenite used to make synthetic rutile in the United States.

Domestic mine output declined in the period from 1973 through 1983, because of the ready availability and relatively low cost of imported concentrates, depletion of some sand deposits in Florida and New Jersey, and greatly decreased production in New York. While the U.S. share of world titanium ore output declined from 19% in 1969 to about 3% in 1983, the United States consumed about 30% of world mine production during that period. To compensate for the shortfall of domestic mine production, the U.S. imports of ilmenite, slag, and rutile increased significantly. Operation of a domestic synthetic rutile plant since 1980 has resulted in significantly lower imports of rutile, but a corresponding increase in ilmenite imports.

Until 1983, a substantial proportion of U.S. ilmenite production came from the NL Industries rock deposit at Tahawus, New York. However, with the shutdown of that mine's major customer, the NL Industries pigment plant at Sayreville, New Jersey, production at Tahawus was sharply reduced. Titanium mineral production in the United States in the future is expected to come predominantly from ilmenite-bearing sands in Florida. Ilmenite output in the rest of the world came mostly from one mine in Canada, one in Norway, eight major producers in Australia, one in the Republic of South Africa, and an unknown number in the U.S.S.R.

Most of the world's rutile production was from Australia, mainly from the east coast. Virtually all of the rutile production of Australia was exported, chiefly to the United States, the United Kingdom, Japan, the Netherlands, the Federal Republic of Germany, and the U.S.S.R. Since 1978, the Republic of South Africa has been an important producer of rutile and high-titanium slag made from ilmenite. In 1979, production of rutile was resumed in Sierra Leone.

Chapter 2

Introduction

IMPORTANT Ti CHARACTERISTICS

Titanium is a low-density element (approximately 60% of the density of steel) which can be strengthened greatly by alloying and deformation processing. (Characteristic properties of elemental titanium are given in Table 2.1.) Titanium is nonmagnetic and has good heat-transfer properties. Its coefficient of thermal expansion is somewhat lower than that of steel's and less than half that of aluminum. Titanium and its alloys have melting points higher than those of steel's, but maximum useful temperatures for structural applications generally range from 427° to 538°C (800° to 1000°F). (Beta-processed, near-alpha alloys, and titanium aluminide alloys show promise for applications above these temperatures.)

Titanium has the ability to passivate and thereby to exhibit a high degree of immunity against attack by most mineral acids and chlorides. Pure titanium is nontoxic; commercially pure titanium and some titanium alloys generally are biologically compatible with human tissues and bones.

The excellent corrosion resistance and biocompatibility coupled with good strengths make titanium and its alloys useful in chemical and petrochemical applications, marine environments, and biomaterials applications.

Furthermore, titanium and titanium alloys are produced in a wide variety of product forms, as suggested in Figure 2.1. Mill products include:

- Ingot
- Billet
- Bar
- Plate
- Sheet
- Strip
- Tube

Nonmill products include:

- Sponge
- Fines
- Powder
- Customized product forms: cast, wrought, powder for P/M

9

(a) strip   (b) slab   (c) billet

(d) wire   (e) sponge

(f) tube   (g) plate

Photos courtesy of Teledyne Wah Chang Albany

Figure 2.1 - Ti and Ti alloys in various product forms

Table 2.1 Physical and mechanical properties of elemental titanium

Atomic number	22
Atomic weight	47.90
Atomic volume	10.6 W/D
Covalent radius	1.32 Å
First ionization energy	158 k-cal/g-mole
Thermal neutron absorption cross-section	5.6 barns/atom
Crystal structure	· Alpha: close-packed, hexagonal $\leq$882.5°C (1620°F) · Beta: body-centered, cubic $\geq$882.5°C (1620°F)
Color	Dark grey
Density	4.51 g/cm^3 (0.163 lb/in.^{3})
Melting point	1668 $\pm$10°C (3035°F)
Solidus/liquidus	1725°C
Boiling point	3260°C (5900°F)
Specific heat (at 25°C)	0.518 J/kg °K (0.124 BTU/lb °F)
Thermal conductivity	9.0 BTU/hr ft^2 °F
Heat of fusion	440 kJ/kg (estimated)
Heat of vaporization	9.83 MJ/kg
Specific gravity	4.5
Hardness	HRB 70 to 74
Tensile strength	35 ksi min
Modulus of elasticity	14.9 x 10^6 psi
Young's modulus of elasticity	116 x 10^9 N/m^2 16.8 x 10^6 lb f/in.^{2} 102.7 GPa
Poisson's ratio	0.41
Coefficient of friction	0.8 at 40 m/min (125 ft/min) 0.68 at 300 m/min (1000 ft/min)
Specific resistance	554 uohm-mm
Coefficient of thermal expansion	8.64 x 10^{-6}/°C
Electrical conductivity	3% IACS (copper 100%)
Electrical resistivity	47.8 uohm-cm
Electronegativity	1.5 Pauling's
Temperature coefficient of electrical resistance	0.0026/°C
Magnetic susceptibility	1.25 x 10^{-6} 3.17 emu/g
Machinability rating	40

TITANIUM ALLOY SELECTION

Titanium and its alloys are used primarily in two areas of application where the unique characteristics of these metals justify their selection:

- Corrosion-resistant service

- Strength-efficient structures

For these two diverse areas, selection criteria differ markedly. Corrosion applications normally utilize low-strength "unalloyed" titanium mill products fabricated into tanks, heat exchangers, or reactor vessels for chemical-processing, desalination, or power-generation plants. In contrast, high-performance applications typically utilize high-strength titanium alloys in a very selective manner, depending on factors such as thermal environment, loading parameters, available product forms, fabrication characteristics, and inspection and/or reliability requirements. (See Figure 2.2.) As a result of their specialized usage, alloys for high-performance applications normally are processed to more stringent and costly requirements than "unalloyed" titanium for corrosion service.

(a) offshore drilling rig components

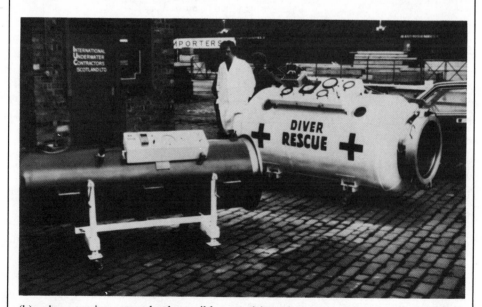

(b) subsea equipment and submersibles requiring ultra-strength

Photos courtesy of Titanium Industries

Figure 2.2 - Typical types of primary areas of applications

Desired mechanical properties such as yield or ultimate strength/ density (strength efficiency), fatigue-crack growth rate, and fracture toughness, plus manufacturing considerations such as welding and forming requirements, normally provide the criteria that determine the alloy composition, structure (alpha, alpha-beta or beta), heat treatment (some variant of either annealing or solution treating and aging), and level of process control selected or prescribed for structural titanium alloy applications. A summary of many commercial and semicommercial titanium grades and alloys is given in Table 2.2.

For lightly loaded structures, where titanium normally is selected because it offers greater resistance to the effects of temperature than aluminum, commercial availability of required mill products along with ease of fabrication may dictate selection. Here, one of the grades of unalloyed titanium usually is chosen. In some cases, corrosion resistance, not strength or temperature resistance, may be the major factor in selection of a titanium alloy.

Selection for Corrosion Resistance

Economic considerations normally determine whether titanium alloys will be used for corrosion service. Capital expenditures for titanium equipment generally are higher than for equipment fabricated from competing materials such as stainless steel, brass, bronze, copper nickel, or carbon steel. As a result, titanium equipment must yield lower operating costs, longer life, or reduced maintenance to justify selection, which most frequently is made on a lower total-life-cycle cost basis. Commercially pure titanium (frequently written CP) satisfies the basic requirements for corrosion service. Unalloyed titanium normally is produced to requirements of ASTM's "standard specifications" B 265, B 338, or B 367 in Grades 1, 2, 3 and 4 in the United States. These grades vary in oxygen and iron content, which control strength level and corrosion behavior, respectively. For certain corrosion applications, Ti-0.2Pd (ASTM Grades 7, 8, and 11) may be preferred over unalloyed Grades 1, 2, 3, and 4. (Refer to Appendix I, "Standards and Specifications," for more information on these grades.)

Due to its unique corrosion behavior, titanium is used extensively in prosthetic devices such as heart-valve parts and load-bearing leg-bone replacements or splints. In general, body fluids are chloride brines that have pH values from 7.4 into the acidic range and that also contain a variety of organic acids and other components—media to which titanium is totally immune. Of the Grades available, ASTM Grade 2 normally is used for low-stress applications, whereas Ti-6Al-4V normally is employed for applications requiring higher strength.

Selection for Strength Efficiency

Historically, wrought titanium alloys have been used widely instead of iron or nickel alloys in aerospace applications, because titanium saves weight in highly loaded components that operate at low-to-moderately elevated temperatures. Many titanium alloys have been custom designed to have optimum tensile, compressive, and/or creep strength at selected temperatures, and at the same time to have sufficient workability to be fabricated into mill products suitable for a specific application.

The most common and widely used titanium alloy is Ti-6Al-4V which accounts for about 50% of the total weight of all titanium alloys shipped. During the life of the titanium industry, various compositions have had transient usage: Ti-4Al-3Mo-1V, Ti-7Al-

Table 2.2 Summary of commercial and semicommercial Ti grades and alloys

Designation	Tensile strength (min) MPa	ksi	0.2% yield strength (min) MPa	ksi	Impurity limits, wt % N (max)	C (max)	H (max)	Fe (max)	O (max)	Nominal composition, wt % Al	Sn	Zr	Mo	Others
Unalloyed grades														
ASTM Grade 1	240	35	170	25	0.03	0.10	0.015	0.20	0.18	...	...	...	...	...
ASTM Grade 2	340	50	280	40	0.03	0.10	0.015	0.30	0.25	...	...	...	...	...
ASTM Grade 3	450	65	380	55	0.05	0.10	0.015	0.30	0.35	...	...	...	...	...
ASTM Grade 4	550	80	480	70	0.05	0.10	0.015	0.50	0.40	...	...	...	...	...
ASTM Grade 7	340	50	280	40	0.03	0.10	0.015	0.30	0.25	...	...	...	...	0.2 Pd
Alpha and near-alpha alloys														
Ti-0.3Mo-0.8Ni	480	70	380	55	0.03	0.10	0.015	0.30	0.25	...	...	...	0.3	0.8 Ni
Ti-5Al-2.5Sn	790	115	760	110	0.05	0.08	0.02	0.50	0.20	5	2.5	...	...	...
Ti-5Al-2.5Sn-ELI	690	100	620	90	0.07	0.08	0.0125	0.25	0.12	5	2.5	...	...	...
Ti-8Al-1Mo-1V	900	130	830	120	0.05	0.08	0.015	0.30	0.12	8	...	...	1	1 V
Ti-6Al-2Sn-4Zr-2Mo	900	130	830	120	0.05	0.05	0.0125	0.25	0.15	6	2	4	2	...
Ti-6Al-2Nb-1Ta-0.8Mo	790	115	690	100	0.02	0.03	0.0125	0.12	0.10	6	...	...	1	2 Nb, 1 Ta
Ti-2.25Al-11Sn-5Zr-1Mo	1000	145	900	130	0.04	0.04	0.008	0.12	0.17	2.25	11.0	5.0	1.0	0.2 Si
Ti-5Al-5Sn-2Zr-2Mo(a)	900	130	830	120	0.03	0.05	0.0125	0.15	0.13	5	5	2	2	0.25 Si
Alpha-beta alloys														
Ti-6Al-4V(b)	900	130	830	120	0.05	0.10	0.0125	0.30	0.20	6.0	...	...	...	4.0 V
Ti-6Al-4V-ELI(b)	830	120	760	110	0.05	0.08	0.0125	0.25	0.13	6.0	...	...	...	4.0 V
Ti-6Al-6V-2Sn(b)	1030	150	970	140	0.04	0.05	0.015	1.0	0.20	6.0	2.0	...	...	0.75 Cu, 6.0 V
Ti-8Mn(b)	860	125	760	110	0.05	0.08	0.015	0.50	0.20	...	...	...	...	8.0 Mn
Ti-7Al-4Mo(b)	1030	150	970	140	0.05	0.10	0.013	0.30	0.20	7.0	...	...	4.0	...
Ti-6Al-2Sn-4Zr-6Mo(c)	1170	170	1100	160	0.04	0.04	0.0125	0.15	0.15	6.0	2.0	4.0	6.0	...
Ti-5Al-2Sn-2Zr-4Mo-4Cr(a)(c)	1125	163	1055	153	0.04	0.05	0.0125	0.30	0.13	5.0	2.0	2.0	4.0	4.0 Cr
Ti-6Al-2Sn-2Zr-2Mo-2Cr(a)(b)	1030	150	970	140	0.03	0.05	0.0125	0.25	0.14	5.7	2.0	2.0	2.0	2.0 Cr, 0.25 Si
Ti-3Al-2.5V(d)	620	90	520	75	0.015	0.05	0.015	0.30	0.12	3.0	...	...	...	2.5 V
Beta alloys														
Ti-10V-2Fe-3Al(a)(c)	1170	170	1100	160	0.05	0.05	0.015	2.5	0.16	3.0	...	...	...	10.0 V
Ti-13V-11Cr-3Al(c)	1170	170	1100	160	0.05	0.05	0.025	0.35	0.17	3.0	...	...	...	11.0 Cr, 13.0 V
Ti-8Mo-8V-2Fe-3Al(a)(c)	1170	170	1100	160	0.05	0.05	0.015	2.5	0.17	3.0	...	...	8.0	8.0 V
Ti-3Al-8V-6Cr-4Mo-4Zr(a)(b)	900	130	830	120	0.03	0.05	0.020	0.25	0.12	3.0	...	4.0	4.0	6.0 Cr, 8.0 V
Ti-11.5Mo-6Zr-4.5Sn(b)	690	100	620	90	0.05	0.10	0.020	0.35	0.18	...	4.5	6.0	11.5	...

(a) Semicommercial alloy; mechanical properties and composition limits subject to negotiation with suppliers. (b) Mechanical properties given for annealed condition; may be solution treated and aged to increase strength. (c) Mechanical properties given for solution treated and aged condition; alloy not normally applied in annealed condition. Properties may be sensitive to section size and processing. (d) Primarily a tubing alloy; may be cold drawn to increase strength.

4Mo, and Ti-8Mn, for example. Ti-6Al-4V is unique in that it combines: attractive properties with inherent workability (which allows it to be produced in all types of mill products, in both large and small sizes); good shop fabricability (which allows the mill products to be made into complex hardware); and the production experience and commercial availability that lead to reliable and economic usage. Thus wrought Ti-6Al-4V has become the standard alloy against which other alloys must be compared when selecting a titanium alloy (or custom designing one) for a specific application. Ti-6Al-4V also is the standard alloy selected for castings that must exhibit superior strength. It has even been evaluated for powder metal processing.

During the approximately 37 years that titanium has been commercially available, many other alloys have been developed, but none match the 50% market share that Ti-6Al-4V enjoys. In addition to the use of Ti-6Al-4V, Pratt & Whitney has used Ti-8Al-1Mo-1V, Ti-5Al-2.5Sn, Ti-6Al-2Sn-4Zr-2Mo, and Ti-6Al-2Sn-4Zr-6Mo in its gas turbine engines. General Electric has used Ti-4Al-4Mn, Ti-1.5Fe-2.7Cr, and Ti-17 among other alloys in addition to the Ti-6Al-4V alloy. Rolls Royce has used IMI 550, IMI 679, IMI 685, and IMI 829 alloys as well as Ti-6Al-4V (IMI 318) in its engines. (IMI Titanium, Ltd. is a British producer-manufacturer.) Some of these mentioned alloys have found use in airframes. Other alloys used in aerospace, missile and space, and high-performance applications have included Ti-6Al-6V-2Sn, Ti-10V-2Fe-3Al, Ti-13V-11Cr-3Al. (The last alloy is also called B120VCA, the first of a line of metastable beta alloys, although it is now considered somewhat obsolete when compared to most contemporary alloys.) Chemical processing operations have been concerned principally with the unalloyed grades, Pd-containing pure grades, and Ti-6Al-4V. Ti-3Al-8V-6Cr-4Zr-4Mo (also called Beta C) has been approved for use in deep, sour well technology. Other alloys are in various stages of use.

(Refer to Appendix J, "Designations, Applications, Properties," for more specific comments on the types of alloys available and their possible applications.)

Rotating components such as jet-engine blades and gas turbine parts require titanium alloys that maximize strength efficiency and metallurgical stability at elevated temperatures. These alloys also must exhibit low creep rates along with predictable behavior with respect to stress rupture with low-cycle fatigue. To reproducibly provide these properties, stringent user requirements are specified to ensure controlled, homogeneous microstructures and total freedom from melting imperfections such as alpha segregation, high-density or low-density tramp inclusions, and unhealed ingot porosity or pipe. The greater the control, however, the greater the cost.

Aerospace pressure vessels similarly require optimized strength efficiency, although at lower temperatures. Required auxiliary properties include weldability and predictable fracture toughness at cryogenic-to-moderately elevated temperatures. To provide this combination of properties, stringent user specifications require controlled microstructures and freedom from melting imperfections. For cryogenic applications, the interstitial elements oxygen, nitrogen, and carbon are carefully controlled to improve ductility and fracture toughness. Alloys with such controlled interstitial element levels are designated ELI (extra low interstitials); for example, Ti-6Al-4V-ELI.

Aircraft structural applications along with high-performance automotive and marine applications also require high-strength efficiency, which normally is achieved by judicious alloy selection combined with close control of mill processing. However, when the design includes redundant structures, when operating environments are not severe, when there are constraints on the fabrication methods that can be used for specific components, or when there are low operational risks, selection of the appropriate alloy and process must take these factors into account.

There are instances of less highly loaded structures when titanium normally is selected because it offers greater resistance to temperature effects than aluminum or greater corrosion resistance than brass, bronze, and stainless steel alloys. In such cases commercial availability of required mill products and ease of fabrication customarily dictate selection. Here, one of the grades of unalloyed titanium usually is chosen. Formability (as with tubes) frequently is a characteristic required of this class of applications.

TITANIUM ALLOY SYSTEMS' AVAILABILITY

In the United States, 70 to 80% of the demand for titanium is from the aerospace industries; 20 to 30% is from industrial applications. Several dozen common alloys are readily available. To a large extent, aerospace applications are the prime cause of titanium alloy and process development and, thus, material availability. The industry has been cyclical in nature and has operated at peak capacity only a few times in the nearly four decades since titanium was introduced as a commercial material. Consequently, producers have been aggressive in their pricing policy because of the less-than-full utilization of capacity.

While total titanium availability has remained relatively flat for some years, the availability of castings has begun to rise. Casting shipments have been increasing at a near 20% per year rate for the past decade and a half. Intricate castings, precision forgings (including near-net shape forgings), and superplastic forming/forging offer promise for continued extensive application of titanium alloys. Titanium castings represent 6% of the weight of some current aircraft gas turbines and may account for substantially more by the mid 1990s. (See Figure 2.3.) Powder parts may be available in limited quantities, yet they are currently and principally restricted to somewhat more exotic alloys and/or applications.

Figure 2.3 - Typical titanium alloy castings

Titanium usage is expected to increase for advanced gas turbines. Titanium usage for airframes has been increasing steadily, as seen in Figure 2.4. Tables 2.3 and 2.4 show the military airframe and/or engine titanium requirements as well as the buy weights for commercial and military airframes. It was not until about 1965 that nonaerospace usage accounted for a significant fraction of the titanium production. Continued modest growth has been taking place since then in many areas: biomedical engineering, marine and chemical applications, automotive, among others.

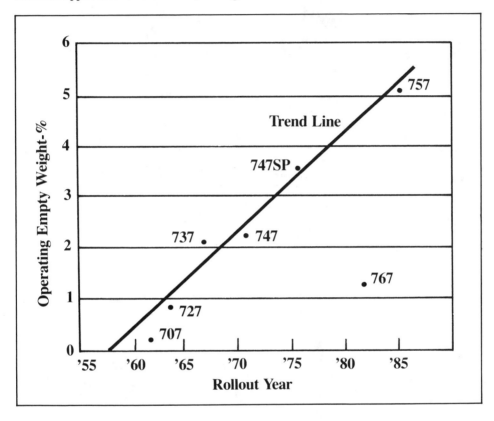

Figure 2.4 - Titanium usage in Boeing aircraft, 1955-85

PROCESSING'S ROLE

Titanium alloys are particularly sensitive to the processing conditions which precede their use in service applications. Processing denotes the wrought, cast, or powder methods used to produce the alloy in the appropriate condition for the intended application, as well as the heat treatments which are applied to the alloy. Heat treatment of alpha-beta alloys seems to produce microstructures which are substantially the same as structures produced, for example, by forging the same alloy in the same general temperature region of the phase diagram as that where the heat treatment is carried out. However, the properties of wrought stock produced by deformation of the alloy at a high temperature generally seem to be better than those produced by heat treatment alone to effect the desired structure. Furthermore, the degree of work placed into the alloy seems to be a controlling factor in the attainment of optimum properties. (Barstock does not have the same properties as a forged disk.)

Table 2.3 Military aircraft (including engines) titanium requirements

| | Titanium buy weight | |
Aircraft/Engine (a)	kg	lb
A-10/(2) TF-34	1 814	4 000
F-5E/(1) J85	635	1 400
F-5G/(1) F404	1 089	2 400
F-14/(2) TF-30	24 630	54 300
F-15/(2) F-100	29 030	64 000
F-16/(1) F-100	3 085	6 800
F-18/(2) F-404	7 620	16 800
C-130/(4) T-56	499	1 100
C-5B/(4) TF-39	24 812	54 700
B-1B/(4)	90 402	199 300
F101-GE-102		
KG-10/CF-6-50	32 206	71 000
CH-53E/(3) T-64	8 800	19 400
CH-60/(2) T-700	2 041	4 500
S-76/(2) All.250	544	1 200
AH-64/(2) T-700	635	1 400

(a) Typical uses are A-10 ballistic armament; structural forgings and wing skins for F-14 and F-15 aircraft; rotor parts for helicopter blade systems; B-1B fracture-critical forgings and wing carry-thru section; and rotor discs, blades, and compressor cases on various engines.

Table 2.4 Titanium buy weights for commercial and military aircraft

| | Titanium buy weight | |
Aircraft/engine (a)	kg	lb
Fairchild A-10	862	1 900
Northrop F-5	408	900
Grumman F-14	18 870	41 600
McDonnell Douglas F-15	24 494	54 000
General Dynamics F-16	861	1 800
McDonnell Douglas F-18	6 214	13 700
Lockheed C-130	454	1 000
Lockheed C-5B	6 804	15 000
Rockwell B-1B	82 646	182 200
707/(4) JT3	4 445	9 800
727/(3) JT8	4 309	9 500
737-200/(2) JT8	3 810	8 400
737/300/(3) CFM-56	3 810	8 400
747/(4) JT-9	42 593	93 900
757/(2) PW2037	12 746	28 100
757/(2) RB211/535	12 973	28 600
767/(2) JT-9	17 554	38 700
767/(2) CF-6	11 703	25 800
MD-80 (2) JT8-217	6 260	13 800
DC-10/(3) CF-6	32 387	71 400
A300/(2) CF-6	6 350	14 000
A310/(2) CF-6	6 350	14 000

(a) Airframe only; slight variations by specific model. Product forms purchased include sheet, plate, bar, billet, and extrusions.

Once the alloy composition is selected, the properties of titanium alloys are linked inextricably to the nature of the processing applied to them. One of the more considerable recent processing challenges has been to develop satisfactory heat treatment procedures for optimizing the properties and the microstructure of cast titanium alloys after they have been hot isostatically pressed. Heat treatments and fabrication conditions to consolidate titanium powder represent ongoing challenges to the process technology involving titanium.

PROPERTY DATA

Properties of commercially pure and alloyed titanium may vary from the data presented in Table 2.1. For specific information on many of the commonly used Ti CP grades and alloys, refer to "Properties of Titanium and Titanium Alloys," Vol. 3 of the 9th edition of *Metals Handbook*.

Chapter 3

Understanding Ti's Metallurgy

CRYSTAL STRUCTURE AND ALLOY TYPES

Titanium is an allotropic element; that is, it exists in more than one crystallographic form. At room temperature, titanium has a hexagonal close-packed (hcp) crystal structure, which is referred to as "alpha" phase. This structure transforms to a body-centered cubic (bcc) crystal structure, called "beta" phase, at 883°C (1621°F). (See Figure 3.1.)

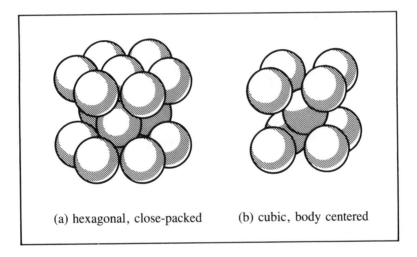

(a) hexagonal, close-packed   (b) cubic, body centered

Figure 3.1 - Hard sphere model of basic crystal structure; atom positions indicated by location of spheres

In discussing the metallurgy of titanium, it is common to separate the alloys into three categories, referring to the common phases present. The alloy categories are:

- α, or alpha

- $\alpha + \beta$, or alpha plus beta

- β, or beta

These categories describe the origin of the microstructure in terms of the basic crystal structure favored by an alloy composition. Crystal structure and grain structure (that is, microstructure) are not synonymous terms. The important fact to keep in mind is that basic crystal structure changes from α to β and back again play the major role in defining titanium properties, as will be discussed in a later chapter.

Some experts have defined five, not three, classes of titanium alloys, depending on the microstructures and chemistries involved. These are:

- Alpha
- Near-alpha (superalpha)
- Alpha-beta

- Near-beta
- Beta

These classes denote the general type of microstructure after processing. An alpha alloy forms no beta phase. A near-alpha (or superalpha) alloy forms only limited beta phase on heating, and so it may appear microstructurally similar to an alpha alloy at lower temperatures. An alpha-beta alloy consists of alpha and retained or transformed beta. A (metastable) near-beta or beta alloy tends to retain the beta phase on initial cooling to room temperature, but it precipitates secondary phases during heat treatment. Typical titanium microstructures are shown in Figure 3.2. How the microstructure of an alloy—

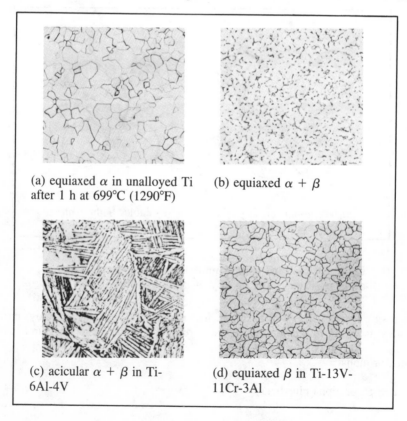

(a) equiaxed α in unalloyed Ti after 1 h at 699°C (1290°F)

(b) equiaxed $\alpha + \beta$

(c) acicular $\alpha + \beta$ in Ti-6Al-4V

(d) equiaxed β in Ti-13V-11Cr-3Al

Figure 3.2 - Typical microstructures of α, $\alpha + \beta$, and β-Ti alloys

in this case a near-alpha alloy—varies with prior heating or working temperature as shown in Figure 3.3.

EFFECTS OF ALLOYING ELEMENTS

Alloying elements generally can be classified as α or β stabilizers. Alpha stabilizers, such as aluminum and oxygen, increase the temperature at which the α phase is stable.

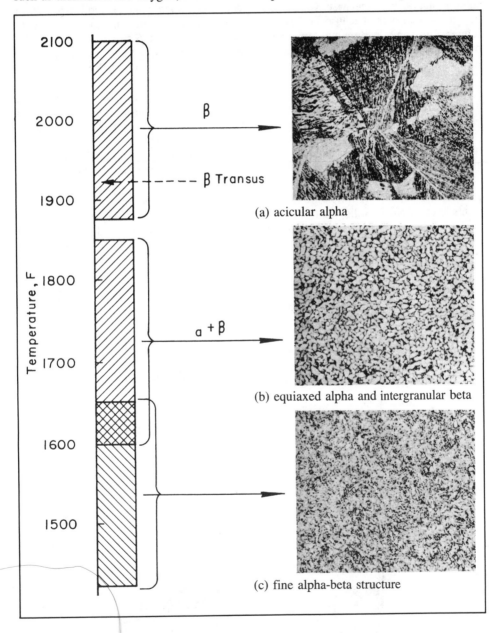

(a) acicular alpha

(b) equiaxed alpha and intergranular beta

(c) fine alpha-beta structure

Figure 3.3 - Room temperature microstructures of annealed Ti-8Al-1Mo-1V bar fabricated at various temperatures, as indicated

Beta stabilizers, such as vanadium and molybdenum, result in stability of the β phase at lower temperatures. This transformation temperature from $\alpha + \beta$ or from α to all β is known as the β transus temperature. The β transus is defined as the lowest equilibrium temperature at which the material is 100% β. The β transus is critical in deformation processing and in heat treatment, as described below.

Below the β transus temperature, titanium is a mixture of $\alpha + \beta$ if the material contains some β stabilizers; otherwise it is all α if it contains limited or no β stabilizers. The β transus is important, because processing and heat treatment often are carried out with reference to some incremental temperature above or below the β transus. Alloying elements that favor the α crystal structure and stabilize it by raising the β transus temperature include aluminum, gallium, germanium, carbon, oxygen, and nitrogen.

Two groups of elements stabilize the β crystal structure by lowering the transformation temperature. The β isomorphous group consists of elements that are miscible in the β phase, including molybdenum, vanadium, tantalum, and columbium. The other group forms eutectoid systems with titanium, having eutectoid temperatures as much as 333°C (600°F) below the transformation temperature of unalloyed titanium. The eutectoid group includes manganese, iron, chromium, cobalt, nickel, copper, and silicon.

Two other elements that often are alloyed in titanium are tin and zirconium. These elements have extensive solid solubilities in α and β phases. Although they do not strongly promote phase stability, they retard the rates of transformation and are useful as strengthening agents. The effects and ranges of some alloying elements used in titanium are indicated in Table 3.1.

Table 3.1 Ranges and effects of some alloying elements used in titanium

Alloying element	Range (approx) wt%	Effect on structure
Aluminum	2 to 7	Alpha stabilizer
Tin	2 to 6	Alpha stabilizer
Vanadium	2 to 20	Beta stabilizer
Molybdenum	2 to 20	Beta stabilizer
Chromium	2 to 12	Beta stabilizer
Copper	2 to 6	Beta stabilizer
Zirconium	2 to 8	Alpha and beta strengthener
Silicon	0.2 to 1	Improves creep resistance

In summary, the transformation temperature (beta transus: completion of transformation to beta on heating) is strongly influenced by:

- The interstitial elements oxygen, nitrogen, and carbon (alpha stabilizers), which raise the transformation temperature

- Hydrogen (beta stabilizer), which lowers the transformation temperature

- Metallic impurity or alloying elements, which may either raise or lower the transformation temperature

The role of the interstitial elements oxygen, nitrogen, and carbon was described above. The substitutional alloying elements also play an important role in controlling the microstructure and properties of titanium alloys.

Tantalum, vanadium, and columbium are beta isomorphous (i.e., have similar phase relations) with body-centered cubic titanium. Titanium does not form intermetallic compounds with the beta isomorphous elements. Eutectoid systems are formed with chromium, iron, copper, nickel, palladium, cobalt, manganese, and certain other transition metals. These elements have low solubility in alpha titanium and decrease the transformation temperature. They usually are added to alloys in combination with one or more of the beta isomorphous elements to stabilize the beta phase and prevent or minimize formation of intermetallic compounds which may occur during service at elevated temperature.

Zirconium and hafnium are unique in that they are isomorphous with both the alpha and beta phases of titanium. Tin and aluminum have significant solubility in both alpha and beta phases. Aluminum increases the transformation temperature significantly whereas tin lowers it slightly. Aluminum, tin, and zirconium commonly are used together in alpha and near-alpha alloys. In alpha-beta alloys, these elements are distributed approximately equally between the alpha and beta phases. Almost all commercial titanium alloys contain one or more of these three elements because they are soluble in both alpha and beta phases, and particularly because they improve creep strength in the alpha phase.

Many more elements are soluble in beta titanium than in alpha. Beta isomorphous alloying elements are preferred as additions because they do not form intermetallic compounds. However, iron, chromium, manganese, and other compound formers sometimes are used in beta-rich alpha-beta alloys or in beta alloys because they are strong beta stabilizers and improve hardenability and response to heat treatment. Nickel, molybdenum, and palladium improve corrosion resistance of unalloyed titanium in certain media.

TRANSFORMATIONS AND SECONDARY PHASE FORMATION

Intermetallic compounds and other secondary phases are formed in titanium alloy systems. The more important phases, historically, have been omega and alpha-2 (α_2), chemically written as Ti_3Al. Omega (ω) phase has not been proved to be a constituent in commercial systems using present-day processing practice. Alpha-2 has been considered to be a factor in some cases of stress-corrosion cracking. (Most present interest in α_2 centers on its use as a matrix for a high-temperature titanium alloy.)

Beta phase decomposes, usually by martensitic transformations, in the alpha-beta alloys. Martensitic reactions are fast, diffusionless (no composition change) transformations in crystal structure and microstructure. There are several martensite types formed in titanium alloys, as discussed below. The beta-to-martensite transition is responsible for the acicular (platelike) structure in quenched and/or quenched and aged titanium alloys. (See Figure 3.4.) *Hardenability of a titanium alloy* is a phrase that refers to its ability to permit full transformation of the alloy to martensitic phases or to retain beta to room temperature. Alpha prime (α') and alpha double prime (α'') martensites are brought out by cooling and decompose, on subsequent aging, to alpha and beta phases.

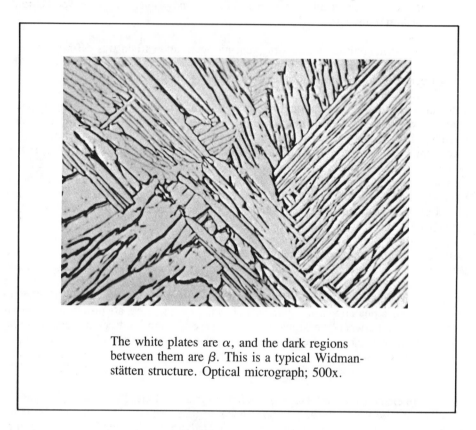

The white plates are α, and the dark regions between them are β. This is a typical Widmanstätten structure. Optical micrograph; 500x.

Figure 3.4 - Microstructure of Ti-6Al-4V alloy after cooling slowly from above the β transus

A graphic illustration of the microstructural transformations which can occur in an alpha-beta alloy, here Ti-6Al-4V, is shown in Figure 3.5. As can be seen, the morphology (shape/location) of the phase changes with prior treatments. The α phase may remain relatively globular (equiaxed), but the transformed beta (martensites or alpha) may be very acicular or elongated. The amount of equiaxed alpha and the coarseness or fineness of the transformed beta products will affect titanium's alloy properties.

Metastable beta can show more variety in decomposition than does the supersaturated alpha or martensitic alpha structure. The omega phase can form, as can alpha phase and a low-solute-content beta phase. Other intermetallic compounds also may form, and, under certain circumstances, ordering of the beta phase can occur. (Ordering removes the randomness in atom location which normally exists and puts atoms in specific locations.)

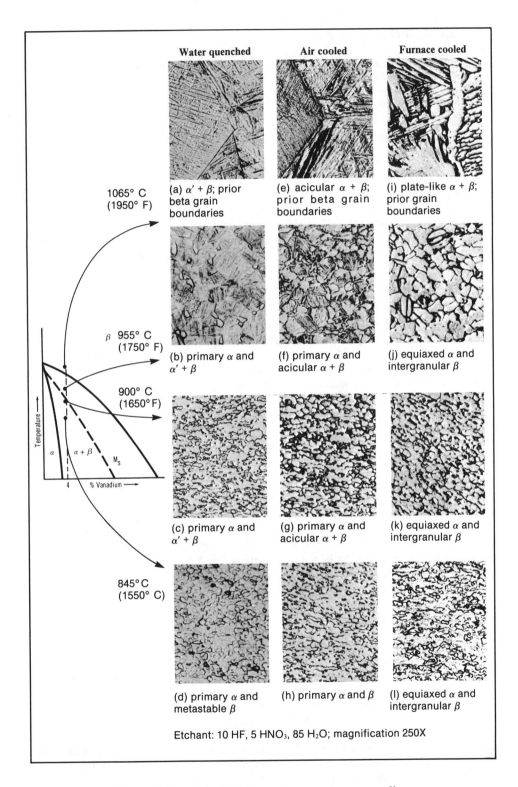

Figure 3.5 - Ti-6Al-4V formation process on cooling

TITANIUM GROUPINGS

When considering unalloyed and alloyed titanium, it is common to group the materials as:

- Unalloyed (CP)

- Alpha and near-alpha

- Alpha-beta

- Metastable beta

Table 2.2, located in Chapter 2, lists some commercial and semicommercial titanium grades (that is, kinds) and alloys currently available. These subdivisions mirror the four titanium groupings listed above. Each of these is discussed separately, below.

Unalloyed Ti

Within the unalloyed titanium system, there are several grades (kinds). The primary difference between grades is oxygen and iron content. (Refer to Appendix I, ASTM specifications table, for more information on "grades.") Grades of higher purity (lower interstitial content) are lower in strength, hardness, and transformation temperature than those higher in interstitial content. The high solubility of the interstitial elements oxygen and nitrogen makes titanium unique among metals and also creates problems not of concern in most other metals. For example, heating titanium in air at high temperature results not only in oxidation but also in solid-solution hardening of the surface as a result of inward diffusion of oxygen (and nitrogen). A surface-hardened zone of "alpha-case" (or "air-contamination layer") is formed. Normally, this layer is removed by machining, chemical milling, or other mechanical means prior to placing a part in service because the presence of alpha-case reduces fatigue strength and ductility.

Unalloyed titanium usually is selected for its excellent corrosion resistance, especially in applications where high strength is not required. Yield strengths of unalloyed (commercially pure) grades vary from 170 MPa (25 ksi) to 480 MPa (70 ksi) simply as a result of variations in the interstitial and impurity levels: strength increases with increasing oxygen and iron contents. (Refer to Table 2.2.)

Alpha, Near-Alpha Alloys

Within the alpha and near-alpha system, those alpha alloys that contain aluminum, tin, and/or zirconium are preferred for high-temperature as well as cryogenic applications. Alpha-rich alloys generally are more resistant to creep at high temperature than alpha-beta or beta alloys. The extra-low-interstitial alpha alloys (ELI grades) retain ductility and toughness at cryogenic temperatures. Ti-5Al-2.5Sn-ELI has been used extensively in such applications.

Unlike alpha-beta and beta alloys, alpha alloys cannot be strengthened by heat treatment. Generally, alpha alloys are annealed or recrystallized to remove residual stresses induced by cold working. Alpha alloys have good weldability because they are insensitive to heat treatment. They generally have poorer forgeability and narrower forging-temperature ranges than alpha-beta or beta alloys, particularly at temperatures below the beta transus. This poorer forgeability is manifested by a greater tendency for center bursts or surface cracks to occur which means that small reduction steps and frequent reheats must be incorporated in forging schedules.

Alpha alloys that contain small additions of beta stabilizers (Ti-8Al-1Mo-1V or Ti-6Al-2Nb-1Ta-0.8Mo, for example) sometimes have been classed as "superalpha" or "near-alpha" alloys. Although they contain some retained beta phase, these alloys consist primarily of alpha and may behave more like conventional alpha alloys than alpha-beta alloys.

Alpha-Beta Alloys

The alloys within the alpha-beta system contain one or more alpha stabilizers or alpha-soluble elements plus one or more beta stabilizers. These alloys retain more beta phase after solution treatment than do near-alpha alloys, the specific amount depending on the quantity of beta stabilizers present and on heat treatment.

Alpha-beta alloys can be strengthened by solution treating and aging. Solution treating usually is done at a temperature high in the two-phase alpha-beta field and is followed by quenching in water, oil, or other soluble quenchant. As a result of quenching, the beta phase present at the solution-treating temperature may be retained or may be partly transformed during cooling by either martensitic transformation or nucleation and growth. The specific response depends on alloy composition, solution-treating temperature (beta-phase composition at the solution temperature), cooling rate and section size. Solution treatment is followed by aging, normally at 480° to 650°C (900° to 1200°F), to precipitate alpha and produce a fine mixture of alpha and beta in the retained or transformed beta phase.

Transformation kinetics, transformation products, and specific response of a given alloy can be quite complex; a detailed review of the subject is beyond the scope of this article. However, Figure 3.6 shows microstructures of the standard alpha-beta alloy, Ti-6Al-4V, in six representative conditions of transformation. Chapters 4 and 5 contain additional discussions of the transformations and microstructures in titanium, as affected by wrought alloy deformation processing and heat treatment. Also refer to the discussion of microstructural development in Ti alloys, below.

Solution treating and aging can increase the strength of alpha-beta alloys from 30 to 50%, or more, over the annealed or over-aged condition. Response to solution treating and aging depends on section size; alloys relatively low in beta stabilizers (Ti-6Al-4V, for example) have poor hardenability and must be quenched rapidly to achieve significant strengthening. For Ti-6Al-4V, the cooling rate of a water quench is not rapid enough to significantly harden sections thicker than about 25 mm (1 in.). As the content of beta stabilizers increases, hardenability increases; Ti-5Al-2Sn-2Zr-4Mo-4Cr, for example, can be through-hardened with relatively uniform response throughout sections up to 150 mm (6 in.) thick. For some alloys of intermediate beta-stabilizer content, the surface of a relatively thick section can be strengthened, but the core may be 10 to 20% lower in hardness and strength. The strength that can be achieved by heat treatment is also a function of the volume fraction of beta phase present at the solution-treating temperature. Alloy composition, solution temperature, and aging conditions must be carefully selected and balanced to produce the desired mechanical properties in the final product.

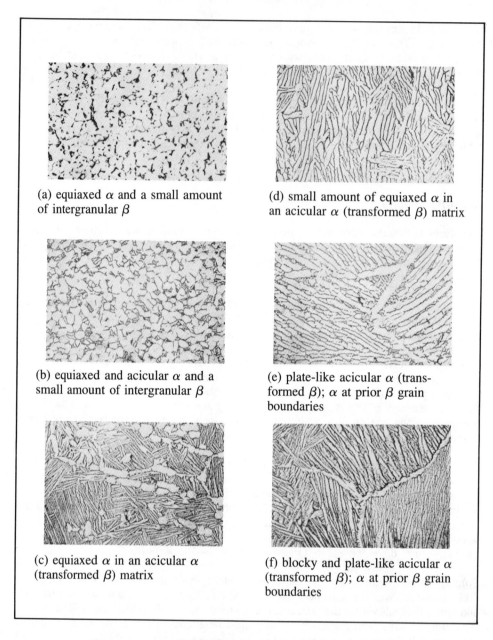

(a) equiaxed α and a small amount of intergranular β

(d) small amount of equiaxed α in an acicular α (transformed β) matrix

(b) equiaxed and acicular α and a small amount of intergranular β

(e) plate-like acicular α (transformed β); α at prior β grain boundaries

(c) equiaxed α in an acicular α (transformed β) matrix

(f) blocky and plate-like acicular α (transformed β); α at prior β grain boundaries

Figure 3.6 - Optical microstructures of Ti-6Al-4V in six representative metallurgical conditions

Although the ability of alpha-beta alloys to be precipitation hardened has been studied in laboratory programs since the early days of the titanium industry, there were relatively few production applications of solution-treated and precipitation- (age) hardened alloys. This situation appears to have changed, because alloys such as Ti-6Al-2Sn-4Zr-6Mo, Ti-5Al-2Sn-2Zr-4Mo-4Cr, and certain high-hardenability beta alloys have been developed specifically to be age hardened for improved strength—about 30 to 40% above that of annealed alloys.

Metastable Beta Alloys

The beta alloys of the metastable beta system are richer in beta stabilizers and leaner in alpha stabilizers than alpha-beta alloys. They are characterized by high hardenability, with the metastable beta phase completely retained on air cooling of thin sections or water quenching of thick sections. Beta alloys have excellent forgeability; in sheet form they can be cold formed more readily than high-strength alpha-beta or alpha alloys. Beta phase usually is metastable and has a tendency to transform to the equilibrium alpha plus beta structure. After solution treating, metastable beta alloys are aged at temperatures of 450° to 650°C (850° to 1200°F) to partially transform the beta phase to alpha. The alpha forms as finely dispersed particles in the retained beta, and strength levels comparable or superior to those of aged alpha-beta alloys can be attained. The chief disadvantages of beta alloys in comparison with alpha-beta alloys are higher density, lower creep strength, and lower tensile ductility in the aged condition. Although tensile ductility is lower, the fracture toughness of an aged beta alloy generally is higher than that of an aged alpha-beta alloy of comparable yield strength. Very high yield strengths (about 170 ksi) with excellent toughness ($K_{Ic} \approx 40$ ksi $\sqrt{in}$.) have been claimed for the beta alloy Ti-10V-2Fe-3Al.

In the solution-treated condition (100% retained beta), beta alloys have good ductility and toughness, relatively low strength and excellent formability. Solution-treated beta alloys begin to precipitate alpha phase at slightly elevated temperatures and thus are unsuitable for elevated-temperature service without prior stabilization or over-aging treatment.

Beta alloys, despite the name, actually are metastable, because cold work at ambient temperature or heating to a slightly elevated temperature can cause partial transformation to alpha. The principal advantages of beta alloys are that they have high hardenability, excellent forgeability, and good cold formability in the solution-treated condition and can be hardened to fairly high strength levels.

MICROSTRUCTURAL DEVELOPMENT IN Ti ALLOYS

Microstructures in titanium alloys usually are developed by heat treatment or processing (wrought/cast/powder metallurgy) followed by heat treatment. Structural changes invariably are achieved through production of beta phase in some amount and the subsequent changes as beta transforms. In all-alpha alloys, microstructural change is limited to grain refinement and, possibly, to grain shape. One illustration each of a typical alpha-beta and beta alloy microstructural development is given in the following discussions.

Ti-6Al-4V Microstructure

One of the most widely used titanium alloys is an alpha-beta type containing 6% Al and 4% V. It is, of course, Ti-6Al-4V. It has an excellent combination of strength and toughness along with excellent corrosion resistance. Typical uses include aerospace applications, pressure vessels, aircraft turbine and compressor blades and disks, surgical implants, etc.

The properties of this alloy are developed by relying on the refinement of the grains upon cooling from the β region, or the $\alpha + \beta$ region, and subsequent low-temperature aging to decompose martensite formed upon quenching. When this alloy is slowly cooled from the β region, α begins to form below the β transus, which is about 980°C (1796°F). The α forms in plates, with a crystallographic relationship to the β in which it forms. The α plates form with their basal (close-packed) plane parallel to a special plane in the β phase. Upon slow cooling, a nucleus of α forms and, because of the close atomic matching along this common plane, the α phase thickens relatively slowly perpendicular to this plane but grows faster along the plane. Thus, plates are developed. Since in a given β grain there are six sets of nonparallel growth planes, then a structure of α plates is formed consisting of six nonparallel sets. The Widmanstätten microstructure developed is illustrated in Figure 3.4.

The formation process is shown schematically in Figure 3.7. It uses a constant-composition phase diagram section at 6% Al to illustrate the formation of α upon cooling. The darker regions are the β phase, left between the α plates which have formed. The microstructure consists of parallel plates of α delineated by the β between them. Where α plates formed parallel to one specific plane of β meet plates formed on another plane, a high-angle grain boundary exists between the α crystals and etches to reveal a line separating them. This microstructural morphology, consisting of these sets of parallel plates which have formed with a crystallographic relationship to the phase from which they formed, is called a Widmanstätten structure.

Upon cooling rapidly, β may decompose by a martensite reaction, similar to that for pure Ti, and form a Widmanstätten pattern. The structure present after quenching to 25°C (77°F) depends upon the annealing temperature. Different types of martensite may form, depending upon the alloy chemistry and the quenching temperature. These are designated α' and α''. Upon quenching from above the β transus (about 980°C or 1796°F), the structure is all martensitic (α' or α'') with a small amount of β (although in some alloys the β has not been observed).

The presence of some β in the structure after quenching from above the β transus is due to the fact that the temperature for the end of the martensite transformation, M_f, is below room temperature (25°C or 77°F) for this alloy. That is because vanadium is a β stabilizer, and the addition of 4% V to a Ti-6% Al alloy is sufficient to place the M_f below 25°C (77°F). Thus, upon quenching to 25°C (77°F), not all of the β is converted to α' or α''.

For the commercial Ti-6Al-4V alloy, there are some commonly used heat treatments. For each of these, the following descriptions are typical of the temperatures and times used. The actual practice varies with alloy producer and user.

To place the alloy in a soft, relatively machinable condition, the alloy is heated to about 730°C (1346°F) in the lower range of the $\alpha + \beta$ region, held for 4 h, then furnace cooled to 25°C (77°F). This treatment, called mill annealing, produces a microstructure of globular crystals of β in an α matrix. A typical microstructure is shown in Figure 3.8.

Another annealing treatment is duplex annealing. Several variants of this treatment are used. Typically, the alloy is heated to 955°C (1751°F) for 10 min, then air cooled. It then is heated to 675°C (1247°F) for 4 h and air cooled to 25°C (77°F).

With the aging treatment called solution treating and aging, typically the alloy is heated at 955°C (1751°F) for 10 min, water quenched, then aged for 4 h at a temperature between 540° and 675°C (1004° to 1247°F), followed by air cooling to 25°C (77°F).

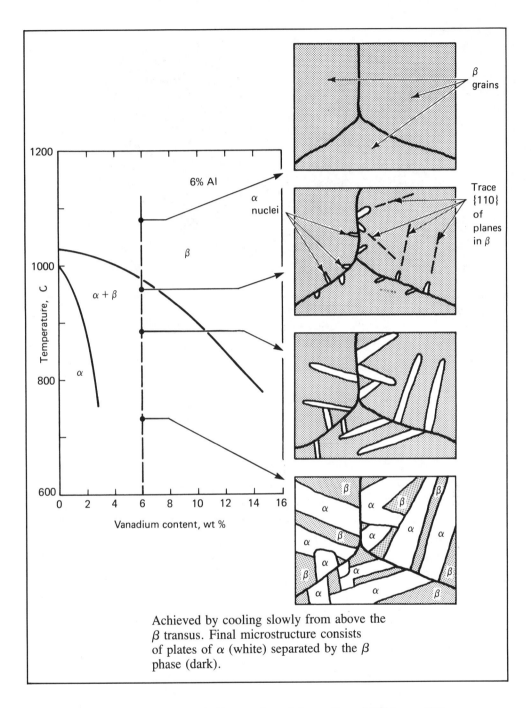

Achieved by cooling slowly from above the β transus. Final microstructure consists of plates of α (white) separated by the β phase (dark).

Figure 3.7 - Schematic illustration of formation of Widmanstätten structure in a Ti-6Al-4V alloy

Typical tensile mechanical properties for the three treatments are compared in Table 3.2. The strongest alloy is the one that has been solution treated and aged. The mill-annealed condition is stronger than the duplex-annealed condition, but the difference is slight.

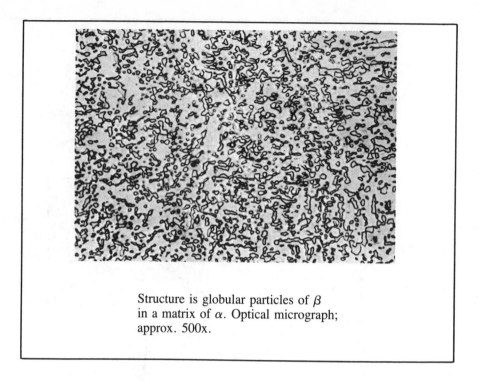

Structure is globular particles of β
in a matrix of α. Optical micrograph;
approx. 500x.

**Figure 3.8 - Microstructure of Ti-6Al-4V alloy, mill annealed
4 h at 750°C (1382°F)**

Table 3.2 Ti-6Al-4V tensile mechanical properties (a)

Condition	Yield strength		Tensile strength		Elongation at fracture, %
	MPa	ksi	MPa	ksi	
Mill annealed	945	137	1069	155	10
Duplex annealed	917	133	965	140	18
Solution treated and aged	1103	160	1151	167	13

(a)   At 25°C (77°F) in the mill-annealed, duplex-annealed, and solution-treated and
       aged conditions

Ti-13V-11Cr-3Al Microstructure

A second alpha-beta alloy type is Ti-13V-11Cr-3Al. Before it is discussed in detail,
there are some basic concepts that should be understood.

The effect of the body-centered cubic elements in stabilizing the β phase raises the possibility of the development of an eutectoid alloy which forms a martensite structure upon
rapid cooling from β and becomes strong and hard, as is developed in steels. In fact,

some elements form an eutectoid region with titanium with a martensite structure formed upon rapid cooling. The M_s (martensite start) and M_f (martensite finish) temperatures are decreased as the alloy content increases, similar to the influence of carbon on iron. Sufficient alloy content allows retained β to be present at 25°C (77°F) after quenching.

However, the martensite is the α' structure and is not particularly strong. Thus, the expectation that this heat treatment can be used to strengthen titanium alloys is not realized. When heated below the eutectoid temperature, the martensite decomposes into a structure of α and other phases, depending upon the exact alloy. This process was expected to provide for more strengthening than the martensitic structure. However, significant strengthening is not obtained, so this latter approach is not used to provide strengthening.

Nevertheless these beta alloys can be used because of the relative formability of the body-centered cubic beta structure compared to the hexagonal close-packed alpha structure. Sufficient alloy content allows retention of a metastable β structure after quenching from the β region to 25°C (77°F). In this condition the alloy can be fabricated by plastic deformation. Then the component can be reheated below the eutectoid temperature to decompose the retained β to a multiphase structure of α and other phases which depend upon the exact alloy, providing considerable strengthening over that of the retained β. It is this approach which is the basis of a few commercial alloys, and, in this discussion, the physical metallurgy of one of these, Ti-13V-11Cr-3Al, is examined.

The addition of Cr should maintain the desirable corrosion- and oxidation-resistance characteristic of titanium alloys. Table 3.3 shows the effect of several elements on the eutectoid temperature and composition and the required alloy content to lower the M_f to 25°C (77°F). Note that Cr is relatively effective in retaining β.

Table 3.3 Effect of alloying elements in several titanium binary alloys on eutectoid temperature and composition, and content needed to retain beta

Alloying element	Eutectoid temperature °C	Eutectoid temperature °F	Eutectoid composition, wt %	Alloy content (a)
Manganese	550	1022	20	6.5
Iron	600	1112	15	4
Chromium	675	1247	15	8
Cobalt	585	1085	9	7
Nickel	770	1418	7	8
Copper	790	1454	7	13
Silicon	860	1580	0.9	–

(a) Needed to retain beta after quenching at 25°C (77°F)

Data for Ti-V and Ti-Cr alloys show that β can be retained upon quenching from the β region. In the Ti-V alloys, hardening occurs upon aging because of the formation of α in the β and the appearance of the intermediate omega phase. In the Ti-Cr alloys, hardening is associated with the formation of α in the β, and also subsequently $TiCr_2$. Thus, in the alloy Ti-Cr-V it is expected that β can be retained upon quenching from the β region to 25°C (77°F), and that, upon aging, hardening associated with the formation of α and

TiCr$_2$, and perhaps the intermediate omega phase, occur. These two latter phases are metastable and disappear upon prolonged aging.

The recommended commercial heat treatment for Ti-13V-11Cr-4Al is to solution heat treat in the β region from about 760° to 815°C (1400° to 1499°F) for 0.2 to 1 h, then air cool or quench (depending upon the size of the part) to retain the β structure. Subsequent aging to precipitate α phase is accomplished around 480°C (896°F) for a time usually between 2 and 100 h, depending upon the properties desired. The use of aging temperatures around 480°C (896°F) is based on data which show that this is the optimum range to use for maximum strength for aging times up to 100 h. Below this temperature, the rate of formation of α is too low to give appreciable hardening. This is seen in the time-temperature transformation (TTT) diagram, shown in Figure 3.9. Above this temperature, the structure is too coarse to attain maximum strengthening.

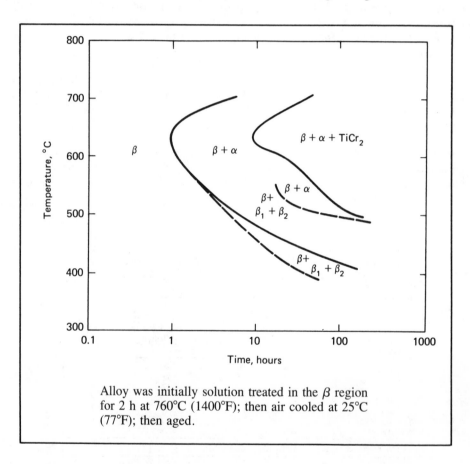

Alloy was initially solution treated in the β region for 2 h at 760°C (1400°F); then air cooled at 25°C (77°F); then aged.

Figure 3.9 - TTT diagram (approx.) for a Ti-13V-11Cr-4Al alloy

Chapter 4

Wrought Alloy Processing

GENERAL ASPECTS OF INGOT METALLURGY

Titanium metal passes through four major steps during processing from ore to finished product. These are:

- Reduction of titanium ore to a porous form of titanium metal called "sponge"

- Melting of sponge or sponge plus a master alloy to form an ingot

- Primary fabrication, in which ingots are converted into general mill products

- Secondary fabrication of finished shapes from mill products

An overall view of the major aspects of a variety of operations performed on titanium is shown in Figure 4.1.

At each of these steps, mechanical and physical properties of titanium in the finished shape may be affected by any of several factors, or by a combination of factors. Among the most important are:

- Amounts of specific alloying elements and impurities

- Melting process used to make ingot

- Method for mechanically working ingots into mill products

- Fabrication or heat treatment, the final step employed in working

Because the properties of titanium are so readily influenced by processing, great care must be exercised in controlling the conditions under which the processing is carried out. At the same time, this characteristic of titanium makes it possible for the titanium industry to serve a wide range of applications with a minimum number of grades or alloys. By varying thermal or mechanical processing, or both, a broad range of special properties can be produced in commercially pure titanium and titanium alloys.

Control of raw materials is extremely important in producing titanium and its alloys because there are many elements of which even small amounts can produce major and at

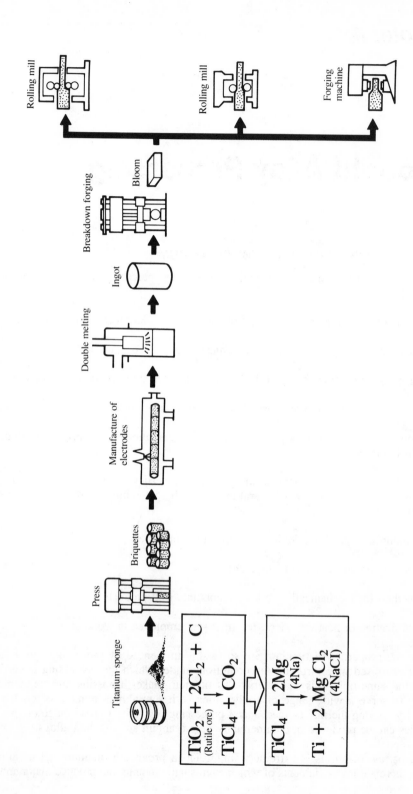

Figure 4.1 - Overview of titanium production cycle

(Continued)

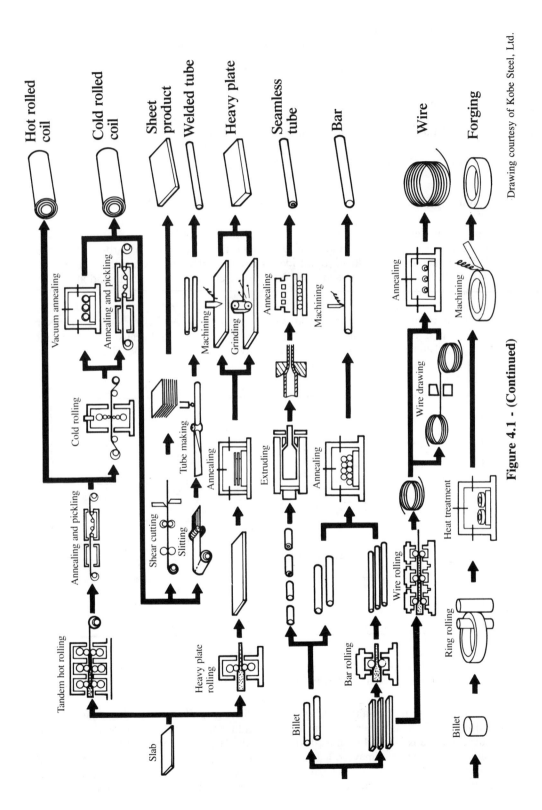

Figure 4.1 - (Continued)

Drawing courtesy of Kobe Steel, Ltd.

times undesirable, effects on the properties of these metals in finished form. The raw materials commonly used in producing titanium and its alloys are:

- Titanium in the form of sponge metal

- Alloying elements

- Reclaimed titanium scrap (usually called "revert")

Each of these materials is discussed below.

Titanium Sponge

Titanium sponge must meet stringent specifications in order to assure control of an ingot's composition. Most importantly, sponge must not contain hard, brittle, and refractory titanium oxide, titanium nitride, or complex titanium oxynitride particles that, if retained through subsequent melting operations, could act as crack-initiation sites in the final product.

Carbon, nitrogen, oxygen, silicon, and iron commonly are found as residual elements in sponge. These elements must be held to acceptably low levels because they raise the strength and lower the ductility of the final product. (See Figure 4.2 and Chapter 11, "Relationships of Processing and Properties.")

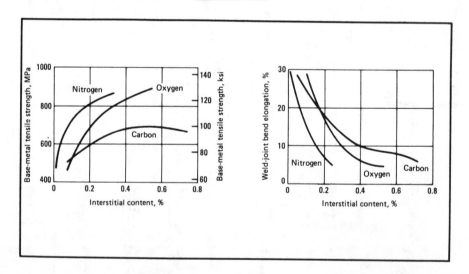

Figure 4.2 - Effects of interstitial-element content on strength and ductility of unalloyed titanium

Titanium sponge is manufactured by first chlorinating rutile ore and then reducing the resulting $TiCl_4$ with either sodium or magnesium metal. (See Figure 4.1.) Sodium-reduced sponge is leached with acid to remove the NaCl by-product of reduction. Magnesium-reduced sponge may be leached, inert-gas swept, or vacuum distilled to remove the excess $MgCl_2$ by-product. Vacuum distilling results in lower residual levels of magnesium, hydrogen, and chlorine, but increases cost. Modern melting techniques remove volatile substances from sponge, so that ingot of high quality can be produced regardless of which method is used for production of sponge.

Alloying Elements

Purity of the alloying elements added to titanium during melting is as important as the purity of sponge at an earlier stage. It must be controlled with the same degree of care to avoid undesirable residual elements — especially those that can form refractory or high-density inclusions in the titanium matrix.

Basically, oxygen and iron contents determine strength levels of commercially pure (CP) titanium (ASTM and ASME Grades 1, 2, 3, and 4) and, also, the differences in mechanical properties between extra-low-interstitial (ELI) grades and standard grades of titanium alloys. (Table 4.1 illustrates this effect.) In higher strength grades, oxygen and iron are intentionally added to the residual amounts already in the sponge to provide extra strength. On the other hand, carbon and nitrogen usually are held to minimum residual levels to avoid embrittlement.

Table 4.1 Tensile properties of annealed titanium sheet as influenced by oxygen and iron contents

Material	Maximum impurity content, %		Minimum tensile strength		Minimum yield strength (a)	
	Oxygen	Iron	MPa	ksi	MPA	ksi
Unalloyed Ti, Grade 1	0.18	0.20	240	35	170	25
Unalloyed Ti, Grade 2	0.25	0.30	345	50	275	40
Unalloyed Ti, Grade 3	0.35	0.30	450	65	380	55
Unalloyed Ti, Grade 4	0.40	0.50	655	95	485	70
Ti-6Al-4V	0.20	0.30	925	134	870	126
Ti-6Al-4V-ELI	0.13	0.25	900	130	830	120
Ti-5Al-2.5Sn	0.20	0.50	830	120	780	113
Ti-5Al-2.5Sn-ELI	0.12	0.25	690	100	655	95
(a) At 0.2% offset.						

Reclaimed Scrap (Revert)

The addition of scrap makes production of ingot titanium more economical than if only sponge were used. If properly controlled, addition of scrap is fully acceptable, and it can be used even in materials for critical structural applications, such as rotating components for jet engines.

All forms of scrap can be remelted — machining chips, cut sheet, trim stock, and chunks. To be utilized properly, scrap must be thoroughly cleaned and carefully sorted by alloy and by purity before being remelted. During cleaning, surface scale must be removed, because adding titanium scale to the melt could produce refractory inclusions or excessive porosity in the ingot. Machining chips from fabricators who use carbide tools are acceptable for remelting only if all carbide particles adhering to the chips are removed; otherwise, hard, high-density inclusions could result. Improper segregation of alloy revert would produce off-composition alloys and could potentially degrade the properties of the resulting metal.

Titanium Ingot Production

In some production methods granules of titanium are mixed, and, if required, alloying elements are blended in with the mix. (See Figure 4.3.) At times the granules are pressed into a form called either a "compact," or "briquette." A number of these compacts may be welded together to form a 4.5 m- (15 ft-) long electrode, which is lowered into a furnace for melting. (Alternatively, carefully selected titanium revert may be welded to form the electrode. See Figure 4.3.)

A process called double melting is considered necessary for all applications to ensure an acceptable degree of homogeneity in the resulting product. Thus the first melt is used as the electrode for the second melt, which frequently produces the final ingot. Triple melting is used to achieve even better uniformity. Triple melting also reduces oxygen-rich or nitrogen-rich inclusions in the microstructure to a very low level by providing an additional melting operation to dissolve them.

Melting Practice

Most titanium and titanium alloy ingot is melted twice in an electric-arc furnace under a vacuum — a procedure known as the "double consumable-electrode vacuum-melting process." (For certain critical applications, a third, or "triple," melting step has been specified at times.) In the two-stage process, titanium sponge, revert, and alloy additions are initially mechanically consolidated and then are melted together to form ingot. (See Figure 4.3.) Ingots from the first melt are used as the consumable electrodes for second-stage melting. (Processes other than consumable-electrode arc melting are used in some instances for first-stage melting of ingot for noncritical applications.) Usually, all melting is done under vacuum, but in any event the final stage of melting must be done by the consumable-electrode vacuum-arc process.

Segregation and other compositional variations directly affect the final properties of mill products. Melting technique alone does not account for all segregation and compositional variations and thus cannot be correlated with final properties.

Melting in a vacuum reduces the hydrogen content of titanium and essentially removes other volatiles. This tends to result in high purity in the cast ingot. However, anomalous operating factors such as air leaks, water leaks, arc-outs, or even large variations in power level affect both soundness and homogeneity of the final product.

Still another factor is ingot size. Normally, ingots are 650 to 900 mm (26 to 36 in.) in diameter and weigh 3600 to 6800 kg (8000 to 15 000 lb). Larger ingots are economically advantageous to use and are important in obtaining refined macrostructures and microstructures in very large sections, such as billets with diameters of 400 mm (16 in.) or greater. Ingots up to 1000 mm (40 in.) in diameter and weighing more than 9000 kg (20 000 lb) have been melted successfully, but there appear to be limitations on the improvements that can be achieved by producing large ingots due to increasing tendency for segregation with increasing ingot size.

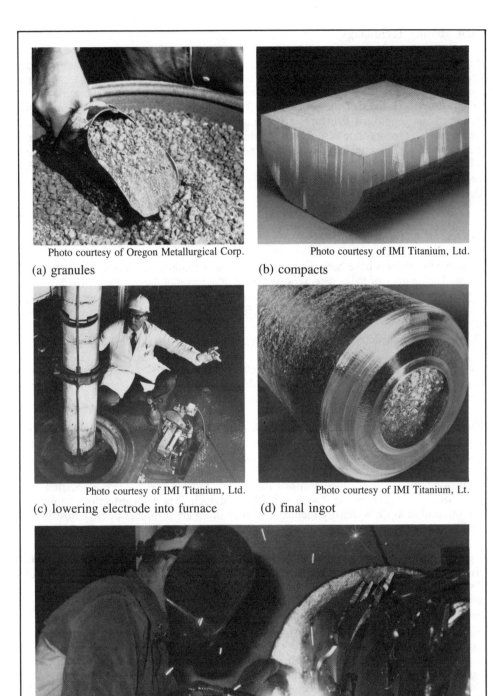

Photo courtesy of Oregon Metallurgical Corp.

(a) granules

Photo courtesy of IMI Titanium, Ltd.

(b) compacts

Photo courtesy of IMI Titanium, Ltd.

(c) lowering electrode into furnace

Photo courtesy of IMI Titanium, Lt.

(d) final ingot

Photo courtesy of Howmet Turbine Components Corp.

(e) welding revert

Figure 4.3 - Ti ingot production

New Melting Technology

Significant effort has been devoted to the development of improved titanium melting technology relative to reduction of impurities and improvements in structure. Electron beam technology now is available for the melting of titanium alloys or the remelting of scrap. The advantages of the process are the much lower levels of gaseous impurities achieved, the improved structure, and the sharply reduced inclusion level. Efforts are being made to incorporate plasma arc melting as a lower cost alternative to the electron beam process. Other techniques for melting titanium alloys continue to be evaluated. Electroslag melting, induction slag melting, induction skull melting, and other processes are being evaluated as techniques that may produce improved properties for the most demanding titanium applications.

Defects in Titanium Ingots

Historically, defects have been a troublesome part of titanium ingot metallurgy development. The sources of these defects were varied, the types of defects have now been recognized. The resulting strict process controls agreed upon jointly by metal suppliers and customers alike have done much to attain either reduced-defect or defect-free materials.

A prime source of defects in alloys is segregation.

Segregation in titanium ingot must be controlled because it leads to several different types of imperfections that cannot be readily eliminated by homogenizing heat treatments or combinations of heat treatment and primary mill processing.

Type I imperfections, usually called "high interstitial defects," are regions of interstitially stabilized alpha phase that have substantially higher hardness and lower ductility than the surrounding material. These regions also exhibit a higher beta transus temperature. They arise from very high nitrogen or oxygen concentrations in sponge, master alloy, or revert. Type I imperfections frequently, but not always, are associated with voids or cracks. (See Figure 4.4(a) and (b).) Although Type I imperfections sometimes are referred to as "low-density inclusions," they often are of higher density than is normal for the alloy.

Type II imperfections, sometimes called "high-aluminum defects," are abnormally stabilized alpha-phase areas that may extend across several beta grains. (See Figure 4.4(c).) Type II imperfections are caused by segregation of metallic alpha stabilizers, such as aluminum. They contain an excessively high proportion of primary alpha having a microhardness only slightly higher than that of the adjacent matrix. Type II imperfections sometimes are accompanied by adjacent stringers of beta—areas low in both aluminum content and hardness. This condition, shown in Figure 4.4(d), is generally associated with closed-solidification pipe into which alloy constituents of high vapor pressure migrate, only to be incorporated into the microstructure during primary mill fabrication. Stringers normally occur in the top portions of ingots and can be detected by macroetching or blue etch anodizing. Material containing stringers usually must undergo metallographic review to ensure that the indications revealed by etching are not artifacts.

Beta flecks, another type of imperfection, are small regions of stabilized beta in material that has been alpha-beta processed and heat treated. In size, they are equal to, or greater than, prior beta grains. (See Figure 4.4(e).) Beta flecks are either devoid of primary alpha or contain less than some specified minimum level of primary alpha. They are

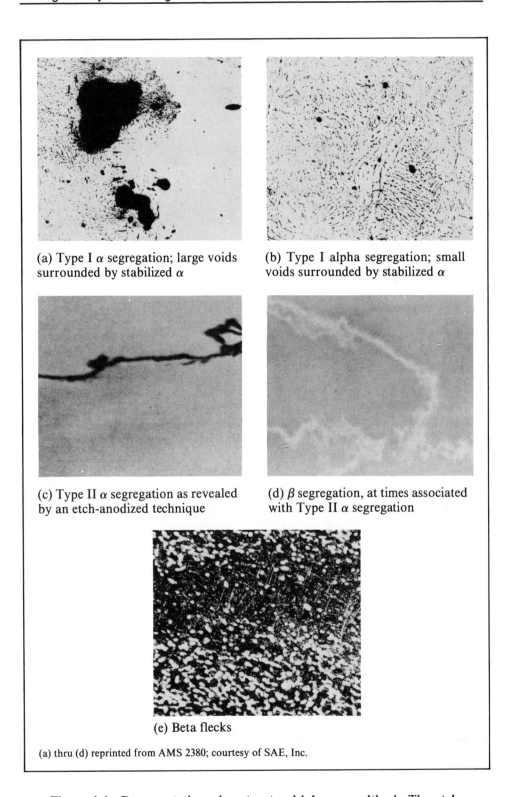

(a) Type I α segregation; large voids surrounded by stabilized α

(b) Type I alpha segregation; small voids surrounded by stabilized α

(c) Type II α segregation as revealed by an etch-anodized technique

(d) β segregation, at times associated with Type II α segregation

(e) Beta flecks

(a) thru (d) reprinted from AMS 2380; courtesy of SAE, Inc.

Figure 4.4 - Representative microstructural inhomogeneities in Ti metals

caused by localized regions either abnormally high in beta-stabilizer content or abnormally low in alpha-stabilizer content. Beta flecks are attributed to microsegregation during solidification of ingots of alloys that contain strong beta stabilizers. They are most often found in products made from large-diameter ingots. Beta flecks also may be found in beta-lean alloys such as Ti-6Al-4V that have been heated to a temperature near the beta transus during processing.

Type I and Type II imperfections are not acceptable in aircraft-grade titanium because they degrade critical design properties. Beta flecks are not considered harmful in alloys lean in beta stabilizers if they are to be used in the annealed condition. On the other hand, they constitute regions that incompletely respond to heat treatment, and for this reason microstructural standards have been established for allowable limits on beta flecks in various alpha-beta alloys. Beta flecks are more objectionable in beta-rich alpha-beta alloys than in leaner alloys.

PRIMARY FABRICATION

General Aspects

Primary fabrication includes all operations that convert ingot into general mill products—billet, bar, plate, sheet, strip, extrusions, tube, and wire. These mill products can readily be utilized in secondary manufacture of parts and structures. (See Figure 4.1.)

Primary fabrication is very important in establishing final properties, because many secondary fabrication operations may have little or no effect on metallurgical characteristics. Some secondary fabrication processes, such as forging and ring rolling, do impart sufficient reduction to play the major role in establishing material properties.

Reduction to Billet

Generally, the first breakdown of production ingot is a press cogging operation done in the beta temperature range. Modern processes utilize substantial amounts of working below the beta transus to produce billets with refined structures. These processes are carried out at temperatures high in the alpha region to allow greater reduction and improved grain refinement with a minimum of surface rupturing. Where maximum fracture toughness is required, beta processing (or alpha-beta processing followed by beta heat treatment) generally is preferred. Table 4.2 gives standard forging-temperature ranges for the manufacture of billet stock.

Some billets intended for further forging, rolling, or extrusion go through a grain-refinement process. This technique utilizes titanium's characteristic to recrystallize when it is heated above the beta transus. By starting with grain-refined billet, secondary fabricators may be able to produce forgings that meet strict requirements with respect to macrostructure, microstructure, and mechanical properties, without extensive hot working below the beta transus.

Final tensile properties of alpha-beta alloys are strongly influenced by the amount of processing in the alpha-beta field—both below the beta transus temperature and after recrystallization. Such processing increases the strength of high-alpha grades in large section sizes. With modern processing techniques, billet and forged sections readily meet specified tensile properties prior to final forging. Table 4.3 shows how billet and

Table 4.2 Standard forging temperatures for several titanium metals

Alloy	Beta transus		Ingot Breakdown	
	°C	°F	°C	°F
CP Ti:				
Grades 1 to 4	900-955	1650-1750	955-980	1750-1800
Alpha or near-alpha alloys:				
Ti-5Al-2.5Sn	1030	1890	1120-1175	2050-2150
Ti-6Al-2Sn-4Zr-2Mo-0.08Si	995	1820	1095-1150	2000-2100
Ti-8Al-1Mo-1V	1040	1900	1120-1175	2050-2150
Alpha-beta alloys:				
Ti-8Mn	800	1475	925-980	1700-1800
Ti-6Al-4V	995	1820	1095-1150	2000-2100
Ti-6Al-6V-2Sn	945	1735	1040-1095	1900-2000
Ti-7Al-4Mo	1005	1840	1120-1175	2050-2150
Beta alloy:				
Ti-13V-11Cr-3Al	720	1325	1120-1175	2050-2150
	Intermediate		**Finish**	
Alloy	°C	°F	°C	°F
CP Ti:				
Grades 1 to 4	900-925	1650-1700	815-900	1500-1650
Alpha or near-alpha alloys:				
Ti-5Al-2.5Sn	1065-1095	1950-2000	1010-1040	1850-1900
Ti-6Al-2Sn-4Zr-2Mo-0.08Si	1010-1065	1850-1950	955-980	1750-1800
Ti-8Al-1Mo-1V	1065-1095	1950-2000	1010-104	1850-1900
Alpha-beta alloys:				
Ti-8Mn	845-900	1550-1650	815-845	1500-1550
Ti-6Al-4V	980-1040	1800-1900	925-980	1700-1800
Ti-6Al-6V-2Sn	955-1010	1750-1850	870-940	1600-1725
Ti-7Al-4Mo	1010-1065	1850-1950	955-980	1750-1800
Beta alloy:				
Ti-13V-11Cr-3Al	1010-1065	1325	925-980	1700-1800

forging section size affects room-temperature tensile properties of various titanium alloys.

Rolling of Bar, Plate, Sheet

Roll cogging and hot roll finishing of bar, plate, and sheet are now standard operations, and special rolling and auxiliary equipment have been installed by the larger titanium producers to allow close control of all rolling operations. Rolling processes used by each manufacturer are proprietary and in some respects unique, but because all techniques must produce the same specified structures and mechanical properties, a high degree of similarity exists among the processes of all manufacturers.

A representative range of temperatures used for hot rolling of titanium metals is presented in Table 4.4. Rolling at these temperatures produces end products with the desired grain structures.

Bars up to about 100 mm (4 in.) in diameter are unidirectionally rolled, and their properties commonly reflect total reduction in the alpha-beta range. For example, a round bar 50 mm (2 in.) in diameter rolled from a Ti-6Al-4V billet 100 mm square typically is 140 to 170 MPa (20 to 25 ksi) lower in tensile strength than rod 7.8 mm (5/16 in.) in diameter rolled on a rod mill from a billet of the same size at the same rolling temperatures. For bars about 50 to 100 mm (2 to 4 in.) in diameter, strength does not decrease with section size, but transverse ductility and notched stress-rupture strength at room temperature do become lower. In diameters greater than about 75 to 100 mm (3 to 4 in.), annealed Ti-6Al-4V bars usually do not meet prescribed limits for stress-rupture at room temperature — 1170 MPa (170 ksi), min, to cause rupture of a notched specimen in 5 h — unless the material is given a special duplex anneal. Transverse ductility is lower in bars about 65 to 100 mm (2.5 to 4 in.) in diameter because it is not possible to obtain the preferred texture throughout bars of this size.

Table 4.3 Variation of typical room-temperature tensile properties with section size for four titanium alloys

Section size (a)		Tensile strength		Yield strength		Elongation (b), %	Reduction in area, %
mm	in.	MPa	ksi	MPa	ksi		
6Al-4V (c):							
25-50	1-2	1015	147	965	140	14	36
102	4	1000	145	930	135	12	25
205	8	965	140	895	130	11	23
330	13	930	135	860	125	10	20
6Al-4V-ELI (c):							
25-50	1-2	950	138	885	128	14	36
102	4	885	128	827	120	12	28
205	8	885	128	820	119	10	27
330	13	870	126	795	115	10	22
6Al-6V-2Sn (c):							
25-50	1-2	1105	160	1035	150	15	40
102	4	1070	155	965	145	13	35
205	8	1000	145	930	135	12	25
8Al-1Mo-1V:							
25-50	1-2(d)	985	143	905	131	15	36
102	4(e)	910	132	840	122	17	35
205	8(f)	1000	145	895	130	12	23
6Al-2Sn-4Zr-2Mo+Si (g):							
25-50	1-2	1000	145	930	135	14	33
102	4	1000	145	930	135	12	30
205	8	1035	150	940	136	12	28
330	13	1000	145	825	120	11	21

(a) Properties are in longitudinal direction for sections 50 mm (2 in.) or less, and in transverse direction for sections 100 mm (4 in.) or more, in section size.
(b) In 50 mm or 2 in.
(c) Annealed 2 h at 700°C (1300°F) and air cooled.
(d) Annealed 1 h at 900°C (1650°F), air cooled, then heated 8 h at 600°C (1100°F) and air cooled.
(e) Annealed 1 h at 1010°C (1850°F), air cooled, then heated to 566°C (1050°F).
(f) Annealed 1 h at 1010°C (1850°F) and oil quenched.
(g) Annealed 1 h at 954°C (1750°F), air cooled, then heated 8 h to 600°C (1100°F) and air cooled.

Table 4.4 Typical rolling temperatures for several titanium metals

Alloy	Bar °C	Bar °F	Plate °C	Plate °F	Sheet °C	Sheet °F
	Rolling temperatures					
CP Ti Grades 1 to 4	760-815	1400-1500	760-790	1400-1450	705-760	1300-1400
Alpha, near-alpha alloys: Ti-5Al-2.5Sn	1010-1065	1850-1950	980-1040	1800-1900	980-1010	1800-1850
Ti-6Al-2Sn-4Zr-2Mo	955-1010	1750-1850	955-980	1750-1800	925-980	1700-1800
Ti-8Al-1Mo-1V	1010-1040	1850-1900	980-1040	1800-1900	980-1040	1800-1900
Alpha-beta alloys: Ti-8Mn	-----	---	705-760	1300-1400	705-760	1300-1400
Ti-4Al-3Mo-1V	925-955	1700-1750	900-925	1650-1700	900-925	1650-1700
Ti-6Al-4V	955-1010	1750-1850	925-980	1700-1800	900-925	1650-1700
Ti-6Al-6V-2Sn	900-955	1650-1750	870-925	1600-1700	870-900	1600-1650
Ti-7Al-4Mo	955-1010	1750-1850	925-955	1700-1750	925-955	1700-1750
Beta alloy: Ti-13V-11Cr-3Al	955-1065	1750-1950	980-1040	1800-1900	730-900	1350-1650

Plate and sheet may exhibit the same tensile properties in both the transverse and longitudinal directions relative to the final rolling direction. With the precise control systems (or types of titanium) now available, texturing and directionality can be obtained in alpha-beta sheet by unidirectional rolling. These characteristics favorably affect tensile properties of Ti-6Al-4V sheet in various gages. (See Table 4.5.) Other properties, such as fatigue resistance, also are improved by this type of rolling.

Directionality in properties generally is observed only as a slight drop in transverse ductility of plate greater than 25 mm (1 in.) thick. Military (MIL spec), AMS (Aerospace Material Specifications), and customer specifications all prescribe lower minimum ten-

Table 4.5 Tensile properties of unidirectionally rolled Ti-6Al-4V sheet

Gage mm	Gage in.	Tensile strength MPa	Tensile strength ksi	Yield strength MPa	Yield strength ksi	Elongation (a), %	Tensile modulus GPa	Tensile modulus 10^6 psi
Longitudinal direction:								
0.737	0.029	945	137	870	126	7.0	100	14.5
1.016	0.040	970	141	855	124	6.5	106	15.4
1.168	0.046	915	133	860	125	6.5	105	15.2
1.524	0.060	985	143	925	134	6.5	104	15.1
1.778	0.070	995	144	915	133	8.0	105	15.3
Transverse direction:								
0.737	0.029	1105	160	1061	154	7.5	130	18.8
1.016	0.040	1195	173	1105	160	7.5	145	21.1
1.168	0.046	1225	178	1165	169	7.5	140	20.2
1.524	0.060	1125	163	1090	158	8.0	125	18.2
1.778	0.070	1095	159	1055	153	9.5	135	19.5
(a) In 50 mm or 2 in.								

sile and yield strengths as plate thickness increases. For forming applications, some customers specify a maximum allowable difference between tensile strengths in the transverse and longitudinal directions.

SECONDARY FABRICATION

General Aspects

Secondary fabrication refers to manufacturing processes such as die forging, extrusion, hot and cold forming, machining, chemical milling, and joining, all of which are used for producing finished parts from mill products. Each of these processes may strongly influence properties of titanium and its alloys, either alone or by interacting with effects of processes to which the metal has previously been subjected.

Die Forging

One of the main purposes of die forging, in addition to shape control, is to obtain a combination of mechanical properties that generally does not exist in bar or billet. Tensile strength, creep resistance, fatigue strength, and toughness all may be better in forgings than in bar or other forms.

Forging is a common method of producing wrought titanium alloy articles. Forging sequences and subsequent heat treatment can be used to control the microstructure and resulting properties of the product. Forging is more than just a shapemaking process. The key to successful forging and heat treatment is the beta transus temperature. Figure 4.5 shows, schematically, the possible locations for temperature of forging and/or heat treatment of a typical alpha-beta alloy such as Ti-6Al-4V. The higher the processing temperature in the alpha plus beta region, the more beta is available to transform on cooling. On quenching from above the beta transus, a completely transformed, acicular structure arises. The exact form of the globular (equiaxed) alpha and the transformed beta structures produced by processing depends on the exact location of the beta transus, which varies from heat to heat of a given alloy, and also on the degree and nature of deformation produced.

Section size is important, and the number of working operations can be significant. Conventional forging may require two or three operations, whereas isothermal forging may require only one. A schematic representation of a conventional forging and subsequent heat treatment sequence is shown in Figure 4.6. The solution heat treatment offers a chance to modify, or tune, the as-forged microstructure, while the aging cycle modifies the transformed beta structures to an optimum dispersion.

Microstructural control is basic to successful processing of titanium alloys. Undesirable structures (grain-boundary alpha, beta fleck, "spaghetti" or elongated alpha) can interfere with optimum property development. Titanium ingot structures can carry over to affect the forged product. Beta processing, despite its adverse effects on some mechanical properties, can reduce forging costs, while isothermal forging offers a means of reducing forging pressures and/or improving die fill and part detail. Isothermal beta forging is finding use in the production of more creep-resistant components of titanium alloys.

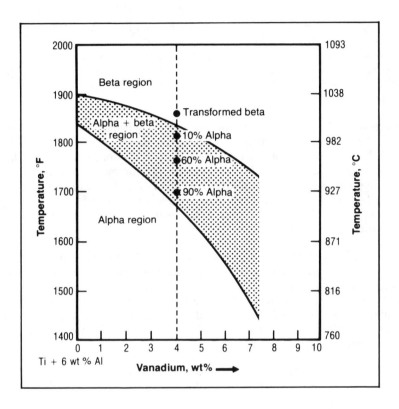

Figure 4.5 - Phase diagram that predicts results of forging or heat treatment practice

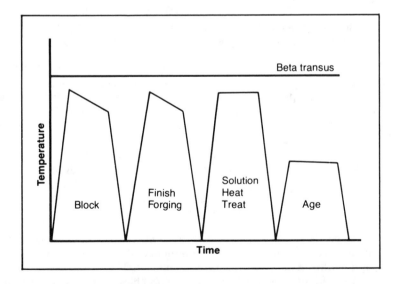

Figure 4.6 - Schematic diagram of a conventional forging and subsequent heat treatment sequence for producing an alpha-beta structure

Reheating of alpha-beta titanium alloys after hot working can substantially alter the microstructure. Careful attention must be paid to the development of microstructure right through the heat treatment steps. Superficially similar microstructures may not produce the same levels of mechanical properties. Solution heat treatment and aging of non-worked or insufficiently worked structures will not produce optimum strengths or toughness in titanium alloys.

The effects of different thermomechanical processing schedules on the mechanical properties and the corresponding structures of three titanium alloys (Ti-6Al-4V, Ti-6Al-6V-2Sn, and Ti-6Al-2Sn-4Zr-6Mo) may be considered to illustrate the effects of processing schedules on properties. Table 4.6 summarizes four thermomechanical schedules that produced optimum combinations of properties in Ti-6Al-4V test forgings: excellent tensile strength; good-to-excellent notch fatigue strength; low-cycle fatigue strength; and fracture toughness. Also included in the table are three schedules that produced subnormal properties.

The microstructures of Ti-6Al-4V shown in Figure 4.7 correspond to two of the schedules that produced good combinations of properties and two that produced inferior combinations. Note the substantial difference in microstructure in the same final product—which, in combination with the resulting properties, demonstrates that control of thermomechanical processing can control the microstructures and corresponding final properties of forgings. Read the description in Figure 4.7's footnote carefully to understand this statement.

Figure 4.8 summarizes the results of an extensive study of alpha-beta forging versus beta forging for several titanium alloys. Although yield strength after beta forging was not always as high as that after alpha-beta forging, values of notch tensile strength and fracture toughness were consistently higher for the beta-forged material.

The beta-forged alloys tend to show a transformed beta or acicular microstructure, whereas alpha-beta forged alloys show a more equiaxed structure. This latter structure generally is as shown in Figure 4.7(a). Tradeoffs are required for each structural type (acicular vs. equiaxed) since each structure has unique capabilities. Table 4.7 shows the relative advantages of equiaxed and acicular microstructures.

A titanium die forging alternative procedure involves the use of precision isothermal (sometimes superplastic) forging techniques. Isothermal forging is a process in which the material being forged is held at essentially constant temperature conditions, and thus does not undergo the thermal fluctuations of heat-up and cool-down which are experienced several times in a conventional forging sequence. Identical forging presses may be used, although there may also be a press dedicated to isothermal forging. There is, however, a difference in the dies used. Die block materials have been a significant concern in the development of the isothermal forging processes. With the advent of more readily forgeable near-beta alloys such as Ti-10V-2Fe-3Al, it is thought that isothermal precision forging of titanium alloys to very near net shape is going to be a common procedure, for suitable alloys and at a premium cost, in the future.

The near-net-shape (NNS) concept is the motivational basis of the isothermal forging process. This implies, of course, a desire to minimize the amount of costly machining that must be done to produce the finished component. It also implies a desire not to absorb costs of material that will only be scrapped as chips. Alpha-beta alloy isothermal forging is technically feasible, although high process and tooling costs, catastrophic die failures, and other engineering problems associated with very high process temperatures combine to minimize its use on conventional alpha-beta alloys.

Table 4.6 Thermochemical schedules for producing various combinations of properties in Ti-6Al-4V forgings

Initial microstructure	Blocker forging temperature range	Finish forging temperature range	Finish forging reduction	Cooling after forging	Heat treated condition	Final microstructure
Best combinations of properties						
...	Alpha-beta	Alpha-beta	...	Air cooled	Annealed	6% equiaxed alpha plus fine platelet alpha
Grain-boundary alpha	Alpha-beta	Alpha-beta	...	Air cooled	Annealed	26% elongated partly broken up grain-boundary primary alpha plus fine platelet alpha
Grain-boundary alpha	Alpha-beta	Alpha-beta	...	Water quenched	Annealed	23% elongated partly broken up primary alpha plus very fine platelet alpha
...	Beta	Alpha-beta	10%	Air cooled	Annealed	63% fine elongated primary alpha plus fine platelet alpha
Subnormal properties						
Spaghetti alpha	Alpha	Alpha	...	Air cooled	STOA	25% blocky primary alpha plates plus very fine platelet alpha
...	Beta	Alpha-beta	10%	Water quenched	STOA	43% coarse elongated primary alpha plates plus very fine platelet alpha
...	Beta	Beta	...	Slow cooled	Annealed	92% alpha basket-weave structure

(a) 6% equiaxed primary alpha plus fine platelet alpha in Ti-6Al-4V alpha-beta forged, then annealed 2 h at 705°C (1300°F) and air cooled.

(b) 23% elongated, partly broken up primary alpha plus grain-boundary alpha in Ti-6Al-4V, alpha-beta forged and water quenched, then annealed 2 h at 705°C and air cooled.

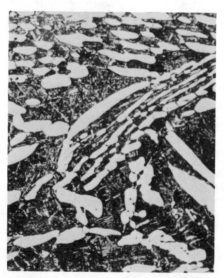

(c) 25% blocky (spaghetti) alpha plates plus very fine platelet alpha in Ti-6Al-4V alpha-beta forged from a spaghetti-alpha starting structure, then solution treated 1 h at 955°C (1750°F) and reannealed 2 h at 705°C

(d) 92% alpha basket-weave structure in Ti-6Al-4V beta forged and slow cooled, then annealed 2 h at 705°C.

Figure 4.7 - Microstructures corresponding to different combinations of properties in Ti-6Al-4V forgings

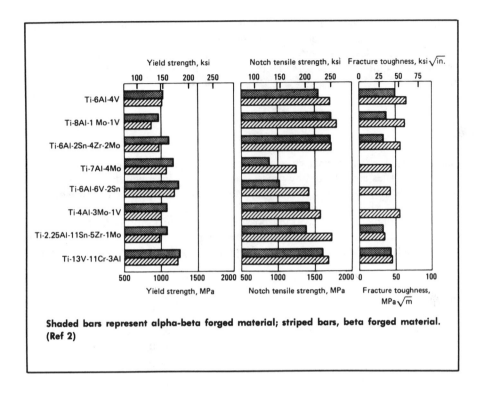

Figure 4.8 - Comparison of mechanical properties of alpha-beta forged and of beta forged titanium alloys

Table 4.7 Relative advantages of equiaxed and acicular microstructures

Equiaxed:

* Higher ductility and formability
* Higher threshold stress for hot salt stress corrosion
* Higher strength (for equivalent heat treatment)
* Better hydrogen tolerance
* Better low-cycle fatigue (initiation) properties

Acicular:

* Superior creep properties
* Higher fracture-toughness values

Isothermal forging can be performed in the alpha-beta or the beta phase field. Microstructures can be controlled quite accurately. Thus property uniformity should be better than that which is achieved by means of conventional forging. Although initially isothermal forging was applied to billet material, the technology can be applied to P/M ingot or to a preform. (A preform is a P/M consolidation which has a shape designed to be somewhat like the final shape to maximize die-filling capability in the forging process and to minimize pressures and excess material. Refer to Chapter 8, "Powder Metallurgy," for more details.)

According to some sources, isothermal forging of titanium has been a production deformation process available for quite some time. As its capabilities become better recognized, and as alloy development proceeds, or, alternatively, as property, manufacturing, and economic requirements become more stringent, there probably will be increased use of this technique.

Extrusion

Extrusion is used as an alternative to rolling as a mill process in order to make rodlike products. Properties are affected by processing conditions in much the same way as they are for rolled or forged products. The properties of extruded products, however, are not identical to those of die-forged structures. Even where similar microstructures are produced, the thermomechanical working possible in open- and closed-die forging permits much more control over the resultant properties. One of the more unusual applications of extrusion has been in the production of tapered wing spars for a military aircraft.

Forming

Titanium and titanium alloy sheet and plate are strain hardened by cold forming. This normally increases tensile and yield strengths, and causes a slight drop in ductility. Beta alloys generally are easier to form than are alpha and alpha-beta alloys.

Titanium metals exhibit a high degree of springback in cold forming. To overcome this characteristic, titanium must be extensively overformed or, as is done most frequently, hot sized after cold forming.

Hot forming does not greatly affect final properties. Forming at temperatures from 595° to 815°C (1100° to 1500°F) allows the material to deform more readily, and, simultaneously, it stress relieves the deformed material. It also minimizes the degree of springback. The true net effect in any forming operation depends on total deformation and actual temperature during forming. Titanium metals also tend to creep at elevated temperatures; holding under load at the forming temperature (creep forming) is another alternative for achieving the desired shape without having to compensate for extensive springback.

In all forming operations, titanium and its alloys are susceptible to the Bauschinger effect. This is a drop in compressive yield strength in one loading direction accompanied by an increase in tensile strength in another direction due to strain hardening. The Bauschinger effect is most pronounced at room temperature: plastic deformation (1 to 5% tensile elongation) at room temperature always introduces a significant loss in compressive yield strength, regardless of the initial heat treatment or strength of the alloys. At 2% tensile strain, for instance, the compressive yield strengths of Ti-4Al-3Mo-1V and Ti-6Al-4V drop to less than half the values for solution-treated material. Increasing the temperature reduces the Bauschinger effect; subsequent full thermal stress relieving completely removes it.

Temperatures as low as the aging temperature remove most of the Bauschinger effect in solution-treated titanium alloys. Heating or plastic deformation at temperatures above the normal aging temperature for solution-treated Ti-6Al-4V causes overaging to occur and, as a result, all mechanical properties decrease.

Chapter 5

Heat Treating

GENERAL BACKGROUND

Titanium and titanium alloys are heat treated for the following purposes:

- To reduce residual stresses developed during fabrication, a process called stress relieving

- To produce an acceptable combination of ductility, machinability, and dimensional and structural stability, especially in alpha-beta alloys under less stringent processing rules than those used to generate optimum strength or special property combinations, a process called annealing

- To increase strength by means of a process that combines solution treatment and aging

- To optimize special properties such as fracture toughness, fatigue strength, and high-temperature creep strength

The response of titanium and titanium alloys to heat treatment depends on the composition of the metal and, by extension, upon the metal's characterization as alpha, alpha-beta, or beta. (Remember: near-alpha and/or superalpha, along with near-beta are also used as designations. Refer to Chapter 3, "Understanding Ti's Metallurgy.")

Alpha and near-alpha titanium alloys can be stress relieved and annealed, but high strength cannot be developed in the alpha alloys by heat treatment. Depending upon the exact definition of near-alpha in mind, some of the near-alpha alloys such as Ti-8Al-1Mo-1V can be solution treated and aged in order to develop higher strengths. Solution treatment plus aging is used to produce maximum strengths in alpha-beta alloys. In commercial (metastable) beta alloys, stress-relieving and aging treatments can be combined and, also, annealing and solution treating may be identical operations.

Not all heat-treating cycles are applicable to all titanium alloys, because the various alloys are designed for different purposes. For example:

- Alloys Ti-5Al-2Sn-2Zr-4Mo-4Cr (commonly called Ti-17) and Ti-6Al-2Sn-4Zr-6Mo are designed for strength in heavy sections.

• Ti-6Al-2Sn-4Zr-2Mo is designed for creep resistance.

• Ti-6Al-2Cb-1Ta-1Mo and Ti-6Al-4V-ELI are designed both to resist stress corrosion in aqueous salt solutions and for high fracture toughness.

• Ti-5Al-2.5Sn is designed for weldability.

• Ti-6Al-6V-2Sn, Ti-6Al-4V, and Ti-10V-2Fe-3Al are designed for high strength at low-to-moderate temperatures.

(Refer to Appendix J, "Designations, Applications, Properties," for more details.)

Actual heat-treat cycles for each alloy vary. Any heat treatment at temperatures about 427°C (800°F) should provide the titanium with an atmospheric protection that prevents pickup of oxygen or nitrogen and alpha case formation. The protection also obviates the possibility of undesirable scale formation. (See later discussions in this chapter concerning contamination during heat treatment.)

STRESS RELIEVING

Titanium and titanium alloys can be stress relieved without adversely affecting strength or ductility. Stress-relieving treatments decrease the undesirable residual stresses that result from:

• Nonuniform hot forging deformation from cold forming and straightening

• Asymmetric machining of plate (hogouts) or forgings

• Welding of wrought or cast articles and cooling of castings

Removal of such stresses helps maintain shape stability and eliminates unfavorable conditions such as the loss of compressive yield strength, commonly known as the Bauschinger effect.

Separate stress relieving may be omitted when the manufacturing sequence can be adjusted to employ annealing or hardening as the stress-relieving process. For example, forging stresses may be relieved by annealing prior to machining. Large, thin rings have been effectively processed with minimum distortion by rough machining in the annealed state. This is followed by solution treating, quenching, partial aging, finish machining, and final aging. Partial aging relieves quenching stresses, and final aging relieves stresses developed during finish machining.

Table 5-1 presents combinations of time and temperature that are used for stress relieving titanium and titanium alloys. The ranges in both time and temperature indicate that more than one combination may yield satisfactory results. For effective stress relief, the high temperatures usually are used with shorter times and the lower temperatures with longer times. During stress relief of solution-treated and aged titanium alloys, care should be taken to prevent overaging which would cause lower strength. This usually involves selection of a time-temperature combination that provides partial stress relief. The parts, in bulk or in fixtures, may be charged directly into a furnace operating at the stress-relief temperature. If a part is mounted in a massive fixture, a thermocouple should be attached to the largest part of the fixture.

Table 5.1 Selected stress-relief treatments for titanium and titanium alloys (a)

Alloy	°C	°F	Time, h
Commercially pure Ti (All Grades)	480-595	900-1100	1/4-4
Alpha or near-alpha titanium alloys:			
Ti-5Al-2.5Sn	540-650	1000-1200	1/4-4
Ti-8Al-1Mo-1V	595-705	1100-1300	1/4-4
Ti-6Al-2Sn-4Zr-2Mo	595-705	1100-1300	1/4-4
Ti-6Al-2Cb-1Ta-0.8Mo	595-650	1100-1200	1/4-2
Ti-0.3Mo-0.8Ni (Ti Code 12)	480-595	900-1100	1/4-4
Alpha-beta titanium alloys:			
Ti-6Al-4V	480-650	900-1200	1-4
Ti-6Al-6V-2Sn (Cu + Fe)	480-650	900-1200	1-4
Ti-3Al-2.5V	540-650	1000-1200	1/2-2
Ti-6Al-2Sn-4Zr-6Mo	595-705	1100-1300	1/4-4
Ti-5Al-2Sn-4Mo-2Zr-4Cr (Ti-17)	480-650	900-1200	1-4
Ti-7Al-4Mo	480-705	900-1300	1-8
Ti-6Al-2Sn-2Zr-2Mo-2Cr-0.25Si	480-650	900-1200	1-4
Ti-8Mn	480-595	900-1100	1/4-2
Beta or near-beta titanium alloys:			
Ti-13V-11Cr-3Al	705-730	1300-1350	1/12-1/4
Ti-11.5Mo-6Zr-4.5Sn (Beta III)	720-730	1325-1350	1/12-1/4
Ti-3Al-8V-6Cr-4Zr-4Mo (Beta C)	705-760	1300-1400	1/6-1/2
Ti-10V-2Fe-3Al	675-705	1250-1300	1/2-2
Ti-15V-3Al-3Cr-3Sn	790-815	1450-1500	1/12-1/4

(a) Parts can be cooled from stress relief by either air cooling or slow cooling.

The effects of stress relieving Ti-6Al-4V at five temperatures ranging from 260° to 620°C (500° to 1150°F) for periods of time ranging from 5 min to 50 h are illustrated in Figure 5.1.

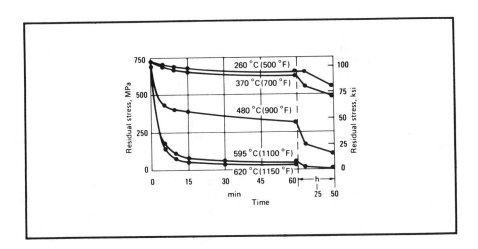

Figure 5.1 - Relation between time and relief of residual stress at various temperatures for Ti-6Al-4V

The rate of cooling from the stress-relieving temperature is not critical. Uniformity of cooling, however, is critical, particularly in the temperature range from 480° to 315°C (900° to 600°F). Oil or water quenching should not be used to accelerate cooling, however, because this can induce residual stress by unequal cooling. Either furnace or air cooling is acceptable.

Stress-relieving treatments must be based on the metallurgical response of the alloy involved. Generally, this requires holding at a temperature sufficiently high to relieve stresses without causing an undesirable amount of precipitation or strain aging in alpha-beta and beta alloys, or without producing undesirable recrystallization in single-phase alloys that rely on cold work for strength.

Stress relieving of the more highly alloyed alpha-beta compositions, and of beta alloys, should be performed using a thermal exposure that is compatible with annealing, solution-treating, stabilization, or aging processes.

There are no nondestructive testing methods that can measure the efficiency of a stress-relief cycle other than direct measurement of residual stresses by X-ray diffraction. No significant changes in microstructure due to stress-relieving heat treatments can be detected by optical microscopy.

The temperatures used for stress relieving complex weldments of alpha or alpha-beta alloys should be near the high ends of the ranges given in Table 5.1. Complex weldments may be defined as those having multiple welds in complex configurations, ones possibly involving combinations of machine and manual welding. In complex weldments made with commercially pure titanium, Ti-5Al-2.5Sn alloy, or Ti-6Al-4V alloy, more than 70% of the residual stress is relieved during the first hour at temperature. (Simple weldments of commercially pure titanium often are used without stress relief.)

PROCESS ANNEALING

General

The term "annealing" is so generic that solution treatment is frequently considered an "annealing" process as are mill annealing and recyrstallization annealing. Even stress relief frequently is called stress-relief annealing. For the purposes of description, the techniques which serve primarily to increase toughness, ductility at room temperature, dimensional and thermal stability, and, sometimes, creep resistance can be described as process annealing. In this chapter, the term is frequently shortened to annealing.

Since process annealing treatments usually are less closely controlled, more property variability, or "scatter," is found in annealed titanium alloys than in solution-treated and aged alloys.

Many titanium alloys are placed in service in the annealed state. Because improvement in one or more properties generally is obtained at the expense of some other property, the annealing cycle should be selected according to the objective of the treatment. Common process annealing treatments are:

- Mill annealing
- Recrystallization annealing
- Beta annealing

At times multiple anneals are performed. Such instances are termed duplex and triplex annealing. They frequently are used in the context of solution treatment and aging. Duplex heat treatment usually does not occur in the area of process annealing. (For more details refer to the solution treating and aging discussion, below.)

Mill annealing is a general-purpose treatment given to all mill products. It is not a full anneal and may leave traces of cold or warm working in the microstructures of heavily worked products (particularly sheet). Recrystallization treatments are used to improve toughness. The alloy is heated into the upper end of the alpha-beta range, held for a time, and then very slowly cooled. Beta annealing uses a slow cool from temperatures above the beta transus of the alloy being annealed. Some representative annealing treatments for titanium and titanium alloys are given in Table 5.2.

Table 5.2 Selected annealing treatments for titanium and titanium alloys

	Temperature			
Alloy	°C	°F	Time, h	Cooling method
Commercially pure Ti (All Grades)	650-760	1200-1400	1/10-2	Air
Alpha or near-alpha titanium alloys:				
Ti-5Al-2.4Sn	720-845	1325-1550	1/6-4	Air
Ti-8Al-1Mo-1V	790 (a)	1450 (a)	1-8	Air or furnace
Ti-6Al-2Sn-4Zr-2Mo	900 (b)	1650 (b)	1/2-1	Air
Ti-6Al-2Cb-1Ta-0.8Mo	790-900	1450-1650	1-4	Air
Alpha-beta titanium alloys:				
Ti-6Al-4V	705-790	1300-1450	1-4	Air or furnace
Ti-6Al-6V-2Sn (Cu + Fe)	705-815	1300-1500	3/4-4	Air or furnace
Ti-3Al-2.5V	650-760	1200-1400	1/2-2	Air
Ti-6Al-2Sn-4Zr-6Mo	(c)	(c)	-	-
Ti-5Al-2Sn-4Mo-2Zr-4Cr (Ti-17)	(c)	(c)	-	-
Ti-7Al-4Mo	705-790	1300-1450	1-8	Air
Ti-6Al-2Sn-2Zr-2Mo-2Cr-0.25Si	705-815	1300-1500	1-2	Air
Ti-8Mn	650-760	1200-1400	1/2-1	(d)
Beta or near-beta titanium alloys:				
Ti-13V-11Cr-3Al	705-790	1300-1450	1/6-1	Air or water
Ti-11.5Mo-6Zr-4.5Sn (Beta III)	690-760	1275-1400	1/6-1	Air or water
Ti-3Al-8V-6Cr-4Zr-4Mo (Beta C)	790-815	1450-1500	1/4-1	Air or water
Ti-10V-2Fe-3Al	(c)	(c)	-	-
Ti-15V-3Al-3Cr-3Sn	790-815	1450-1500	1/12-1/4	Air

(a) For sheet and plate, follow by 1/4 h at 790°C (1450°F), then air cool.
(b) For sheet, follow by 1/4 h at 790°C (1450°F), then air cool (plus 2 h at 595°C or 1100°F, then air cool, in certain applications). For plate, follow by 8 h at 595°C (1100°F), then air cool.
(c) Not normally supplied or used in annealed condition.
(d) Furnace or slow cool to 540°C (1000°F), then air cool.

Straightening, sizing, and flattening may be combined with annealing (or stress relief) by use of appropriate fixtures. The parts, in bulk or in fixtures, may be charged directly into a furnace operating at the annealing temperature.

Either air or furnace cooling may be used, but the two methods may result in different levels of tensile properties. For example, air cooling of Ti-6Al-6V-2Sn from the mill-annealing temperature results in lower tensile strength than that obtained by furnace cooling. If distortion is a problem, the cooling rate should be uniform down to 315°C (600°F).

Stability

In alpha-beta titanium alloys, thermal stability is a function of beta-phase transformations. During cooling from the annealing temperature, beta may transform and may, under certain conditions and in certain alloys, form the (brittle) intermediate phase omega. A stabilization annealing treatment is designed to produce a stable beta phase capable of resisting further transformation when exposed to elevated temperatures in service. Alpha-beta alloys that are lean in beta, such as Ti-6Al-4V, can be air cooled from the annealing temperature without impairing their stability. Furnace (slow) cooling may promote formation of Ti_3Al, a reaction that can degrade resistance to stress corrosion.

In the case of alloys which are solution treated and then aged, the aging treatment may be regarded, in some cases, as a stabilization heat treatment.

Straightening During Annealing

It may be difficult to prevent distortion of close-tolerance, thin sections during annealing. Straightening of bar to close tolerances, along with flattening of sheet, present major problems for titanium producers and fabricators. Because of springback and resistance to straightening at room temperature, it is necessary to employ elevated-temperature forming. At annealing temperatures, many titanium alloys have creep resistance low enough to permit straightening during annealing. With proper fixturing, and in some instances with judicious weighting, sheet-metal fabrications and thin, complex forgings have been straightened with satisfactory results. Again, uniformity of cooling to below 315°C (600°F) can improve results.

Various jigs and processing techniques have been proposed for annealing titanium in a manner that yields a flat product. "Creep flattening" and "vacuum creep flattening" are two such techniques. Creep flattening consists of heating titanium sheet between two clean, flat sheets of steel in a furnace containing an oxidizing or inert atmosphere. Vacuum creep flattening is used to produce stress-free flat plate for subsequent machining. The plate is placed on a large, flat, ceramic bed that has integral electric heating elements. Insulation is placed on top of the plate, and a plastic sheet is sealed to the frame. The bed is slowly heated to the annealing temperature while a vacuum is pulled under the plastic. Atmospheric pressure is used to creep flatten the plate.

SOLUTION TREATING AND AGING

General

A wide range of strength levels can be obtained in alpha-beta or beta alloys by solution treating and aging. The origin of heat-treating responses of titanium alloys lies in the instability of the high-temperature beta phase at lower temperatures. Solution treatment and aging (stabilization) usually, but not always, follow working operations to generate optimum and mechanical properties. Heating an alpha-beta alloy to the solution-treating temperature produces a higher ratio of beta phase. This partitioning of phases is maintained by quenching; on subsequent aging, decomposition of the unstable beta phase and of the martensite (if any) occurs. Commercial beta alloys, generally supplied in the solution-treated condition, need only be aged to achieve properties.

After being cleaned, titanium components should be loaded into fixtures or racks that permit free access to the heating and quenching media. Thick and thin components of the same alloy may be solution treated together, but the time at temperature (soaking time) is determined by the thickest section. To determine the required temperature for most alloys, the rule is 20 to 30 min/in. of thickness. This is followed by the required soak time.

Time-temperature combinations for solution treating are given in Table 5.3. A load may be charged directly into a furnace operating at the solution-treating temperature. Although preheating is not essential, it may be used to minimize distortion of complex parts.

Table 5.3 Selected solution treating and aging (stabilizing) treatments for titanium alloys

Alloy	Solution temperature °C	Solution temperature °F	Solution time, h	Cooling method
Alpha or near-alpha alloys:				
Ti-8Al-1Mo-1V	980-1010(a)	1800-1850(a)	1	Oil or water
Ti-6Al-2Sn-4Zr-2Mo	955-980	1750-1800	1	Air
Alpha-beta alloys:				
Ti-6Al-4V	955-970(b)(c)	1750-1775(b)(c)	1	Water
	955-970	1750-1775	1	Water
Ti-6Al-6V-2Sn (Cu + Fe)	885-910	1625-1675	1	Water
Ti-6Al-2Sn-4Zr-6Mo	845-890	1550-1650	1	Air
Ti-5Al-2Sn-2Zr-4Mo-4Cr	845-870	1550-1600	1	Air
Ti-6Al-2Sn-2Zr-2Mo-2Cr-0.25Si	870-925	1600-1700	1	Water
Beta or near-beta alloys:				
Ti-13V-11Cr-3Al	775-800	1425-1475	1/4-1	Air or water
Ti-11.5Mo-6Zr-4.5Sn (Beta III)	690-790	1275-1450	1/8-1	Air or water
Ti-3Al-8V-6Cr-4Mo-4Zr (Beta C)	815-925	1500-1700	1	Water
Ti-10V-2Fe-3Al	760-780	1400-1435	1	Water
Ti-15V-3Al-3Cr-3Sn	790-815	1450-1500	1/4	Air

Alloy	Aging temperature °C	Aging temperature °F	Aging time, h
Alpha or near-alpha alloys:			
Ti-8Al-1Mo-1V	565-595	1050-1100	-
Ti-6Al-2Sn-4Zr-2Mo	595	1100	8
Alpha-beta alloys:			
Ti-6Al-4V	480-595	900-1100	4-8
	705-760	1300-1400	2-4
Ti-6Al-6V-2Sn (Cu + Fe)	480-595	900-1100	4-8
Ti-6Al-2Sn-4Zr-6Mo	580-605	1075-1125	4-8
Ti-5Al-2Sn-2Zr-4Mo-4Cr	580-605	1075-1125	4-8
Ti-6Al-2Sn-2Zr-2Mo-2Cr-0.25Si	480-595	900-1100	4-8
Beta or near-beta alloys:			
Ti-13V-11Cr-3Al	425-480	800-900	4-100
Ti-11.5Mo-6Zr-4.5Sn (Beta III)	480-595	900-1100	8-32
Ti-3Al-8V-6Cr-4Mo-4Zr (Beta C)	455-540	850-1000	8-24
Ti-10V-2Fe-3Al	495-525	925-975	8
Ti-15V-3Al-3Cr-3Sn	510-595	950-1100	8-24

(a) For certain products, use solution temperature of 890°C (1650°F) for 1 h, then air cool or faster.
(b) For thin plate or sheet, solution temperature can be used down to 890°C (1650°F) for 6 to 30 min, then water quench.
(c) This treatment is used to develop maximum tensile properties in this alloy.

Solution Treating

To obtain high strength with adequate ductility, it generally is necessary to solution treat at a temperature high in the alpha-beta field, normally 28° to 83°C (50° to 150°F) below the beta transus of the alloy. If higher fracture toughness or improved resistance to stress corrosion is required, beta annealing or beta solution treating may be desirable. A change in the solution-treating temperature of alpha-beta alloys alters the amount of beta phase and, consequently, changes the response to aging. (See Table 5.4.) Selection of solution-treating temperature usually is based upon practical considerations such as the desired level of tensile properties and the amount of ductility to be obtained after aging.

Table 5.4 Variation of tensile properties of Ti-6Al-4V bar stock with solution-treating temperature

Solution-treating temperature		Room-temperature tensile properties (a)				Elongation in 4D,
		Tensile strength		Yield strength (b)		
°C	°F	MPa	ksi	MPa	ksi	%
845	1550	1025	149	980	142	18
870	1600	1060	154	985	143	17
900	1650	1095	159	995	144	16
925	1700	1110	161	1000	145	16
940	1725	1140	165	1055	153	16

(a) Properties determined on 13-mm (1/2-in.) bar after soltution treating, quenching, and aging. Aging treatment 8 h at 480°C (900°F), air cool.
(b) At 0.2% offset

Because alpha-beta solution treating involves heating to temperatures only slightly below the beta transus, proper control of temperature is essential. If the beta transus is exceeded, tensile properties (especially ductility) are reduced and cannot be fully restored by subsequent thermal treatment. The beta transus temperatures for commercial alloys are listed in Table 5.5.

Solution-treating temperatures for beta alloys may be above the beta transus. Beta alloys normally are obtained from producers in the solution-treated condition. If reheating is required, soak times should be only as long as necessary to obtain complete solutioning because grain growth can proceed rapidly under these conditions, since no second phase is present. For lean beta content (near-beta) alloys, solution heat treatment may have to be carried out below the beta transus (alpha-beta anneal). Such a solution-treated product contains globular alpha plus retained beta. The final aged product would contain a bimodal alpha distribution (primary alpha plus alpha from aging).

Quenching

The rate of cooling from the solution-treating temperature for alpha-beta alloys has an important effect on strength. If the rate is too low, appreciable diffusion may occur during cooling, and decomposition of the altered beta phase during aging may not provide effective strengthening.

Table 5.5 Beta transformation temperatures of titanium alloys

Alloy	Beta transus °C(±15°)	°F(±25°)
Commercially pure Ti, 0.25 max O	910	1675
Commercially pure Ti, 0.40 max O	945	1735
Alpha or near-alpha alloys:		
Ti-5Al-2.5Sn	1050	1925
Ti-8Al-1Mo-1V	1040	1900
Ti-6Al-2Sn-4Zr-2Mo	995	1820
Ti-6Al-2Cb-1Ta-0.8Mo	1015	1860
Ti-0.3Mo-0.8Ni (Ti Code 12)	880	1615
Alpha-beta alloys:		
Ti-6Al-4V	1000 (a)	1830 (b)
Ti-6Al-6V-2Sn (Cu + Fe)	945	1735
Ti-3Al-2.5V	935	1715
Ti-6Al-2Sn-4Zr-6Mo	940	1720
Ti-5Al-2Sn-4Mo-2Zr-4Cr (Ti-17)	900	1650
Ti-7Al-4Mo	1000	1840
Ti-6Al-2Sn-2Zr-2Mo-2Cr-0.25Si	970	1780
Ti-8Mn	800 (c)	1475 (d)
Beta or near-beta alloys:		
Ti-13V-11Cr-3Al	720	1330
Ti-11.5Mo-6Zr-4.5Sn (Beta III)	760	1400
Ti-3Al-8V-6Cr-4Zr-4Mo (Beta C)	795	1460
Ti-10V-2Fe-3Al	805	1480
Ti-15V-3Al-3Cr-3Sn	760	1400

(a) ±20°
(b) ±30°
(c) ±35°
(d) ±50°

For alloys relatively high in beta-stabilizer content and for products of small section size air or fan cooling may be adequate; such slow cooling, where allowed by specified mechanical properties, is preferred because it minimizes distortion. Beta alloys generally are air quenched from the solution-treated temperature.

Water, or a 5% brine, or a caustic soda solution is preferred for quenching alpha-beta alloys, because these quenchants provide cooling rates necessary to prevent decomposition of the beta phase obtained by solution treating, as well as to provide maximum response to aging. The need for rapid quenching is further emphasized by short quench-delay-time requirements. Depending on the mass of the sections being heat treated, some alpha-beta alloys can only tolerate a maximum delay of 7 sec, whereas more highly beta-stabilized alloys can tolerate quench delay times of up to 20 sec. For example, the effect of quench delays on Ti-6Al-4V bar is shown in Figure 5.2.

Less sensitive to delayed quenching are alloys such as Ti-6Al-2Sn-4Zr-6Mo, and Ti-5Al-2Sn-2Zr-4Mo-4Cr in which fan air cooling develops good strength through 100-mm (4-in.) sections.

Section size influences effectiveness of quenching and, in turn, response to aging. The amount and type of beta stabilizer in the alloy determine depth of hardening or strengthening. Unless the alloy is highly alloyed with beta stabilizers, thick sections exhibit

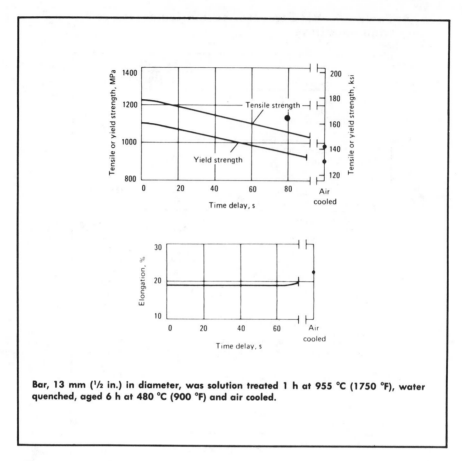

Figure 5.2 - Effects of quench delay on tensile properties of Ti-6Al-4V bar

lower tensile properties. The practical significance of section size for some alloys is shown in Table 5.6. The effects of quenched section size on the tensile properties of Ti-6Al-4V alloy are illustrated in Figure 5.3.

Aging

The final step in heat treating titanium alloys to high strength consists of reheating to an aging temperature between 425° and 650°C (800° and 1200°F). Aging causes decomposition of the supersaturated beta phase retained on quenching and the transformation of any martensite to alpha. A summary of aging times and temperatures is presented in Table 5.3. The time-temperature combination selected for a specific alloy composition depends on required strength.

Aging above the standard aging temperature for an alloy, yet still several hundred degrees below the beta transus temperature, results in overaging. This condition, called solution treated and overaged (STOA), sometimes is used to obtain modest increases in strength while maintaining satisfactory toughness and dimensional stability.

Table 5.6 Effect of section size on tensile strength of solution-treated and aged titanium alloys

Alloy	Tensile strength of square bar in section size of:											
	13 mm (1/2 in.)		25 mm (1 in.)		50 mm (2 in.)		75 mm (3 in.)		100 mm (4 in.)		150 mm (6 in.)	
	MPa	ksi	MPa	ksi	MPa	ksi	MPA	ksi	MPa	ksi	MPa	ksi
Ti-6Al-4V	1105	160	1070	155	1000	145	930	135	-	-	-	-
Ti-6Al-6V-2Sn (Cu+Fe)	1205	175	1205	175	1070	155	1035	150	-	-	-	-
Ti-6Al-2Sn-4Zr-6Mo	1170	170	1170	170	1170	170	1140	165	1105	160	-	-
Ti-5Al-2Sn-2Zr-4Mo-4Cr (Ti-17)	1170	170	1170	170	1170	170	1105	160	1105	160	1105	160
Ti-10V-2Fe-3Al	1240	180	1240	180	1240	180	1240	180	1170	170	1170	170
Ti-13V-11Cr-3Al	1310	190	1310	190	1310	190	1310	190	1310	190	1310	190
Ti-11.5Mo-6Zr-4.5Sn (Beta III)	1310	190	1310	190	1310	190	1310	190	1310	190	-	-
Ti-3Al-8V-6Cr-4Zr-4Mo (Beta C)	1310	190	1310	190	1240	180	1240	180	1170	170	1170	170

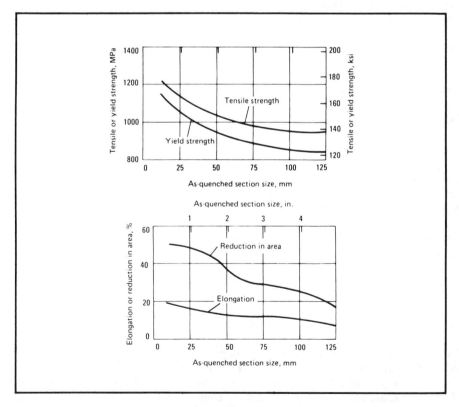

Figure 5.3 - Effects of quenched section size on tensile properties of Ti-6Al-4V

Although the aged condition is not necessarily one of equilibrium, proper aging produces high strength with adequate ductility and metallurgical stability. Heat treatment of alpha-beta alloys for high strength frequently involves a series of compromises and modifications, depending on the type of service and on special properties that are required, such as ductility and suitability for fabrication. This has become especially true where fracture toughness is important in design and where strength is lowered to improve design life.

During aging of some highly beta-stabilized alpha-beta alloys, beta transforms first to a metastable transition phase referred to as omega phase before the equilibrium alpha phase is produced. Retained omega phase, which produces brittleness unacceptable in alloys heat treated for service, can be avoided by severe quenching and rapid reheating to aging temperatures above 425°C (800°F). Because a coarse alpha phase forms, however, this treatment might not produce optimum strength properties. An aging practice that ensures that aging time and temperature are adequate to carry out any omega reaction to completion usually is employed. Aging above 425°C (800°F) generally is adequate to complete the reaction.

The metastable beta alloys usually do not require solution treatment. Final hot working, followed by air cooling, leaves these alloys in a condition comparable to a solution-treated state. In some instances, however, solution treating at 790°C (1450°F) has produced better uniformity of properties after aging. Aging at 480°C (900°F) for 8 to 60 h produces tensile strengths of 1.10 to 1.38 GPa (160 to 200 ksi). Aging for times longer than 60 h may provide higher strengths, but decrease ductility and fracture toughness in alloys containing chromium, when titanium-chromium compounds are formed. Short aging times can be used on cold-worked material to produce a significant increase in strength over that obtained by cold working.

CONTAMINATION AND ATMOSPHERIC EFFECTS

General

Titanium reacts with the oxygen, water, and carbon dioxide normally found in the oxidizing heat-treating atmospheres. It also reacts with hydrogen formed by decomposition of water vapor. Unless the heat treatment is performed in a vacuum furnace or in an inert atmosphere, and unless surface cleanliness is maintained, there is a direct effect on the titanium's properties. While properties can be recovered by vacuum heat treatment or stock removal, depending on the situation, it is more efficient to prevent or to minimize interactions through the surface, where possible. Even when, for example, coatings are used in forging to protect as well as to lubricate a billet, some oxygen/nitrogen pickup occurs, and stock removal is required. In some cases, surface contamination can render a piece unfit for use. (Refer to the discussion of chlorides, below.)

Contamination During Heat Treatment

Before being subjected to any thermal treatment, titanium components should be cleaned and dried. Caution must be taken not to use ordinary tap water for cleaning such components. Oil, fingerprints, grease, paint, and other foreign matter should be removed from all surfaces. Cleaning is required because the chemical reactivity of titanium at elevated temperatures can lead to its contamination or embrittlement and can increase its susceptibility to stress corrosion. After cleaning, parts should be handled with clean gloves to prevent recontamination. If a component is to be sized, straightened, or heat treated in a fixture, the fixture also should be free of any foreign matter and loosely adhering scale.

Titanium is chemically active at elevated temperatures and will oxidize in air, resulting in the formation of a scale. However, oxidation is not of primary concern in heat treating of titanium, although it may be a problem in sheet-forming operations.

Oxygen (or nitrogen) pickup during heat treatment results in a surface structure composed predominantly of alpha phase; oxygen and nitrogen are alpha stabilizers. Of the two alpha case formers, oxygen is the more potent. Oxygen is absorbed at a much greater rate than nitrogen. Alpha case is detrimental because of the brittle nature of the oxygen-enriched alpha structure, which also is very abrasive to either carbide or high-speed steel machine tools. At 955°C (1750°F), the alpha structure can extend 0.2 to 0.3 mm (0.008 to 0.012 in.) below the surface; it must be removed. (See Figure 7.1 for a microsection showing alpha case.)

An antioxidant spray coating may be applied beforehand to clean sheet metal parts in order to minimize oxygen pickup during heat treatment. Such coatings work effectively at temperatures up to about 760°C (1400°F), but such use does not fully eliminate the need for removing the surface structure after heat treating.

The danger of hydrogen pickup is of greater importance than that of oxidation since hydrogen does not create a visible surface condition which may be used as a check against excess hydrogen. Current specifications limit hydrogen content to a maximum of 125 to 200 ppm, depending on alloy and mill form. Above these limits, hydrogen embrittles some titanium alloys, thereby reducing impact strength and notch tensile strength and causing delayed cracking.

Hydrogen Pickup

With the exceptions of high vacuum, salt baths, and chemically inert gases such as argon, all heat-treating atmospheres contain some hydrogen at temperatures used for annealing titanium. Hydrocarbon fuels produce hydrogen as a by-product of incomplete combustion; electric furnaces with air atmospheres contain hydrogen from breakdown of water vapor. However, because small amounts of hydrogen can be tolerated in titanium and because inert media are expensive, most titanium heat-treating operations are performed in conventional furnaces employing oxidizing atmospheres with at least 5% excess oxygen in the flue gas.

An oxidizing atmosphere serves in two ways to reduce hydrogen pickup: it reduces the partial pressure of hydrogen in the surrounding atmosphere, and it provides the titanium with a protective surface oxide that retards hydrogen pickup.

Oxidation Rates

Titanium alloy oxidation rates vary considerably. A comparison of the scaling rates of commercially pure titanium and titanium alloys in air at temperatures from 650° to 980°C (1200°C to 1800°F) is given in Figure 5.4. Table 5.7 indicates the measurable thickness of oxide formed on commercially pure titanium after one-half hour at various temperatures in air.

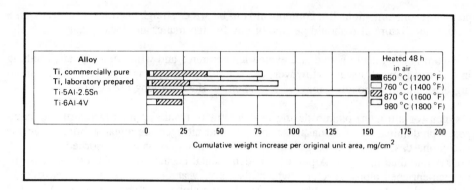

Figure 5.4 - Scaling rates of Ti and Ti alloys in air at various temperatures

Table 5.7 Thickness of oxide on CP Ti (a)

Temperature		Measurable thickness	
°C	°F	mm	in.
315	600	None	None
425	800	None	None
540	1000	None	None
650	1200	<0.005	<0.0002
705	1300	0.005	0.0002
760	1400	0.008	0.0003
815	1500	<0.025	<0.001
870	1600	<0.025	<0.001
925	1700	<0.05	<0.002
980	1800	0.05	0.002
1040	1900	0.10	0.004
1095	2000	0.36	0.014
(a) Heated for one-half hour in air.			

Nitrogen

Nitrogen is absorbed by titanium during heat treatment at a much slower rate than oxygen and thus does not present a serious contamination problem. Dry nitrogen has been used successfully as a lower-cost protective atmosphere for heat treating of titanium forgings that are to be fully machined after treatment. If absorbed in sufficient quantities, however, nitrogen forms a hard, brittle compound.

Carbon Monoxide and Carbon Dioxide

The gases CO and CO_2 decompose in the presence of hot titanium and produce surface oxidation. (Refer to the discussion of contamination during heat treatment, above.)

Chlorides

Titanium alloys are subject to stress corrosion when parts with high residual stress are exposed to chlorides at temperatures above 290°C (550°F). Salt from fingerprints, and the chlorides contained in some degreasing solutions, may cause stress-corrosion cracking at temperatures above 315°C (600°F). Although this phenomenon is readily produced in laboratory testing, and is known to occur during heat treatment, hot-salt cracking in service has not been a significant problem. Care is required during thermal processing to ensure freedom from chloride contamination.

POST HEAT-TREATING REQUIREMENTS AFFECTED BY ATMOSPHERIC INTERACTIONS

Oxygen and nitrogen react with the titanium at the metal's surface and produce an oxygen-enriched layer commonly called "alpha case." This brittle layer must be removed before the component is put into service. It can be removed by machining, but certain machining operations may result in excessive tool wear. The standard practice is to remove alpha case by other mechanical methods or by chemical methods, or by both. Oxidation rates of commercial titanium alloys vary. Table 5.8 can be used as a guide to determine how much metal should be removed. Temperature and total time at temperature must be known.

Table 5.8 Estimated min metal removal required after thermal exposure (in air) of titanium alloys

Heat treating temperature		Time at temperature,	Minimum stock removal per surface (a)	
°C	°F	h	mm	in
480-593	900-1100	Up to 12	0.005	0.0002
594-648	1101-1200	Up to 4	0.008	0.0003
		4 to 12	0.015	0.0006
649-704	1201-1300	Up to 1	0.013	0.0005
		1 to 8	0.020	0.0008
		8 to 12	0.025	0.0010
705-760	1301-1400	Up to 1	0.025	0.0010
		1 to 4	0.036	0.0014
		4 to 8	0.038	0.0015
		8 to 12	0.043	0.0017
761-787	1401-1450	Up to 1	0.030	0.0012
		1 to 2	0.038	0.0015
		2 to 4	0.046	0.0018
		4 to 8	0.051	0.0020
		8 to 12	0.056	0.0022
788-815	1451-1500	Up to 1/2	0.036	0.0014
		1/2 to 1	0.041	0.0016
		1 to 2	0.051	0.0020
816-871	1501-1600	Up to 1/2	0.058	0.0023
		1/2 to 1	0.066	0.0026
		1 to 2	0.076	0.0030
872-898	1601-1650	Up to 1/2	0.058	0.0023
		1/2 to 1	0.081	0.0032
		1 to 2	0.089	0.0035
899-926	1651-1700	Up to 1/2	0.086	0.0034
		1/2 to 1	0.091	0.0036
		1 to 2	0.107	0.0042
927-954	1701-1750	Up to 1/2	0.097	0.0038
		1/2 to 1	0.107	0.0042
		1 to 2	0.122	0.0048

(a) Values shown are typical; actual values may vary with alloy type.

One method to check for complete removal of alpha case is to etch the component with a solution composed of 18 g of ammonium bifluoride per litre of water (2.4 oz/gal). The presence or absence of alpha case is detected by the difference in etching characteristics: light gray shows the presence of alpha case; dark gray indicates its absence. If the component such as a forging has been machined, the ammonium bifluoride treatment must by preceded by etching in a solution consisting nominally of 5% HF, 30% min HNO_3, and the balance water. For other mill products, such as plate, microexamination of representative samples removed from the plate commonly is used.

Small amounts of hydrogen (100 to 200 ppm) can be tolerated in titanium alloys with the specific limiting amount determined by the type of alloy. High hydrogen content can lead to premature failure of a component. Hydrogen pickup occurs not only during heat treatment but also during pickling or chemical cleaning operations used to remove alpha case. The amount of hydrogen pickup can only be determined by chemical analysis. If high hydrogen content is found, vacuum annealing is required. A typical vacuum annealing cycle consists of heating at, or close to, the annealing temperature for 2 to 4 h in a vacuum of not less than 10 μm.

Hardness testing is not recommended as a nondestructive method of checking the efficiency of heat treatment. The correlation between strength and hardness is poor. Whenever verification of a property is required, the appropriate mechanical test should be used.

GROWTH DURING HEAT TREATMENT

Solution treating of large parts requires allowances for growth during heat treatment. The growth due to heating may be retained after cooling, and this growth may be increased either by longer holding times at solution temperature or by lower heating rates. Table 5.9 gives examples of net growth of Ti-6Al-4V specimens heated to 955°C (1750°F).

Table 5.9 Effect of heating rate and time at 955 °C (1750°F) on growth of Ti-6Al-4V (a)

Mill heat (b)	Heating rate °C/min	°F/min	Holding time (c), h	Net growth (d), %
A	3.3	6	0	0.27
B	3.3	6	0	0.22
A	3.3	6	1	0.60
B	3.3	6	1	0.49
A	3.3	6	2	1.00
B	3.3	6	2	0.90 (e)
B	10	18	1	0.32
B (f)	10	18	1	0.35

(a) Test conditions: 50-mm (2-in.) specimens were taken in the longitudinal direction (except where otherwise indicated) from material annealed 2 h at 705°C (1300°F) and air cooled. No growth was observed in specimens tested during annealing.
(b) Beta transus temperatures (determined metallographically) were 990°C (1810°F) for heat A and 1015°C (1860°F) for heat B.
(c) All specimens water quenched after holding for time indicated.
(d) As determined by Leitz-Wetzler dilatometer.
(e) Calculated from curve.
(f) Specimen taken in transverse direction.

HOT ISOSTATIC PRESSING

Hot isostatic pressing (HIP) has become an accepted method for the closure of solidification or gas porosity in titanium castings. Since thermal conditions between 899° and 954°C (1650° and 1750°F) are used for 2 to 4 h, HIP clearly functions as a heat treatment. HIP is conducted in a heated, argon-filled pressure vessel, usually at pressures of 10 to 15 ksi. The temperatures used are in the high end of the alpha-beta range for the few alloys (principally Ti-6Al-4V) which are being cast plus HIPed. Heat treatment after HIP generally is close to, yet below, the beta transus. Cooling rate from HIP can affect subsequent post-heat treatment properties of titanium alloys. (See Chapter 7, "Castings," for more information on the effects of HIP.)

Chapter 6

Machining

INTRODUCTION TO MACHINING

Many of titanium's material and component design characteristics make it expensive to machine. A considerable amount of stock must be removed from primary forms such as forgings, plates, bars, etc. In some instances, as much as 50 to 90% of the primary form's weight ends up as chips. (The complexity of some finished parts, such as bulk-heads, makes difficult the use of near-net-shape methods that would minimize chip forming.) Maximum machining efficiency for titanium alloys is required to minimize the costs of stock removal.

Historically, titanium has been perceived as a material that is difficult to machine. Due to titanium's growing acceptance in many industries, along with the experience gained by progressive fabricators, a broad base of titanium machining knowledge now exists. Manufacturers now know that, with proper procedures, titanium can be fabricated using techniques no more difficult than those used for machining 316 stainless steel.

Stories about problems encountered when machining titanium have usually originated in shops working with aircraft alloys. The fact is that commercially pure grades of titanium (ASTM B, Grades 1, 2, 3, and 4) with tensile strengths of 241 to 552 MPa (35 to 80 ksi) machine much easier than the aircraft alloys (i.e. ASTM B, Grade 5: Ti-6Al-4V).

With higher alloy content and hardness, the machinability of titanium alloys by traditional chip-making methods generally decreases. (This is true of most other metals.) At a hardness level over 38 R_C (350 BHN) increased difficulty in operations such as drilling, tapping, milling, and broaching can be expected. In general, however, if the particular characteristics of titanium are taken into account, the machining of titanium and its alloys should not present undue problems.

Machining of titanium alloys requires cutting forces only slightly higher than those needed to machine steels, but these alloys have metallurgical characteristics that make them somewhat more difficult to machine than steels of equivalent hardness. The beta alloys are the most difficult titanium alloys to machine. When machining conditions are selected properly for a specific alloy composition and processing sequence, reasonable production rates of machining can be achieved at acceptable cost levels.

Care must be exercised to avoid loss of surface integrity, especially during grinding; otherwise a dramatic loss in mechanical behavior such as fatigue can result. To date, techniques such as high-speed machining have not improved the machinability of titanium. A breakthrough appears to require the development of new tool materials.

CHARACTERISTICS INFLUENCING MACHINABILITY

The fact that titanium sometimes is classified as difficult to machine by traditional methods in part can be explained by the physical, chemical, and mechanical properties of the metal. For example:

- Titanium is a poor conductor of heat. Heat, generated by the cutting action, does not dissipate quickly. Therefore, most of the heat is concentrated on the cutting edge and the tool face.

- Titanium has a strong alloying tendency or chemical reactivity with materials in the cutting tools at tool operating temperatures. This causes galling, welding, and smearing along with rapid destruction of the cutting tool.

- Titanium has a relatively low modulus of elasticity, thereby having more "springiness" than steel. Work has a tendency to move away from the cutting tool unless heavy cuts are maintained or proper backup is employed. Slender parts tend to deflect under tool pressures, causing chatter, tool rubbing, and tolerance problems. Rigidity of the entire system is consequently very important, as is the use of sharp, properly shaped cutting tools.

- Titanium's fatigue properties are strongly influenced by a tendency to surface damage if certain machining techniques are used. Care must be exercised to avoid the loss of surface integrity, especially during grinding. (This characteristic is described in greater detail below.)

- Titanium's work-hardening characteristics are such that titanium alloys demonstrate a complete absence of "built-up edge." Because of the lack of a stationary mass of metal (built-up edge) ahead of the cutting tool, a high shearing angle is formed. This causes a thin chip to contact a relatively small area on the cutting tool face and results in high bearing loads per unit area. The high bearing force, combined with the friction developed by the chip as it rushes over the bearing area, results in a great increase in heat on a very localized portion of the cutting tool. Furthermore, the combination of high bearing forces and heat produces cratering action close to the cutting edge, resulting in rapid tool breakdown.

With respect to titanium's fatigue properties, briefly noted in the above list, the following details are of interest.

As stated, loss of surface integrity must be avoided. If this precaution is not observed, a dramatic loss of mechanical behavior (such as fatigue) can result. Even proper grinding practices using conventional parameters (wheel speed, downfeed, etc.) may result in appreciably lower fatigue strength due to surface damage. The basic fatigue properties of many titanium alloys rely on a favorable compressive surface stress induced by tool action during machining. Electromechanical removal of material, producing a stress-free surface, can cause a debit from the customary design fatigue strength properties. (These results are similar when mechanical processes such as grinding are involved, although the reasons are different.)

TRADITIONAL MACHINING OF TITANIUM

General

The term "machining" has broad application and refers to all types of metal removal and cutting processes. These include turning, boring, milling, drilling, reaming, tapping, both sawing and gas cutting, broaching, planing, gear hobbing, shaping, shaving, and grinding.

The technology supporting the machining of titanium alloys basically is very similar to that for other alloy systems. Efficient metal machining requires access to data relating the machining parameters of a cutting tool to the work material for the given operation. The important parameters include:

- Tool life

- Forces

- Power requirements

- Cutting tools and fluids

Subsequent paragraphs discuss these parameters in general terms.

Tool Life

Tool-life data have been developed experimentally for a wide variety of titanium alloys. A common way of representing such data is shown in Figure 6.1 where tool life (as time) is plotted against cutting speed (fpm) for a given cutting tool material at a constant feed and depth in relation to Ti-6Al-4V. It can be seen that at a high cutting speed, tool life is extremely short. As the cutting speed decreases, tool life dramatically increases.

Titanium alloys are very sensitive to changes in feed, as in Figure 6.1. Industry generally operates at cutting speeds providing long tool life. Curve fitting of tool life to feed, speed, and other machining parameters is commonly being done by means of computer techniques. However, in cases where no data base exists, certain rules of thumb should be recognized. For example, when cutting titanium, a high shear angle is produced between the workpiece and chip, resulting in a thin chip flowing at high velocity over the tool face. High temperatures develop, and, since titanium has low thermal conductivity, the chips have a tendency to gall and weld themselves to the tool cutting edges. This speeds up tool wear and failure. When dealing with high-fixed-cost machine tools, production output may be much more important than a cutting tool's life! It thus may be wise to work a tool at its maximum capacity, and then to replace it as soon as its cutting efficiency starts to drop off noticeably, thereby maintaining uptime as much as possible.

When machining titanium in circumstances in which production costs are not of paramount concern, it is still unsound practice to allow tools to run to destruction. The other extreme, premature tool changing, may result in a low number of pieces per tool grind, but the lower the tool wear, the less expensive the regrinding.

Ideally, a tool should be permitted to continue cutting as long as possible without risking damage to the tool or the work but with the retention of surface integrity. The only way

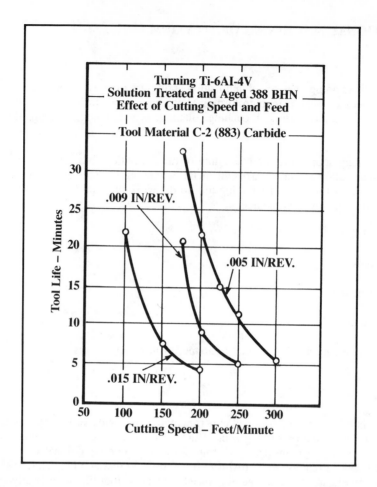

Figure 6.1 - Effects of cutting speed and feed on tool life when turning Ti-6Al-4V

to find a safe stopping point is to check a few runs by counting the pieces produced and inspecting the surface finish, dimensions, and surface integrity. In this manner it can be established how many acceptable pieces can be produced before the tool fails.

Forces and Power Requirements

The forces in machining can be determined with a tool dynamometer. In turning, the tool dynamometer usually measures three components:

- Tangential, or cutting force

- Thrust, or separating force

- Feed, or axial force

The cutting force is important since, when multiplied by the cutting velocity, it determines the power requirements in machining. The thrust, or separating force, determines the accuracy produced on a part.

For general approximations, the power requirements in turning and milling can be obtained by measuring the power input to the machine tool's drive motor during a cutting operation and by subtracting from it the tare, or idle power. A good approximation of the horsepower required in most machining operations can be predicted from unit power requirements. Table 6.1 shows the power requirements for titanium in comparison to other alloys.

Table 6.1 Average unit power requirements for turning, drilling, or milling of titanium compared with other competitive alloy systems

Material	Hardness Bhn (3000 kg)	Unit power for sharp tools (a) $hp/in.^3/min$		
		Turning HSS and carbide tools	Drilling HSS drills	Milling HSS and carbide tools
Steels	35-40 R_c	1.4	1.4	1.5
Titanium alloys	250-375	1.2	1.1	1.1
High-temperature nickel and cobalt base alloys	200-360	2.5	2.0	2.0
Aluminum alloys	30-150 (500 kg)	0.25	0.16	0.32

(a) Power requirements at spindle drive motor, corrected to 80% spindle drive efficiency. Dull tools may require 25% more power.

Cutting Tools

Major improvements in the rate at which workpieces are machined usually result from the development and application of new tool materials. In the past several years, there have been major advancements in the development of cutting tools including coated carbides, ceramics, cermets, cubic boron nitride, silicon nitride, and polycrystalline diamond. These have found useful applications in the machining of cast irons, steels, and high-temperature and aluminum alloys.

Unfortunately, none of these or other new materials has improved the removal rate of titanium alloys. In studies conducted as early as 1950, the straight tungsten carbide (WC) cutting tools, typically C-2 grades, performed best in operations such as turning and face milling, while the high-cobalt, high-speed steels were most applicable in drilling, tapping, and end milling.

Today, the situation is much the same. C-2 carbides are used extensively in engine and airframe manufacturing for turning and face milling operations. In recent years, in the United States as well as in Europe, solid C-2 end mills and end mills with replaceable C-2 carbides are finding applications, particularly in aerospace plants. Today, the M7 and, more frequently, the M42 and M33 high-speed steels are recommended for end milling, drilling, and tapping of titanium alloys.

Cutting Fluids

Cutting fluids used in machining titanium alloys require special consideration because chlorine ions have, under certain circumstances, caused stress-corrosion cracking in laboratory testing of these alloys for mechanical properties. Consequently, chlorine at one time was considered a suspect element regardless of the concentration and specific conditions used in manufacturing operations, such as machining.

When specifying cutting fluids for machining titanium, some companies have practically no restrictions other than using controlled-washing procedures on parts after machining. Other manufacturers do likewise, except that they do not use cutting fluids containing chlorine on parts which are subjected to higher temperatures in welding processes or in service. Also when assemblies are machined, the same restrictions apply because of the difficulty in doing a good cleaning job after machining. Still other organizations in aerospace manufacturing permit no active chlorine in any cutting fluid used for machining titanium alloys.

A program to define the effect of experimental chlorinated and sulfurized cutting fluids on the mechanical properties of the Ti-6Al-4V alloy (annealed, 34 R_C) was performed. Mechanical property evaluations included:

- High-cycle fatigue at both room and elevated temperatures

- Fatigue crack propagation at two cyclic frequencies

- Fracture toughness

- Stress-corrosion/surface-embrittlement exposures

Within the scope of the program, and within the range of variables investigated, the results indicated generally that no degradation of mechanical properties relative to those obtained from neutral cutting fluids occurred. Similar results were obtained by using chlorinated and sulfurized fluids in machining, or by having those cutting fluids present as an environment during testing. The use of chlorine-containing (or halogen-containing) cutting fluids generally is not a recommended practice, despite the above-noted results which pertain to only a single titanium alloy.

There are excellent cutting fluids available which do not contain any halogen compounds. In fact, from extensive test data collected by the Air Force Materials Laboratory, it can be concluded that chlorine-containing cutting fluids do not always provide better tool life. For certain alloys and operations, dry machining is preferred. Usually the heavy chlorine-bearing fluids excel in operations such as drilling, tapping, and broaching. Figure 6.2 shows the effect of various cutting fluids on tool life in drilling Ti-6Al-4V.

MACHINING DATA: SPEEDS AND FEEDS

Cutting speed and feed are two of the most important parameters for all types of machining operations. Extensive testing has developed the tool-life data, as illustrated in Figure 6.1, for turning Ti-6Al-4V. Tool life charts are available, as noted in Appendix C, "Machining Practices." One manufacturer offers the following general guidelines for typical machining operations.

Although the basic machining properties of titanium metal cannot be altered significantly, their effects can be greatly minimized by decreasing temperatures generated at

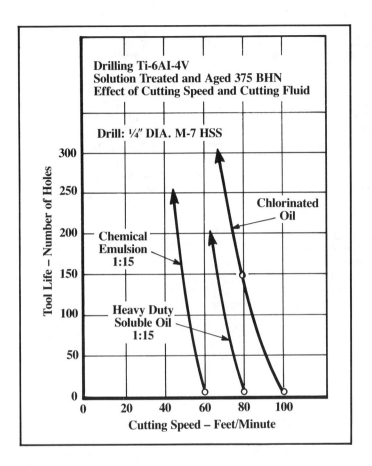

Figure 6.2 - Effects of various cutting fluids and speeds on tool life when drilling Ti-6Al-4V (375 Bhn)

the tool face and cutting edge. Economical production techniques have been developed through application of these basic rules in machining titanium:

- Use low cutting speeds. Tool tip temperatures are affected more by cutting speed than by any other single variable. A change from 6 to 46 meters per min (20 to 150 sfm) with carbide tools results in a temperature change from 427° to 927°C (800° to 1700°F).

- Maintain high feed rates. Temperature is not affected by feed rate so much as by speed, and the highest feed rates consistent with good machining practice should be used. A change from 0.05 to 0.51 mm (0.002 in. to 0.020 in.) per revolution results in a temperature increase of only 149°C (300°F).

- Use generous amounts of cutting fluid. Coolant carries away heat, washes away chips, and reduces cutting forces.

- Use sharp tools and replace them at the first sign of wear, or as determined by production/cost considerations. Tool wear is not linear when cutting titanium. Complete tool failure occurs rather quickly after a small initial amount of wear takes place.

• Never stop feeding while a tool and a workpiece are in moving contact. Permitting a tool to dwell in moving contact causes work hardening and promotes smearing, galling, seizing, and total tool breakdown.

Machining recommendations, such as noted above, may require modification to fit particular circumstances in a given shop. For example, cost, storage, or requirements may make it impractical to accommodate a very large number of different cutting fluids. Savings achieved by making a change in cutting fluid may be offset by the cost of changing fluids. Likewise, it may be uneconomical to inventory cutting tools which may have only infrequent use. Also, the design of parts may limit the rate of metal removal in order to minimize distortion (of thin flanges, for example) and to corner without excessive inertia effects.

An example of typical machining parameters currently used to machine Ti-6Al-4V bulkheads containing deep pockets, thin flanges, and floors at an important United States airframe manufacturer are shown in Table 6.2. A bulkhead frequently contains numerous pockets and some flanges as thin as 0.76 mm (0.030 in.). Typical example bulkhead rough forgings weigh in excess of 450 kg (1000 lb), but the finished part is less than 67.5 kg (150 lb) after machining. Extensive machining is done on gas turbine engine components, just as is done on the larger airframe components. Table 6.3 lists typical parameters for machining Ti-6Al-4V jet engine components such as fan disks, spacers, shafts, and rotating seals.

Table 6.2 Example of typical machining parameters currently used to machine Ti-6Al-4V airframe bulkheads

Operation	Part surface	Cutter description and material	Speed, ft/ min	Feed, in./ tooth
Milling rough/finish	Peripheral ML flanges	2" dia. x 6" flute length, 6 flute, 35° helix, M42	50	0.0066/0.0096
Milling rough	Thin flanges Walls	1-1/4" dia. x 2" flute length, 4 flute, 35° helix, M42	50	0.0062/0.009
Milling finish	Thin flanges	3/4" dia. x 2-1/2" flute length, 4 flute, 35° helix, M42	50	0.0024/0.0034
Milling finish	Pocket floor	1-1/4" dia. x 2" flute length, 4 flute, 35° helix, M42	50	0.0062/0.009

INCREASED PRODUCTIVITY BY SPECIAL TECHNIQUES

The inability to improve cutting-tool performance by developing new cutting-tool materials—coatings in particular—has been very frustrating. Likewise, very little improvement in productivity has been experienced by exploring new combinations of speeds, feeds, and depths. However, developments of interest include specially designed turning tools and milling cuttings along with the use of a special end mill pocketing technique.

In recent years, ceramic tools have been used successfully in machining high-temperature alloy jet-engine components at speeds much higher than those conventionally used. At speeds of 183 to 213 m/min (600 to 700 ft/min), tool life is short (3 to 5 min), but it

Table 6.3 Example of typical parameters for machining Ti-6Al-4V gas turbine components

Operation	Tool Material	Cutting Speed ft/min	Feed	Depth of Cut in.
Turn (Rough)	C-2	150	0.010 in./rev	0.250
Turn (Finish)	C-2	200	0.006-0.008 in./rev	0.010-0.030
Turn (Finish)	C-2	300	0.006-0.008 in./rev	0.010-0.030
End mill (3/4-1" dia.)	M42 HSS (a)	60	0.003 in./tooth	Axial depth: 0.125 Radial depth: up to two-thirds cutter diameter
End mill (3/4-1" dia.)	C-10	200	0.005 in./tooth	Axial depth: 0.150-0.200 Radial depth: up to two-thirds cutter diameter
Drill (1/4-1/2" dia.)	M42 HSS (a)	30	0.005 in./rev	
Drill (1/4-1/2" dia.)	C-2	40	0.004 in./rev	
Ream	M42 HSS (a)	20	0.010 in./rev	
	C-2	35	0.010 in./rev	
Tap	M7 HSS	15	-	
Broach	M3 HSS	12	0.003 in./tooth max	
Spline shape	M42 HSS	12	0.012 in./stroke	

(a) Designates tool material most widely used.

is possible to finish a cut at these speeds and then index the cutting tip for making the next pass. This same technique has potential in machining of titanium with C-2 carbides. Data are needed to determine the speeds at which reproducible and reliable tool life of the order of 3 to 5 min can be obtained, and to determine whether these conditions improve the economics of titanium machining.

One of the practical techniques for increasing productivity is to determine the optimum cost in machining a given titanium part for a specific machining operation. If specific data are available relating tool life to speed, feed, and depth for a given operation and cutter, it is possible to calculate the overall cost and time of machining as a function of the cutting parameters. Some companies are now using computers to perform such cost analyses and to arrive at minimum costs and optimum production rates for specific machining operations.

NONTRADITIONAL MACHINING

The design of titanium alloy components often requires the use of the so-called nontraditional machining methods. Among these electrochemical machining (ECM), chemical milling (CHM), and laser beam torch (LBT) are probably the most widely used. Technical information on procedures and techniques is generally proprietary, however.

Chemical and electrochemical methods of metal removal are expected to be used increasingly in years to come, because of their many favorable features. They are particularly useful for rapid removal of metal from the surface of formed or complex-shaped parts, from thin sections, and from large areas down to shallow depths. These processes have no damaging effect on the mechanical properties of the metal. (See the earlier comments about fatigue properties of stress-free surfaces.) There is no hydrogen entry into the metal to cause embrittlement or loss of ductility.

ECM is the removal of electrically conductive material by anodic dissolution in a rapidly flowing electrolyte which separates the workpiece from a shaped electrode. ECM can generate difficult contours and provide distortion-free, high-quality surfaces. For ECM of titanium alloys, a very common electrolyte is sodium chloride used at concentrations of about 1 lb/gal.

CHM is the controlled dissolution of a workpiece material by contact with a strong chemical reagent. The part being processed is cleaned thoroughly and covered with a stripable, chemically-resistant mask. Areas where chemical action is desired are stripped off the mask, and then the part is submerged in the chemical reagent to dissolve the exposed material.

Another operation usable in the processing of titanium alloys is the LBT method. In this process, material is removed by focusing a laser beam and a gas stream on a workpiece. The laser energy causes localized melting, and an oxygen gas stream promotes an exothermic reaction and purges the molten material from the cut. Titanium alloys are cut at very rapid rates using a continuous wave CO_2 laser with oxygen assist.

SURFACE INTEGRITY

The surface of titanium alloys is thought to be easily damaged during some traditional machining operations. Damage appears in the form of microcracks; built-up edge; plastic deformation; heat-affected zones; and tensile residual stresses. In service, this damage can lead to degraded fatigue strength and stress corrosion resistance. In a study of grinding effects on Ti-6Al-4V alloy, gentle or low-stress grinding parameters displayed no readily identifiable changes at the surface, while conventional and abusive practices altered the surface layer noticeably. There was an appreciable drop in hardness in the gently ground specimen, but very good high-cycle-fatigue values were noted.

Figure 6.3 indicates an endurance limit of 372 MPa (54 ksi) for the gentle grinding and values of 83 and 97 MPa (12 and 14 ksi) for conventional and abusive conditions, respectively. Figure 6.3 also presents values for other machining operations including electrical discharge machining (EDM) and chemical milling (CHM). As can be seen, in operations like end mill cutting or turning, the same sensitivity to abusive conditions was not observed, possibly due to residual surface compressive stresses.

Machinists and companies specializing in the machining of aerospace materials generally will have developed techniques to maximize surface integrity of titanium alloys. Thus optimum properties usually are achieved during the production machining of titanium. In those areas of application where maximum fatigue strength is required, not only are appropriate machining parameters used, but also selected surface areas of components may be glass bead blasted to restore, or to retain, a high level of favorable compressive surface stress.

Note: the text which appears following the paragraph Traditional Machining of Titanium is reprinted with permission of the *Journal of Metals*, a publication of The Metallurgical Society, Warrendale, PA. (Reference Vol. 37, No. 4, 1985, "The Machining if Titanium Alloys," **J. F. Kahles, M. Field, D. Eylon, F. H. Froes**.) Other paragraphs have been added by the editor to amplify the general points originally presented.

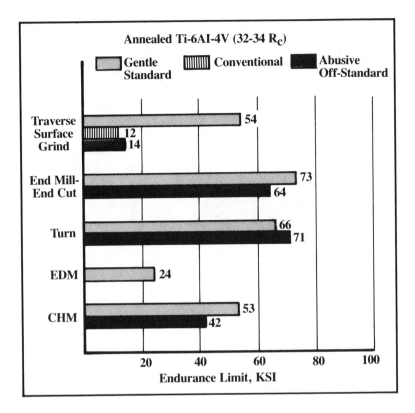

Figure 6.3 - Summary of machining effects on high-cycle-fatigue behavior of Ti-6Al-4V (annealed, 32-34 R_c)

Chapter 7

Cleaning and Finishing

GENERAL

Cleaning and finishing processes for titanium and its alloys are similar in some ways to those for other metals. However, the differences in processes, methods, and cleaning solutions are of major importance to maximize the eventual use of the finished metal and to maintain safe procedures during its working.

Generally, a heavy oxide layer resulting from hot working of a metal is removed by grit blasting or other mechanical means. This is followed by acid pickling to achieve the metallic luster of the material. Heavy grease, oil, and black lubricant coatings resulting from cold working of a metal are removed by alkaline or caustic soaking followed by acid pickling. To remove soil or light oil from a metal, cleaning by detergents, solvent washing or vapor degreasing is generally performed.

Electroplating on reactive metals is not generally practiced due to the difficulty of the process. However, plating by other means for refractory metals, in order to protect the metal from speedy oxidation at elevated service temperatures, is generally mandatory.

Cleaning procedures serve to remove scale, tarnish films, and other contaminants which form or are deposited on the surface of titanium and titanium alloys during hot working, heat treating, and other processing operations. Because their resistance to many corrosive environments is excellent, titanium alloys do not require special surface treatments to improve corrosion resistance; however, several different coatings may be applied to the surface of these materials to develop or to enhance specific properties, such as lubricity and emissivity.

CLEANING AND DESCALING PROBLEMS

General

The metallurgical and chemical properties of titanium create a number of very special cleaning problems. These include:

- Affinity of titanium to common gases

- Galvanic effects caused by discontinuities in scaled surfaces

- Metallurgical restrictions on the temperature of the descaling media

- Variety of scales encountered in titanium descaling

- Protective coatings used in titanium manufacturing

Gas Absorption

The property that causes the most difficulty is titanium's capacity to absorb common gases including oxygen, hydrogen, and nitrogen, all of which tend to embrittle the product. Because tightly packed, hot rolling scale acts as partial protection against additional gas absorption, some mills perform two or three heat treatments over the scale. Each additional heat treatment toughens the scale and compounds descaling difficulties. An additional problem is that heat-treating furnace atmospheres are maintained on the oxidizing side to minimize chances for hydrogen pickup, thereby promoting oxygen absorption and scale formation. A layer of oxygen-rich metal, as shown in Figure 7.1, develops beneath the resulting scale formation. It varies in thickness from 0.05 to 0.07 mm (0.002 to 0.003 in.) in the heat-treated condition to 0.15 to 0.20 mm (0.006 to 0.008 in.) in the hot-rolled condition. This brittle surface is usually removed by acid or electrolytic chemical pickling.

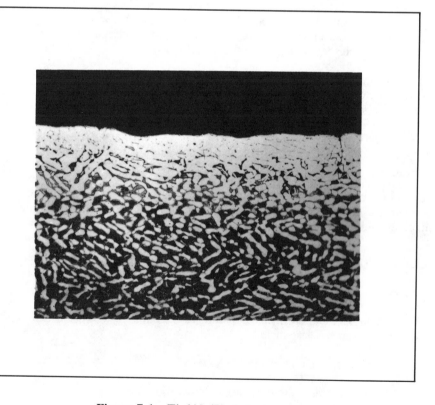

Figure 7.1 - Ti-6Al-4V alpha case

Galvanic Effects and Discontinuities

Galvanic effects and discontinuities in the surface scale are encountered in all types of metal descaling, but they appear to be more pronounced in titanium. Although the exact cause of small pits or cells formed in descaled material is a debatable issue, possibilities include alloy or nonmetallic segregations, scale porosity and, surface contamination. A more severe galvanic attack problem is created by patch scale conditions on titanium surfaces when areas of heavy scale flake away from an apparently uniform surface. The same problem has been observed with superimposed oxides, even though the surface layer may be quite thin and powderlike. Surface contamination with oil, grease, or fingerprints can also create a patch scale condition. All of these factors promote severe localized attack when areas of the basis metal are exposed selectively during descaling. Some producers have considered, as a possible solution, reoxidation of the product during processing.

Metallurgical Restrictions on Descaling

Solution-treated, age-hardenable alloys of titanium are sensitive to time-temperature reactions. The temperatures of descaling media may induce a subsequent aging effect and cause a change in mechanical properties. This is particularly true in thin-gage sheet materials where descaling temperature-time conditions may cause property changes of as much as 70 MPa (10 ksi) in tensile strength. Alloy descaling temperatures normally should not be allowed to exceed 260°C (500°F).

Variety of Scale

Another factor that contributes appreciably to titanium's descaling problems is the wide variety of scale encountered, including scale formed by annealing, forging, solution treating, stress relieving, extruding, rolling, aging, hot forming, or a combination of several of these operations. With processing temperatures ranging from 425° to 1150°C (800° to 2100°F), the scale spectrum for titanium is far broader than for most other difficult-to-descale materials.

Coatings

Protective coatings are often used in titanium manufacturing operations, where they are frequently an asset and a necessity, although they become a liability and a contaminant in cleaning operations. They are soluble and removable if the proper techniques are used. Protective coatings are applied to titanium surfaces during manufacturing operations to:

- Lubricate and aid in metal flow, die contouring, and forming operations

- Act as barrier films, thereby reducing gas contamination during high-temperature forming and heat-treating cycles

- Reduce surface flaws caused by nicking and scratching during manufacturing operations

The gas-protective films are usually applied directly to the titanium's surface. They are silicate-based materials that deposit uniform, fusible films through solvent evaporation. These films form glassy barriers at treatment temperatures up to 815°C (1500°F) and are quite effective in reducing oxygen, hydrogen, and nitrogen contamination. Above about 815°C (1500°F), most of these films are less effective.

Lubricant films or abrasion-protective films are applied over a silica-based coating. This process has the advantage of providing double protection against scratching and scoring. During hot-forming operations and metal surface stretching, some voiding and penetration occurs, creating a titanium oxide on the surface. The contaminant then consists of organic bond or residues, graphite, molybdenum disulfide, silicates, and titanium oxides.

REMOVAL OF SCALE

General

Scale is removed from titanium products by several mechanical methods. Abrasive methods, such as grinding and grit blasting, are preferred for removing heavy scale from large sections. Centerless grinding is used for finishing round bars, and wide-belt grinding is used for finishing sheet and strip. Grinding is usually most efficient when it is performed at low wheel and belt speeds.

Belt Grinding

Dry belt grinding is dangerous because of the hazards of explosion and fire. It is also uneconomical because of poor belt life. When stock is removed during dry grinding, small globules of molten metal and oxide roll along the sheet, causing a type of pitting by burning that is not removed by the grinding. Weld grit scratches and embedded grit result when titanium welds to the dry grit.

A 5% aqueous solution of potassium orthophosphate, K_3PO_4, is widely used as a grinding lubricant. It is applied as a flood at both the entrance and exit side of the contact line.

Titanium should be ground at belt speeds not exceeding rates of 8 m/sec (1500 ft/min). Using a 5% solution of potassium orthophosphate as a lubricant, maximum efficiency is achieved at about 6 m/sec (1100 ft/min). Both billy roll and flat table grinding machines have been successful in grinding titanium. Sheet grinding machines, equipped with feed rolls, sometimes leave a ground line on the sheet. A high degree of grinding uniformity is obtained on machines equipped with a flat table and vacuum chuck. On these machines, the table holding the sheet usually oscillates. Traveling-head machines are available also. The sequence for the belt grinding of Ti-6Al-4V sheet is shown in Figure 7.2.

Abrasive Blasting

Abrasive-blast cleaning techniques, either wet or dry, are convenient for removing scale from a variety of titanium products ranging from massive ingots to small parts. Because it can be used at lower velocities and is less likely to be embedded in the surface, alumina sand is preferred to silica sand.

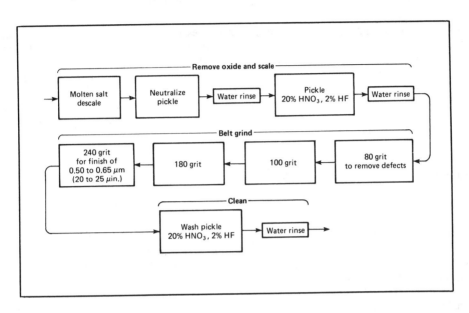

Figure 7.2 - Cleaning and belt-grinding sequences for Ti-6Al-4V sheet

Sheet in thicknesses to about 0.50 mm (0.020 in.) can be descaled without distortion if fine sand and low velocities are used. Mill scale on titanium products can be removed with coarse, high-carbon steel shot or grit, while finished compressor blades can be cleaned with zircon sand of 150 to 200 mesh. The type of product to be cleaned, the cleaning rate, and the cost of the abrasive must be balanced in the selection of a specific blast- cleaning method.

Mineral abrasive particles, such as silica, zircon, or alumina sands, are used more commonly than metal abrasives for blasting finished or semifinished products. Although these abrasives are more expensive, they produce the finer finish that is required in final processing or service. Adequate safety precautions must be observed to avoid inhalation of fine sand particles. Air-circulating and dust-collecting systems must be cleaned frequently. They must be equipped to cope with the fire hazard associated with titanium dust.

A fine dust remains on the titanium from the blasting operation, particularly when mineral abrasives have been used. This is not considered detrimental, although a washing or pickling cycle following the blast is desirable if the part is to be welded subsequently.

The following describes both a wet-blasting and a dry-blasting procedure used for descaling titanium parts:

- **Wet blasting**. Parts are wet-blast cleaned using a slurry that consists of 400-mesh aluminum oxide, 40 vol %, and water. Air pressure of 655 kPa (95 psi) is used to pump the slurry in a steady stream with a pressure of about 34 kPa (5 psi). The descaling rate, normally about 50 min/m² (5 min/ft²), depends on the complexity

of the part. Distortion and the need for planishing are held to a minimum by placing the blast nozzle at a distance of approximately 50 mm (2 in.) from the workpiece, and by using an angle of impingement of 60°.

- **Dry blasting**. Rocket motor case assemblies have been dry blasted after final
  stress relieving at 480° to 540°C (895° to 1005°F). Blasting is accomplished with
  100- to 150-mesh zircon sand at an air pressure of 275 kPa (40 psi). Each assembly is rotated at 2.5 rev/min and is passed at a speed of 65 mm/min (2.5 in./min)
  between two diametrically opposed, fixed-position blasting nozzles. The nozzles
  blast the inside and outside surfaces simultaneously at the same wall location. To
  prevent distortion, each nozzle is placed at the same distance, 300 mm (12 in.),
  from the metal surface.

MOLTEN SALT DESCALING BATHS

General

Molten salt descaling baths are primarily used for descaling bar, sheet products, and
tubing. Even with the most effective barrier films available today, some gas penetration
of titanium surfaces can be expected at the elevated temperatures required for working
and heat treatment. The alpha case or oxygen-enriched layer resulting from this gas
reaction is extremely hard and brittle and must be removed. Bar products used for
machining finished parts must have this hard scale and oxide removed because these are
very abrasive and cause rapid tool wear. Material used for welding or forming must
have these scales removed, or poor and small welds are made; subsequent forming (hot
or cold) is virtually impossible without surface rupture or failure of parts. Removal presents no serious problems since chemical milling techniques have been perfected by the
aircraft industry to effect weight savings. In the case of titanium, the purpose is to
improve the structural soundness of metal, and the solvent materials applied are of a different chemical composition.

One specific problem encountered in alpha case removal is that the titanium oxide
formed is substantially more insoluble in the nitric hydrofluoric etchant than the base
metal. Residues of oxide on the surface develop areas resembling craters on the finished
product. Examination of the artist's conception sketch shown in Figure 7.3 indicates the
surface as a result of a nonuniform cleaning operation.

Where alpha case removal is a required part of a manufacturing operation, salt bath
cleaning is specified because proper cycling practically guarantees a chemically clean
surface. Conditioning salt baths fall into the two basic categories of high-temperature
salt baths and low-temperature salt baths.

High-Temperature Molten Salt

High-temperature salt baths may vary in chemical reaction and effectiveness, depending
on composition. All types operate at a range of 370° to 480°C (700° to 895°F). The temperature range is sufficiently high to produce the most rapid reaction possible for soiled
and oxide films. The range also is sufficiently high to possibly promote metallurgical
changes in some alloys, as, for example, omega phase precipitation in metastable beta
alloys. High-temperature oxidizing salt baths are also capable of reacting chemically
with organic films to destroy them. These baths are also excellent solvents for silicate

barrier films. They do require special fixturing to reduce the strong galvanic effects present at these temperatures, and, for this reason, they are used in cleaning primary forming operation products such as forgings, extrusions, rolled plate, and sheet.

The major advantage of high-temperature oxidizing or reducing salt baths for titanium descaling is their great speed in removing extremely tenacious scale. Although reducing baths have the inherent disadvantage of promoting hydrogen absorption, this can be overcome or minimized by chemical additions. Vacuum degassing is another solution to the hydrogen problem.

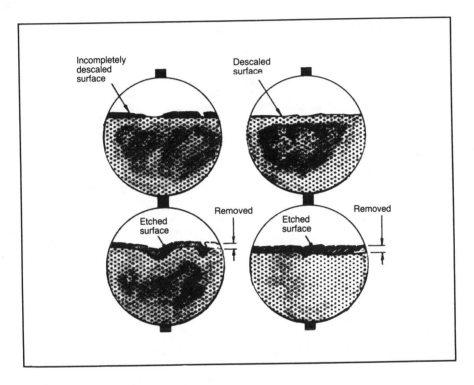

Figure 7.3 - Effect of surface condition on etched metal surface

Low-Temperature Baths

The temperature range used with low-temperature baths for cleaning fabricated parts is 200° to 220°C (390° to 430°F). Descaling systems based on salts in this temperature range eliminate some of the possible problems associated with higher temperature baths including:

- Age hardening

- Dissimilar metal reactions

- Chemical attack

- Metal distortion

• Hydrogen embrittlement

Salts in this range have a very limited composition because of the effect of various compounds on the melting point. Although they contain oxidizing agents, the effect of these materials is not as aggressive as it is in the high-temperature fused salts. Consequently, organic materials are not destroyed, but they are saponified and absorbed. Silicate barrier films and molybdenum disulfide are soluble in these low-temperature salts. The temperature range permits cycling between salt and acid to reduce cleaning times and costs. Examples of salt bath and acid cycle times are given in Table 7.1.

Table 7.1 Low-temperature salt bath and acid bath conditions

Sample composition	Scale formation temperature °C	°F	Salt bath immersion time (a), min	Acid cleaning bath time (b) min	Acid cleaning bath time (c) sec
Ti-6Al-4V	650	1200	2	2	30
Ti-8Al-1Mo-1V	650	1200	2	2	30
Ti-8Al-1Mo-1V	820	1510	5	2	30
Ti-6Al-4V (d)	820	1510	5	5	30
Ti-6Al-4V (e)	950	1745	5	5	60
Ti-8Al-1Mo-1V (f)	950	1745	5	5	60

(a) Salt bath temperature 205°C (400°F).
(b) Bath composition, 30% sulfuric acid.
(c) Bath composition, 30% nitric acid, 3% hydrofluoric acid.
(d) Sample recycled in salt bath for 5 min, in sulfuric acid bath for 5 min, in nitric acid-hydrofluoric acid bath for 30 sec.
(e) Sample recycled in salt bath for 5 min, in sulfuric acid bath for 5 min, in nitric acid-hydrofluoric acid bath for 60 sec.
(f) Sample recycled in salt bath for 5 min, in sulfuric acid bath for 5 min, in nitric acid-hydrofluoric acid bath for 60 sec.

Aqueous Caustic Descaling

Aqueous caustic descaling baths have been developed to remove light scale and tarnish from titanium alloys. Aqueous caustic solutions containing 40 to 50% sodium hydroxide have been used successfully to descale many titanium alloys. One bath containing 40 to 43% sodium hydroxide operates at a temperature near its boiling point, 125°C (260°F). Descaling normally requires from 5 to 30 min. Immersion time is not critical because little weight loss is encountered after the first 5 min. Caustic descaling conditions the scale so that it is removed readily during subsequent acid pickling.

A more effective aqueous solution contains either copper sulfate or sodium sulfate in addition to sodium hydroxide. This bath operates at a lower temperature, 105°C (220°F). A composition of this solution by weight is as follows: 50% sodium hydroxide, 10% copper sulfate pentahydrate ($CuSO_4 5H_2O$), and 40% water. Using immersion times of 10 to 20 min, this bath has proved effective in descaling Ti-6Al-4V and Ti-2.5Al-16V alloys.

PICKLING PROCEDURES FOLLOWING DESCALING

All advantages gained through proper conditioning and handling of titanium parts during cleaning can be lost if the composition of the final pickling acid is not controlled. Cold, spent acid solutions have increased appreciably the time requirements for pickling and the possible quality problems experienced with hydrogen pickup. Highly concentrated hot acids can be overly aggressive, resulting in surface finish problems, such as a rough and pitted surface caused by preferential acid attack. Sulfuric acid, 35 vol % at 65°C (150°F), is recommended for pickling immediately following salt bath conditioning and rinsing to remove molten salt and residual softened scales. An acid of this formula has very little effect on titanium metal. Metal salts in the original and additional acid solutions further minimize these base metal attacks. Table 7.2 gives conditions for corrosion of titanium in various sulfuric acid pickle baths.

Table 7.2 Corrosion of titanium in sulfuric acid pickle baths (a)

Sulfuric acid concentr. %	Acid addition		Bath temperature °C	Bath temperature °F	Corrosion rate μm/ yr	Corrosion rate mils/ yr
30	0.5%	(b)	38	100	100	4.0
30	1%	(b)	38	100	20	0.8
30	10%	(b)	38	100	400	16.0
30	0.25%	(b)	95	205	76	3.0
10	2%	(c)	Boiling point	Boiling point	125	5.0
17	7–8%	(c)	60	140	125	5.0

(a) A nitric-hydrofluoric acid solution, which is the final stage brightening in most alloy cleaning lines, should be maintained at a minimum ratio of 15 parts nitric acid to 1 part hydrofluoric acid to reduce hydrogen pickup effects; the concentration of hydrofluoric acid may vary from 1 to 5%, or even higher as long as the ratio is not exceeded; the activity of these pickle solutions is affected by titanium content, and the acids are frequently discarded at a level of 26 g/L (3 oz/gal); the solution used for final brightening can be used for the required alpha case removal also, assuming a careful watch on the titanium content.
(b) Copper sulfate
(c) Ferrous sulfate

REMOVAL OF TARNISH FILMS

General

Tarnish films are thin oxide films that form on titanium in air temperatures between 315° and 650°C (600° and 1200°F), after exposure at 315°C (600°F). The film is barely perceptible, but with increasing temperature and time at temperature, it becomes thicker and darker. The film acquires a distinct straw yellow color at about 370°C (700°F), and a blue color at 480°C (900°F). At about 650°C (1200°F), it assumes the dull gray appearance of a light scale. Alloying elements and surface contaminants also influence the color and characteristics.

Tarnish films are readily removed by abrasive methods, and all but the heaviest films can be removed by acid pickling. Prolonged exposures at temperatures above about 595°C (1104°F), in combination with surface contaminants, result in heavier surface films that are not removed satisfactorily by acid pickling, but require descaling treatments for their removal.

Acid Pickling

Acid pickling removes a light amount of metal, usually a few tenths of a mil. It is used to remove smeared metal, which could affect penetrant inspection. Titanium and titanium alloys can be satisfactorily pickled by the following procedure:

- Clean thoroughly in alkaline solution to remove all shop soils, soap drawing compounds, and identification inks. If coated with heavy oil, grease, or other petroleum-based compounds, parts may be degreased in trichlorethylene before alkaline cleaning. Degreasing will not be harmful to the part in subsequent processing.

- Rinse thoroughly in clean, running water after alkaline immersion cleaning.

- Pickle for 1 to 5 min in an aqueous nitric-hydrofluoric acid solution containing 15 to 40% nitric acid and 1.0 to 2.0% hydrofluoric acid by weight, and operated at a temperature of 24° to 60°C (75° to 140°F). The ratio of nitric acid to hydrofluoric acid should be at least 15-to-1. The preferred acid content of the pickling solution, particularly for alpha-beta and beta alloys, is usually near the middle of the above ranges. A solution of 33.2% nitric acid and 1.6% hydrofluoric acid has been found effective. When the buildup of titanium in the solution reaches 12 g/L (2 oz/gal), discard the solution.

- Rinse the parts thoroughly in clean water.

- High-pressure-spray wash thoroughly with clean water at 55°C ($\pm$6°) or 130°F ($\pm$10°).

- Rinse in hot water to aid in drying. Allow to dry.

To avoid excessive stock removal, the recommended immersion times for pickling solutions should not be exceeded. It is equally important to maintain the composition and operating temperature of the bath within the limits prescribed to prevent an excessive amount of hydrogen pickup. Gage loss from all-acid pickling after descaling is estimated to be less than 0.025 mm/min (0.001 in./min), as determined by the combination of variables used.

Depending on alloy composition and gage material pickled, hydrogen contamination is estimated to be 0 to 15 ppm/0.025 mm (0.001 in.) of metal removed. Data on hydrogen pickup for an alpha, an alpha-beta and a beta alloy pickled in a 15% nitric acid, 1% hydrofluoric acid bath at 49° to 60°C (120° to 140°F) are given in Table 7.3. Hydrogen contamination can be held to a minimum by maintaining a 10-to-1, or greater, acid ratio of nitric acid to hydrofluoric acid. Hydrogen diffuses more rapidly into the beta phase. Alpha-beta alloys that have $\alpha + \beta$ microstructures, which have been heat treated to complete equilibrium, pick up less hydrogen than microstructures of transformed beta and/or simple mill-annealed structures.

Mass Finishing (Barrel Finishing)

Oxide films formed by heating to temperatures as high as 650°C (1200°F) for 30 min have been effectively removed from Ti-8Mn alloy parts by wet mass finishing. Parts are randomly loaded in the barrel and rotated at relatively low barrel speeds to minimize dis-

Table 7.3 Effect of Ti alloy composition on hydrogen pickup in acid pickling (a)

Alloy	Thickness mm	Thickness in.	Hydrogen pickup (gage removed) ppm/0.0250 mm (ppm/0.001 in.)
Alpha alloy:			
Ti-5Al-2.5Sn	0.50	0.020	0-4
Ti-5Al-2.5Sn	1.00	0.040	0-3
Alpha-beta alloy:			
Ti-6Al-4V	0.50	0.020	4-7
Ti-6Al-4V	1.00	0.040	3-5
Beta alloy:			
Ti-13V-11Cr-3Al	0.50	0.020	10-15
Ti-13V-11Cr-3Al	1.00	0.040	5-8

(a) Pickling bath is an aqueous solution containing 15% nitric acid and 1% hydrofluoric acid by weight; operating temperature is 49 to 60°C (120 to 140°F).

tortion and nicking. Conditions for mass finishing of titanium parts are given in Table 7.4. At barrel speeds of 43 000 to 51 000 mm/min (1700 to 2000 in./min), parts have been cleaned satisfactorily in about 1 h. Barrel loading information for three barrels, ranging from 0.02 to 0.25 m^3 (0.75 to 8.85 ft^3) in capacity is also given in Table 7.4.

When mass finishing titanium parts, the medium-to-parts ratio should be between 10- and 15-to-1, depending on the size of the parts. Proportionately more medium is required as part size increases. Water is used to cover parts and medium. Surface finish is improved when more water is added, but cycle time required to obtain a given finish is increased. The rate of descaling increases directly with barrel speed but is limited by the fragility of the parts being processed.

Table 7.4 Mass finishing conditions for titanium parts

Capacity m^3	Capacity ft^3	Barrel size diameter mm	Barrel size diameter in	Width mm	Width in	Speed rev/min
0.02	0.75	381	15	178	7	36
0.07	2.33	559	22	240	10	28
0.25	8.85	813	32	457	18	20

Part load kg	Part load lb	Medium (a) kg	Medium (a) lb	Water L	Water qt	Abrasive compound (b) kg	Abrasive compound (b) lb	Alkaline cleaner (c) kg	Alkaline cleaner (c) lb
1-2	3-4	18	40	1.2	1.25	0.2	0.5	0.2	0.5
4-5	8-12	54	120	4	4	0.7	1.5	0.34	0.75
14-18	30-40	209	460	14	15	2.3	5	0.5	1

(a) Aluminum oxide nuggets 6.4 to 38 mm (0.25 to 1.5 in) or preformed vitrified chips 4.8 by 9.5 to 7.9 by 28.6 mm (3/16 by 3/8 to 5/16 by 1-1/8 in.)
(b) Dry, mildly alkaline compound
(c) Mild cleaner with high soap content

Aluminum oxide mediums are the most satisfactory. They do not contaminate the work, yet they have a long, useful life. For oxide removal, small, well-worn mediums produce the highest finish. To avoid possible metallic contamination, the medium used for titanium should not be used in processing other metals. Strong acid-forming compounds are to be avoided, principally because they are corrosive and contribute to hydrogen embrittlement. Because of the fire hazard created by fine, dry titanium particles, dry mass finishing of titanium parts is not recommended.

POLISHING AND BUFFING

The polishing and buffing of titanium is accomplished with the same equipment used for other metals. Polishing frequently is done wet with mineral oil lubricants and coolants. Silicon carbide abrasive cloth belts have been effective. It is common to polish in two or more steps. Use a coarser grit initially, such as 60 or 80, to remove gross surface roughness. Follow this with 120 or 150 grit to provide a smooth finish. Titanium tends to wear the sharp edges of the abrasive particles and also to load the belts more rapidly than steel. Frequent belt changes are required for effective cutting. A good flow of coolant improves polishing and extends the life of the abrasives.

Dry polishing is more appropriate than wet for some applications. For these operations, belts or cloth wheels with silicon carbide abrasive may be used. Soaps and proprietary compounds may be applied to the belts to improve polishing and to extend belt life. Abrasive belt materials that incorporate solid stearate lubricants offer improved results for dry polishing operations.

Fine polishing of titanium articles for extremely smooth finishes requires several progressive polishing steps with finer abrasives until pumice or rouge types of abrasive are applied. With the softer grades of titanium, such as unalloyed material, fine polishing requires more time and care to prevent scratching. The harder alloy grades can be polished more readily to a surface of high reflectivity. If a matte finish is desired, wet blasting with a fine slurry may be used after initial polishing.

Titanium alloys can be buffed safely. The purpose of buffing is to improve the surface appearance of the metal and to produce a smooth, tight surface. Buffing is used as a final finishing operation and is particularly adaptable to finishing a localized area of a part. Items such as body prosthesis, pacemakers, and heart valves require a highly buffed, tight surface to prevent entrapment of particles. Close fitting parts for equipment, such as the modern guidance systems and electronics applications, require highly polished surfaces obtained by buffing. In addition, sheet sizes too large to be processed by other abrasive finishing methods, such as mass finishing or wet blasting, can be economically processed by buffing.

The principal limitations of buffing are:

- Distortion, caused by the inducement of localized stress

- Surface burning, resulting from prolonged dwell of the buff

- Inability to process inner or restricted surfaces

- Feathering of holes and edges

Proper care of the buffing wheel is essential. Buffing with insufficient compound or a loaded wheel produces a burning or distortion of the part. After buffing, no further cleaning of parts is required—except degreasing to remove the buffing compound.

WIRE BRUSHING

Wire brushing of titanium alloys is not recommended when other finishing methods, such as buffing, can accomplish the objective. In one instance, in an attempt to remove surface scratches or oxide films, wire brushing was used on titanium; it resulted in serious defects. A stiff-bristled wire brush removed surface scratches and oxide films, but the surface was pitted by the wire tips. To avoid pitting, softer wire bristles were tried. The surface of the titanium acquired a burnished appearance; surface layers were cold worked; and grinding scratches, instead of being removed, were filled with smeared metal.

Wire brushing with a silicon-carbide abrasive grease has been used successfully to remove burrs, break sharp edges from edge radii, and blend chamfers.

REMOVAL OF GREASE AND OTHER SOILS

Removal of grease, oil, and other shop soils from titanium parts normally is accomplished with the same type of equipment and the same cleaning procedures used for stainless steel and high-temperature alloy components. Certain aspects of conventional processing, however, must be modified or omitted when titanium alloys are being cleaned.

Vapor degreasing normally employs either trichlorethylene or perchlorethylene. Under certain conditions, these solvents are known to be a cause of stress-corrosion cracking in titanium alloys. Methylethyl ketone is used as a cleaner in situations where chlorinated solutions are not desired. All titanium parts should be acid pickled after vapor degreasing.

Other cleaning methods use chemicals which, if they are left to dry on the part, may have a harmful effect on the properties of titanium. Among these are:

- Soda ash, borates, silicates, and wetting agents commonly used in alkaline cleaners

- Kerosine and other hydrocarbon solvents used in emulsion cleaners

- Mineral spirits employed in hand-wiping operations

Residues of all these cleaning agents must be completely removed by thorough rinsing. To ensure a surface that is free of contaminants, rinsing is frequently followed by acid pickling.

CHEMICAL CONVERSION COATINGS

Chemical conversion coatings are used on titanium to improve lubricity by acting as a base for the retention of lubricants. Titanium has a severe tendency to gall. Lack of lubricity creates serious problems in applications involving the contact of moving parts in various forming operations.

Conversion coatings are applied by immersing the material in a tank containing the coating solution. Spraying and brushing are alternate methods of application. One coating bath consists of an aqueous solution of sodium orthophosphate, potassium fluoride, and hydrofluoric acid. It can be used with various constituent amounts, immersion times and bath temperatures. The resultant coatings are composed primarily of titanium and potassium fluorides and phosphates. Several solutions are listed in Table 7.5.

Table 7.5 Conversion coating baths for titanium alloys

Bath No.	Bath solution	Composition	Amount g/L	Amount oz/gal	Temperature °C	Temperature °F	pH	Immersion time, min
1	Degreasing solution	$Na_3PO_4 \cdot 12H_2O$ $KF \cdot 2H_2O$ HF solution (a)	50 20 11.5	6.5 2.6 1.5	85	185	5.1-5.2	10
2	Pickling solution	$Na_3PO_4 \cdot 12H_2O$ $KF \cdot 2H_2O$ HF solution (a)	50 20 26	6.5 2.6 3.4	27	81	<1.0	1-2
3	Chemical immersion solution	$Na_2B_4O_7 \cdot 10H_2O$ $KF \cdot 2H_2O$ HF solution (a)	40 18 16	5.2 2.3 2.1	85	185	6.3-6.6	20
(a) Hydrofluoric acid, 50.3% by weight								

Ranging from rapid bubbling to relative dormancy, the appearance of the baths varies widely during the coating reaction. Some coatings rub off when still wet; others are adherent. The solutions produce coatings of approximately the same dark gray or black appearance.

The control of pH and immersion time is important. Dissolved titanium and the active fluoride ion make it impossible to use glass electrodes for pH measurements. Indicator paper and colorimetry are the most satisfactory methods for measuring in the degreasing and chemical immersion baths, which are held in the pH range from 5 to 7. The pickling bath is quite acid, and titrametric analysis offers the most practical method of control. When the bath is in the proper coating range, a 20-mL (0.70-fluid oz) sample in 100 mL (3.4 fluid oz) of water neutralizes 11.8 to 12.0 mL (0.4 to 0.41 fluid oz) of normal sodium hydroxide, when using a phenolphthalein indicator.

Coating thickness depends on immersion time. In all three baths, a specific time is reached after which the coating weight remains essentially constant. In the fluoride-phosphate baths, a maximum coating weight is reached at some time before this equilibrium point. The maximum coating weight is obtained in about 2 min in the low-temperature bath and in about 10 min in the two other baths.

Results of extensive wiredrawing experiments, given in Table 7.6, illustrate the effectiveness of conversion coatings when used with various lubricants. Reciprocating wear tests showed that conversion coatings and oxidized surfaces provided some improvement in wear characteristics, but when conversion-coated samples were also oxidized, a marked improvement was noted. The conversion coating increases the oxidation rate of titanium at about 425°C (800°F) and may increase oxidation rates at temperatures up to 595°C (1104°F). The original coating is retained above the titanium oxide layer. High-speed rotary tests have indicated marked improvement in the wear characteristics of the metal after conversion coating and lubricating with one part of molybdenum disulfide and two parts of thermosetting eponphenolic resin.

Coatings are easily removable without excessive loss of metal by pickling in an aqueous solution containing 20% nitric acid and 2% hydrofluoric acid by weight.

Table 7.6 Comparison of conversion coatings in wiredrawing of titanium

Coating	Drawing compound	Total reduction, %	No. of passes	No. of coats	Final condition
Bare	Molybdenum disulfide with grease	–	0	–	Galled
Bare	Soapy wax	–	0	–	Galled
Degreasing bath	Molybdenum disulfide with grease	85	8	2	Smooth
Pickling bath	Molybdenum disulfide with grease	94	17	7	Smooth
Pickling bath	Soapy wax	68	7	3	Galled
Pickling bath (a)	Molybdenum disulfide with grease	70	7	1	Smooth
Chemical immersion bath	Lacquer molybdenum disulfide	63	8	2	Smooth
Chemical immersion bath	Molybdenum disulfide with grease	63	8	3	Smooth

(a) Coating heated for 1 h at 425°C (795°F).

ELECTROPLATING ON TITANIUM

Copper Plating

The electrodeposition of copper of titanium and titanium alloys provides a basis for subsequent plating. A flowchart outlining the processing sequence for copper plating titanium is shown in Figure 7.4. After cleaning and before plating, the surface of the titanium must be chemically activated by immersion in both an acid dip and a dichromate dip to obtain adequate adhesion of the plated coating. The compositions and operating temperatures of these activating solutions are in the tabular area of Figure 7.4.

Water purity is critical in the composition of activating solutions, although technical grade chemicals are as effective as, and may be substituted for, chemicals of the chemically pure grade. In both the acid and dichromate baths, hydrofluoric acid content is most critical and must be carefully controlled.

After proper activation, titanium may be plated in a standard acid copper sulfate bath. The adhesion of the deposited copper is better than that of 60-40 solder-to-copper, and the deposit successfully withstands the heat of a soldering iron. The normal thickness of the plated deposit is about 25 μm (1 mil).

Copper-plated titanium wire is available commercially. The outstanding property of this material is the lubricity of its copper-plated surface. The wire can be drawn easily and can be threaded on rolls. Such wire has been used in applications that require electrical surface conductivity.

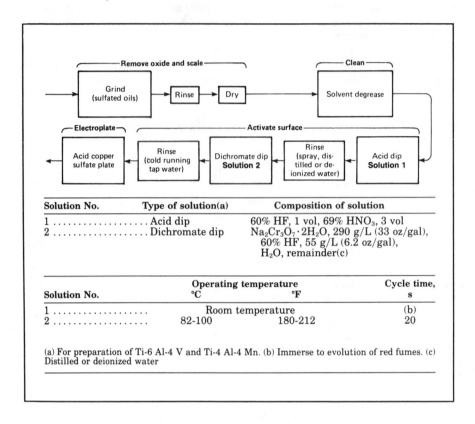

Solution No.	Type of solution(a)	Composition of solution
1	Acid dip	60% HF, 1 vol, 69% HNO_3, 3 vol
2	Dichromate dip	$Na_2Cr_3O_7 \cdot 2H_2O$, 290 g/L (33 oz/gal), 60% HF, 55 g/L (6.2 oz/gal), H_2O, remainder(c)

Solution No.	Operating temperature °C	°F	Cycle time, s
1	Room temperature		(b)
2	82-100	180-212	20

(a) For preparation of Ti-6 Al-4 V and Ti-4 Al-4 Mn. (b) Immerse to evolution of red fumes. (c) Distilled or deionized water

Figure 7.4 - Processing sequence for electroplating copper on titanium alloy parts

The titanium wire is plated continuously at a speed of about 60 m/min (200 ft/min) in a copper fluoborate acid bath at a current density of 7.5 to 12.5 A/dm^2 (75 to 125 A/ft^2). The final copper deposit is a thin flash coating. Higher current densities up to 150 A/dm^2 (1500 A/ft^2) have been tried, but, if the copper coating is too thick, adhesion is poor.

Platinum Plating

Although titanium is not satisfactory as an anode material because of an electrically resistant oxide film that forms on its surface, application of a thin film of platinum to titanium results in a material with excellent electrochemical properties. Theoretically, the thinnest possible film is sufficient to give the highly desirable low overvoltage characteristics of platinum; furthermore, the film need not be continuous or free of defects to be effective.

The greatest immediate use for platinum-coated titanium is for anodes in the chlorine-caustic industry. Some horizontal-type chlorine cells use expanded metal anodes. From

1.3 to 2.5 μm (0.05 to 0.1 mil) of platinum is applied to the anode surface. Replating of the anodes may be required after about 2 years, depending on the operating conditions. The attrition rate for platinum appears to be about 0.6 g/tonne (0.5 g/ton) of chlorine.

Several platinum and electrode suppliers have developed reliable methods for platinum plating of titanium. Most use proprietary solutions. A platinum diamino nitrite bath has been used successfully to apply platinum plates to titanium. In this and other procedures, certain precautionary steps are required to achieve adherent, uniform plates. The surface must be cleaned thoroughly and etched in hydrochloric or hydrofluoric acid to produce a roughened surface. Some procedures also involve a surface activating treatment just before plating. Immersion for 4 min in a solution of glacial acetic acid 895 mL (30 fluid oz) containing hydrofluoric acid 125 mL (4 fluid oz) of 52% hydrofluoric acid, followed by a prompt rinse, appears to be an effective activating treatment if performed immediately before plating. A postplating treatment, consisting of heating to between 400° and 540°C (750° and 1000°F) for a period of 10 to 60 min, stress relieves the plate and improves adhesion. This treatment can be done in an air atmosphere, and a light oxide film forms on unplated areas.

Coatings for Emissivity

Electrodeposits and sprayed coatings of gold on titanium are being used to provide a heat-reflecting surface that reduces the temperature of the base metal. Gold-coated titanium has been used for jet engine components.

The gold coating is applied by spraying a gold-containing liquid on chemically clean titanium sheet. This is followed by a baking treatment. Normal coating thickness is about 25 μm (0.1 mil).

Chapter 8

Castings

GENERAL

Casting is another method for fabricating titanium into shapes. It may be considered a near-net shape technique, although as such it is less "true" than powder metallurgy. For almost a decade large structural castings have been available for aerospace applications, but the method has not always found universal acceptance. Since it is a highly reactive metal, titanium is a challenge to cast. Imperatives within the industry to reduce costs, along with a continually increasing confidence in a combination of cast titanium alloys and hot isostatic pressing (HIP) processing, seem to indicate substantially greater use of the casting process for titanium and its alloys in the near future.

Titanium castings, like wrought titanium products, are used primarily in three areas of application:

- Aerospace products
- Industrial (that is, corrosion) service
- Marine service

Commercially pure titanium, as ASTM Grades 1, 2, or 3, is used for the vast majority of corrosion applications. Ti-6Al-4V is the dominant alloy for aerospace and marine applications. With increasing frequency Ti-6Al-2Sn-4Zr-2Mo-Si is being selected for elevated-temperature service. Castings also have been supplied in alloys Ti-5Al-2.5Sn, Ti-8Al-1Mo-1V, and Ti-6Al-6V-2Sn, as well as several European alloys. However, close to 90% of all castings produced are of Ti-6Al-4V. The majority of the remaining 10% are commercially pure (CP) or other alloys.

All titanium castings have chemical compositions based on those of the common wrought alloys. (There are no commercial titanium alloys developed exclusively for casting applications.) This is unusual, because alloys in other metallic systems have been developed specifically as casting alloys, often to overcome certain problems such as poor castability of a wrought-alloy composition. No unusual problems regarding castability or fluidity have been encountered in any of the titanium metals cast to date.

The major reason for selecting a titanium casting instead of a wrought titanium product is cost. This cost advantage may be attained through increased design flexibility, better utilization of available metal, or reduction in the cost of machining or forming parts.

Titanium castings are unlike castings of many other metals in that they are equal, or nearly equal, in strength to their wrought counterparts. Strength guarantees in most specifications for titanium castings are the same as for wrought forms. Typical ductilities of cast products, as measured by elongation and reduction in area, are lower than typical values for wrought products of the same alloys. Fracture toughness and crack-propagation resistance may equal or exceed those of corresponding wrought material. The fatigue strength of cast titanium is inferior to that of wrought titanium. However, results of ongoing research suggest that fatigue strength of cast titanium can be enhanced by further processing and heat treatment.

CASTING PROCESSES

Castings are produced by melting titanium with a vacuum arc in a copper, water-cooled crucible. The resulting molten metal is poured into either an investment mold or a rammed-graphite mold. Pattern making and investment cast mold techniques are similar to those used with superalloy technology. Although rammed graphite molds are different from investment molds, they are similar to conventional sand molds. Cores are used to cast hollow parts.

(Electron beam or plasma arc melting may find application in the future.)

With respect to titanium and titanium alloy castings, the most significant problem is achieving sufficient levels of superheat in the molten metal to maximize flow and mold-fill characteristics. In many cases either a centrifugal table or mold preheating, or both, must be used to ensure proper mold filling.

A significant difficulty related to large titanium castings is the problem of porosity. Weld repair is used to close gross defects after hot isostatic pressing (HIP). (This type of processing is applied to premium aerospace castings; some applications may not require HIPing.)

A simplified schematic of one type of titanium casting process is shown in Figure 8.1.

Oxygen content is a matter of concern regarding titanium alloy castings since strength increases and ductility (toughness) decreases as oxygen level increases. Control of cast components' oxygen levels at present is achieved mostly by selection of melt stock. An ingot with a low oxygen content generally results in a casting with the lowest oxygen content. The exact role of revert in oxygen content control and in alloy element segregation is not clear. Both revert and virgin ingots are used.

The rammed-graphite method is the oldest mold technique used to produce titanium castings. The method uses a mixture of graphite powder and associated binders and water additions which are rammed against a pattern to form a portion of the mold. Individual mold segments are then fired and assembled for casting.

Most high-performance titanium alloy casting applications rely on the technique of investment casting. Here a wax pattern is produced and "invested" in a ceramic shell. After the mold is completed and dewaxed, it is fired, and is then ready for casting. For small components multiple patterns are used in the same feeding stem in order to create many parts at a single time. This technique may be ideal for biomedical applications, but it is not very practical for large components used in aerospace work.

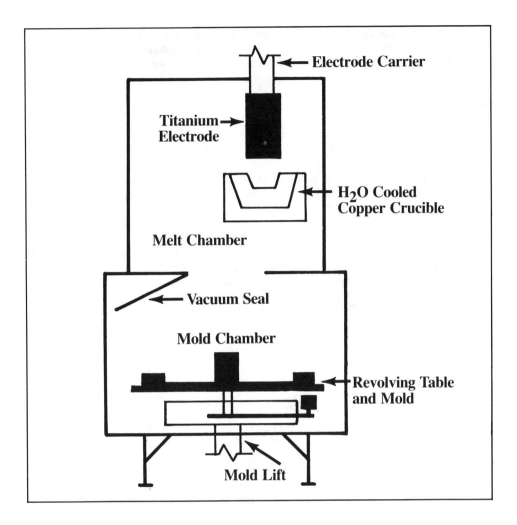

Figure 8.1 - Consumable-electrode titanium vacuum arc-melting furnace with centrifugal casting table

DESIGN CONSIDERATIONS

There are a number of broad guidelines to be followed when designing titanium castings. Briefly stated, they are:

- The supplier and customer must work together to identify the desired properties and to define the intended shape.

- The design must ensure that the mold is filled as completely as possible. (Weld repair, if it is required at all, should be limited as far as possible to noncritical areas.)

- Unnecessary tooling complexity should be avoided.

• Design the casting envelope for smooth transitions between thick and thin areas; to contain generous radii and fillets; and to contain tapered, thin wall sections that promote maxium density (that is minimum shrinkage porosity).

EFFECT OF WELD REPAIR

Weld repair is common foundry practice. However, because titanium can become embrittled due to pickup of oxygen, hydrogen, and other contaminants during welding, weld repair of titanium castings must be carefully executed. Gas tungsten arc welding is used with alloy filler rods to fill pores, cold shuts, or post-HIP depressions in a surface. (These filler rods may be extra-low interstitial ELI) or normal interstitial, that is, regular.) Stress-relief heat treatment may be required after weld repair.

The results of one fatigue study indicate that properly executed weld repair does not degrade the fatigue resistance of cast titanium. Welding also has been evaluated with regard to its effect on creep properties. This study demonstrated that welding does not drastically affect the creep properties of cast Ti-6Al-4V. The rupture times for welded and unwelded bars are similar at 315°C (600°F) and at 650°C (1200°F). Strain rates at 2% creep strain and 650°C (1200°F) are the same for welded and unwelded bars.

HIP

Hot isostatic pressing (HIP) is the application of a hydrostatic pressure to titanium castings in order to effectively sinter nonsurface-connected voids in the castings back together. This process is performed at high temperature in a nonreactive medium such as argon. The reduction in porosity that is achieved by HIPing results in a substantial increment in fatigue strength. (Fatigue strength is, of course, a vital property for typical titanium alloy applications.)

The HIPing of titanium castings became a production reality in the late 1970s. The HIP schedule that has become the industry standard is 2 h at 900°C (1650°F) under argon pressurized to 105 MPa (15 ksi).

Initially, hot isostatic pressing was used with excellent results to salvage parts that had been rejected after radiographic inspection. The effectiveness of the technique gave rise to plans to use HIP for routine parts, but high cost made such plans economically questionable. However, for certain casting configurations, adequate feeding by use of conventional risering is virtually impossible. In order to meet aerospace nondestructive (NDT) inspection standards, shrinkage voids would have to be closed by welding. In such instances, hot isostatic pressing becomes a means of avoiding, or at least minimizing, weld repair and its attendant extra handling and nondestructive test costs.

From a technical viewpoint, hot isostatic pressing is a heat treatment, although some studies have claimed that HIP alone does little, if anything, to enhance mechanical properties of Ti-6Al-4V castings. Properties of hot isostatically pressed alloys are a function of the HIP temperature relative to the beta transus and the post-HIP heat treatment. With castings of marginal-to-substandard quality, hot isostatic pressing raises the lower limit of data scatter and raises the degree of confidence in the reliability of cast products.

HIP is considered by many to be a process that simplifies the problem of defining a standard for internal casting quality. Hot isostatic pressed parts also are aesthetically more acceptable. At the same time, use of HIP ensures that subsurface microporosity will be healed and therefore will not become exposed on a subsequently machined or polished surface to mar the finish or to act as a possible site for fatigue-crack propagation. Fatigue-limited titanium castings always are HIP processed.

HEAT TREATMENT

After casting, but before HIPing, a stress-relief step may be applied. HIP itself is a heat treatment, but it, in turn, is followed by a high-temperature solution treatment at or above the HIP temperature. This temperature is close to, and perhaps above, the beta transus (beta solution heat treatment).

One heat treatment for Ti-6Al-4V consists of beta solution (in vacuum) at 1038°C (1900°F), ±-4°C (±25°F) for 2 to 3 h plus an oil quench. This is followed by overaging in vacuum at 704°C (1300°F), ±-4°C (±25°F) for 2.5 to 3 h. It is then furnace cooled in argon to room temperature. This whole process is called beta solution overaging (beta-STOA).

A more conventional heat treatment for Ti-6Al-4V castings is an alpha-beta solution heated at 954°C (1750°F) for 1 h. The casting is then fan cooled with an inert gas. It is then aged at 621°C (1150°F) for 2 h. (The fan cooling with gas is a relatively new technique.)

Stress-relief heat treatment is carried out below the intended HIP temperature, typically in the range of 704° to 843°C (1300° to 1550°F) for 2 h. This type of treatment has no noticeable effect on the cast structure, although this temperature range would affect the wrought structure. Treatment is carried out in either vacuum or an inert atmosphere to minimize oxidation. Keep in mind that some large castings cool so slowly that no stress relief is needed.

CAST Ti APPLICATIONS

Titanium castings now are used extensively in the aerospace industry and, to lesser but increasing extent, in chemical-process, marine, and other industries.

Current aerospace applications include major structural components weighing over 135 kg (300 lb) each, and small switch guards weighing less than 30 g (1 oz) each. Titanium castings are used for the space shuttle; wings, engine components, brake components, optical-sensor housings, ordnance, and other parts for military aircraft and missiles; and gas turbine engines as well as brake components for commercial passenger aircraft. Additional aerospace applications include rotor hubs for helicopters and flap tracks for fighters.

In the chemical-process industry, components for pumps, valves, and compressors are made of cast titanium. Marine applications include water-jet inducers for hydrofoil propulsion and seawater valve balls for nuclear submarines. Titanium castings are also used in various other industrial applications, such as well-logging hardware for the petroleum industry, special automotive parts, boat deck hardware, and medical implants.

Some of the complex shapes available in cast-plus-HIP titanium alloys are shown in Figure 8.2.

Photo courtesy of Precision Castparts Corp.

Figure 8.2 - Typical complex shapes

COST COMPARISONS

A major aircraft manufacturer made an in-depth study in which costs of precision titanium alloy castings were compared with costs of parts machined from forgings and blocks of both titanium and aluminum. On the average, metal-removal (machining) costs constitute about 60% of total fabrication cost of an airplane. The use of precision investment castings reduces machining cost to about 5% of total part cost compared to as much as 70 to 80% for the same part made from a forging or hogged out of a block.

Figure 8.3 illustrates the relation of cost to number of units produced for a specific design of aircraft fitting. For all quantities, it was least expensive to produce the fitting as an investment casting.

For a series of 16 parts from one model of commercial aircraft, the average cost of a single part was $749 when 200 units were hogged from titanium alloy blocks. The average cost for investment castings in the same production quantity was $227. This represented a savings of $835 100 for each lot of 100 airplanes constructed if the parts were made from castings.

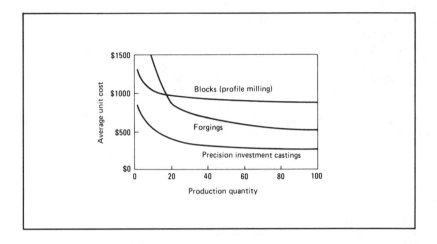

**Figure 8.3 - Cost comparison for one Ti alloy design of aircraft fitting
machined from blocks, forgings, and castings**

For a series of eight parts from another model, average cost was $316 when the parts
were machined from aluminum forgings, but only $179 when the parts were made from
titanium investment castings. This represented a savings of $109 800 for each lot of 100
airplanes.

MECHANICAL PROPERTIES

Cast titanium alloys are equal, or nearly equal, in strength to wrought alloys of the same
compositions. However, typical ductilities are below the typical values for comparable
wrought alloys, yet they are still above the guaranteed minimum values for the wrought
metals. Because castings of Ti-6Al-4V have been used in aerospace applications, the
most extensive data have been developed for this alloy. Typical room-temperature tensile properties are given in Table 8.1 for cast commercially pure titanium and for three
cast titanium alloys.

**Table 8.1 Typical room temperature tensile properties of
several cast titanium alloys**

Alloy	Condition	Tensile strength MPa	ksi	Yield strength (a) MPa	ksi	Elongation % (b)	Reduction in area %
Commercially pure titanium	As-cast or annealed	550	80	450	65	17	32
Ti-6Al-4V	As-cast or annealed	1035	150	890	129	10	19
Ti-6Al-2Sn-4Zr-2Mo	Duplex annealed	1035	150	895	130	8	16
Ti-5Al-2.5Sn-ELI	Annealed	805	117	745	108	11	-

(a) At 0.2% offset.
(b) In 50 mm or 2 in.

These cast property values compare favorably with those of wrought alloys. However fatigue behavior is not always quite so favorable, as may be seen in Table 8.2. HIP helps considerably to raise the fatigue design limits. In addition, claims have been made that heat-treatment modification such as the beta-STOA contribute to a general improvement of fatigue strength to the level of alloy forgings.

Table 8.2 Comparison of cast Ti-6Al-4V; room temperature tensile properties from three sources of test material

Mechanical Property	Separately cast		Cast on Part		Machined from casting	
	MPa	ksi	MPa	ksi	MPa	ksi
Ultimate tensile strength (UTS)	907	113.6	885	128.5	850	123.4
Yield strength (0.2%)	821	119.1	810	117.5	780	113.2
Elongation	10.6%		10.1%		8.6%	
Reduction in area	20.8%		22.2%		15.6%	

(a)   Heat treatment: HIP at 899°C (1650°F), 103 MPa (15 ksi), 2 h plus annealing at 732° C (1350°F), 2 h.

One consideration which should not be overlooked when determining the properties of cast titanium alloys is the origin of test material. Separately cast test bars, bars cast on parts, and bars machined from castings have been used for the determination of properties. Typically, data are reported from separately cast test bars, which are the cheapest source. The least amount of data is derived from test bars machined from castings, which is the most expensive source. (See Table 8.2.)

A second consideration to be kept in mind is that the design for an integrally cast titanium component may not require quite the same strength level as would wrought-ingot metallurgy material. Only after casting trials are run, and properties are determined, should judgments be made concerning the merit(s) of the casting technique.

Chapter 9

Powder Metallurgy

GENERAL

Although titanium, from the standpoint of performance durability, is the best design choice for many systems, far less of it is used in final production than may be anticipated. The cost of titanium alloy components is the "problem," if indeed it can be called that. The initial cost of the alloy, the costs of processing and forging, and the cost of machining at times all go together to make a decision in favor of titanium difficult. As indicated in Table 9.1, a very large difference exists between the planned and realized titanium content of some relatively recent United States Air Force aircraft.

Table 9.1 Titanium content of military airframes

System	Early design, wt%	Final concept, wt%
C-5 (cargo)	24	3
B-1 (bomber)	42	22
F-15 (fighter)	50	34

9

Thus, net shape and near-net-shape technologies are seen as major means of reducing the costs of titanium alloys. These processes include, of course, powder metallurgy (P/M).

Titanium P/M fabrication offers the potential for true net shape capability combined with mechanical properties that are equal to, or exceed, cast and wrought products. This is due to a lack of texture and segregations and to the fine, uniform grain structure inherent in titanium P/M products. Equivalent strength levels along with substantially reduced machining and scrap combine to make P/M titanium products attractive alternatives to conventional ingot metallurgy, involving wrought titanium alloys, and to castings. In order to fully exploit titanium's powder metallurgy potential, substantial additional effort must be made in the reduction of the cost of powder. Furthermore, improved fabrication procedures, coupled with improved nondestructive test (NDI) techniques and realistic NDI standards, are required.

A number of alloy compositions have been produced in powder form. These include:

- Ti-6Al-4V

- Ti-6Al-6V-2Sn

- Ti-10V-2Fe-3Al

- Ti-6Al-2Sn-4Zr-2Mo

- Ti-6Al-2Sn-4Zr-6Mo

Regardless of this variety, well over 90% of the P/M developmental efforts have employed only Ti-6Al-4V. Recently there has been increasing developmental effort centering on the metastable beta alloy Ti-10V-2Fe-3Al. Most references in this discussion, however, involve Ti-6Al-4V.

As with castings, titanium powder metallurgy chemistries have been based on those of common wrought alloys, and so no distinct alloys exist exclusively for P/M. This results in part from ingot metallurgy products' existing and known data base which encourages direct substitution according to product chemistry. A second reason is the availability of barstock and wrought mill products, which are the usual sources of feedstock in powder making. One of the important features of P/M processing, however, is the potential for development of unique, nonstandard chemistries which may not be processable by any other techniques. (Refer to the discussion of rapid solidification rate processing in Chapter 13.)

Ti P/M PRODUCTION PROCESSES

Titanium alloy P/M components are produced by two different processes:

- Blended elemental (BE) method

- Prealloyed (PA) method

Basically, the blended elemental method is a cold press (or cold isostatic press) and sinter process that results in less than fully dense material. As a result, this method provides relatively lower cost products with good tensile strength, fracture toughness, and fatigue-crack growth rates. It also results in somewhat lower fatigue strength. On the other hand, prealloyed powder compacts are fully dense with good mechanical properties, including fatigue strength. Blended elemental components are relatively inexpensive and may be suitable for noncritical applications. The prealloyed method is intended to produce high-performance components in complex shapes.

Prealloyed powder processing is being expanded to include the rapidly solidified alloy powders which offer enhanced alloying opportunities. (Refer to rapid solidification rate, or RSR, in Chapter 13.) RSR is not to be considered a current or near-term titanium alloy technology.

POWDER-MAKING PROCESS

Titanium's inherently high reactivity limits the powder-production process to nontraditional and, therefore, high-cost processes. By contrast to the superalloy industry which managed to adapt gas atomization to powder production, titanium alloys have relied on two principal variations of a centrifugal atomization process. These are:

- Rotating-electrode process (REP)

- Plasma rotating-electrode process (PREP)

The rotating-electrode process uses a tungsten arc to melt the titanium or its alloy from a rotating feedstock. The liquid is rapidly spun off as droplets to cool in the atmosphere. In a hardened form, they collect in the bottom of a chamber.

The plasma rotating-electrode process uses a plasma arc to melt the alloy from a rotating feedstock to produce a powder, much as occurs in REP.

(Both the REP and PREP processes are covered in various powder metallurgy texts if further information is required.)

The principal feature of both REP and PREP is their ability to retain low interstitial element content with minimal contamination. One advantage of the PREP method over the other is the elimination of tungsten inclusions, which in the REP method are produced by the tungsten electrode heat source.

Powder produced by the REP and PREP processes is characteristically coarser than superalloy powders. Also, it is more or less spheroidal in morphology. A sieve and chemical analysis of Ti-6Al-4V REP powder used to make some demonstration aerospace components is given in Table 9.2.

Table 9.2 Sieve and chemical analysis of Ti-6Al-4V REP powder

Sieve size:	35	45	60	80	120	170	230	325	<325
μm:	500	354	250	177	125	88	63	44	<44
% retained:	0.01	1.70	14.95	38.28	33.64	8.95	1.87	0.57	0.03

	%			ppm				
	Al	V	Fe	O	N	C	H	W
Powder batch:	5.97	3.95	0.20	1070	4	(a)	7	36
Typical wrought (MSRR 8618):	5.5-6.75	3.5-4.5	0.3	2000	500	800	100	-
					maxima			

As-HIPed component analyses:

%			ppm			
Al	V	Fe	O	N	C	H
6.22	4.01	0.19	1860-1930	124-176	130-156	40-87

(a) No reliable value available

Thus far this discussion of the powder-making process pertained only to PA powders. Recently, however, much attention has been given to so-called elemental powders. (These may also use a master alloy as one constituent.) Unfortunately the use and future of elemental powder is less clear. Sponge fines of titanium and aluminum-vanadium master alloy powder produced by conventional P/M have been used in the BE process to produce P/M titanium alloy parts.

Alternative methods of producing PA powders have been evaluated, although REP and PREP are currently the only production processes. Those evaluated methods included:

- Powder under vacuum, or pulverization sous vide (PSV)

- Continuous shot casting (CSC)

- Colt-Crucible hydrogenation (C-I) process

- Pendant drop (P-D)

Some effort has been directed to prealloyed powder production by chemical methods, and other entirely different techniques have also been given some consideration. Still, the development of a new technique that produces an inexpensive, high-quality powder is a prerequisite to an expansion of the titanium P/M products market.

For PA powders, a number of processes have been developed to improve the performance of the end product by removing contaminants or adjusting the final microstructure. These include:

- Jet classification

- Electrostatic separation

- Electrodynamic degassing

Characterization of both the as-produced powder particles and the foreign particles which may be present at the loose powder stage can assist in quality control. It has been clearly demonstrated that both the basic microstructure and contaminants present influence mechanical properties, particularly fatigue. A method that is useful in separating out foreign particles for subsequent classification (but not as a cleaning method for titanium) is water elutriation, which distinguishes between particles on a density basis.

CONSOLIDATION

General Processes

Three processes of consolidation are currently being investigated. All of these, it is claimed, are capable of producing fully dense compacts. These processes are:

- Hot isostatic pressing (HIP)

- Press consolidation

- Vacuum hot pressing (VHP)

Of the three procedures, hot isostatic pressing within a heated pressure vessel or autoclave is the most common procedure. By simultaneously applying temperature and pressure, full density in the part is attained by transmittal of the applied pressure.

Press consolidation allows rapid (or lower-temperature) compaction of powder inside a shaped, evacuated can. Very high pressures are attainable and, with certain die designs, close to 100% density is claimed after pressing and sintering. (One particular die design is a fluidized die.)

Finally, powder can be compacted by vacuum hot pressing. In this process, powder is hot compacted in a forge press that is adapted to a vacuum system. Dies press the powder to an essentially 100% density in the required shape. The major disadvantages of this process appear to be the lack of flexibility in shapes that are pressable and, also, the size of parts that can be produced.

With respect to HIP, this process uses the same consolidation equipment as closing casting porosity, although pressure and temperature conditions differ. Cold isostatic compaction (CIP) is similar to HIP, but it is much less elaborate and expensive. It also is performed at room temperature. Shapes can be produced by means of CIP. Molds are much less expensive than those required for HIP. With CIP, vacuum sintering (high-temperature heating for a long period) is required in order to reach 95 to 99% density. HIP is required to reach 100% density.

Most HIP processes use the metal can technology, and fairly complex shapes are produced. The metal can is shaped to the desired configuration by state-of-the-art sheet metal methods. These include brake bending, press forming, spinning, and superplastic forming. Carbon steel is the best suited container material, because it reacts minimally with titanium. It forms titanium carbide, thereby inhibiting further reaction.

Blended Elemental Powders

The basic blended elemental powder compaction methods are:

- Cold isostatic pressing (CIP)

- Press consolidation

These methods may be followed by sintering in order to reach 95% density. They also may be followed by HIP to achieve almost 100% density.

Considerable emphasis recently has been placed on CIP, CIP plus sinter, and CIP plus sinter plus HIP. The last has been designated the CHIP process.

Cost savings realized in BE powder compaction result from reduced raw material requirements and reduced handling-and-process cycle time. BE, of course, assumes that less than 100% density is acceptable.

Powder components for a BE part are blended and consolidated. The titanium powder component comes from the sponge fines, normally surplus material produced during the conversion of ore to ingot. These inexpensive, acicular powders cost only a little more than titanium sponge.

Prealloyed Powders

The principal method of consolidating prealloyed powders appears to be HIP. Fluid die compaction (FDC) and vacuum hot pressing (VHP) also have been evaluated. Because of cost considerations, these methods — and other similar ones — attempt to produce near-net shape (NNS) configurations of the desired component. Limited machining envelopes are incorporated in the design of the cans or dies used, but the requirements of nondestructive testing limit the extent to which the NNS can be approached. Practical considerations of component complexity also affect the final compact volume and geometry.

Direct processing of PA powders to mill products can be accomplished. (This also is possible with BE powders.) Plate and barstock have been produced, although cost effectiveness is questionable. Preforms for subsequent forging also have been considered. (Use of preforms is common in the powder metallurgy field; preconsolidated billets — from HIP or extrusion — are common in the superalloy field. These latter are used as stock for the isothermal forging of aircraft gas turbine components.

Similar process innovations have been evaluated for titanium PA powders. For beta alloy Ti-10V-2Fe-3Al, properties are claimed to be equal to, or exceed, the ingot metallurgy of wrought products. The use of P/M preforms may promote more confidence in property levels of P/M titanium alloy products as many engineers feel that some level of working (that is deformation) enhances inspectability of the product and increases uniformity of products.

SHAPEMAKING

Blended Elemental Powders

Cold isostatic pressing with elastomeric molds can produce extremely complex shapes, such as the impeller shown in Figure 9.1. Because the mold material is elastomeric (as

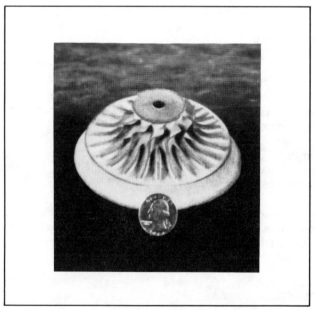

Photo courtesy of Dynamet Technology, Inc.

Figure 9.1 - Impeller made of Ti-6Al-4V elemental blend powder

opposed to hard punch and die tooling), lubricant is not needed, thus eliminating a presintering treatment to remove lubricant, as well as eliminating lubricant contamination. A thick-walled rubber bag and steel mandrel for hydrostatically pressing sponge fines to produce a cold-pressed, commercially pure titanium, thin-walled splined cylinder for a gyro application is shown in Figure 9.2. However, part size is limited to a maximum diameter of 60 cm (24 in.); length is limited by the availability of cold isostatic pressing equipment. Dimensional tolerance as yet has not been determined by production data but appears to be about ±0.02 mm/mm (±0.02 in./in.) length.

Figure 9.2 - Tools to CIP a gyro component

A 15.3 kg (34 lb), 100% dense airframe component made by means of the CHIP method of BE Ti-6Al-4V powder (with extra-low chloride content) is shown in Figure 9.3. By contrast, a 45 kg (100 lb) wrought billet would have been needed for more traditional methods.

Figure 9.3 - Airframe component made of CHIPed blended elemental Ti-6Al-4V extra-low chloride powder

Press consolidation is not limited by size; presses up to 45,000 metric tons (50,000 tons) are available in the United States. Consequently, a part of almost 13,000 cm^2 (2000 in.^{2}) can be produced. Intricate shape capabilities, however, do not approach those possible by cold isostatic pressing. Production data are required to establish dimensional tolerance, but these techniques appear capable of ± 0.01 mm/mm (± 0.01 in./in.) length. A typical part produced from elemental blended Ti-6Al-4V powder by the BE powder consolidation method is shown in Figure 9.4.

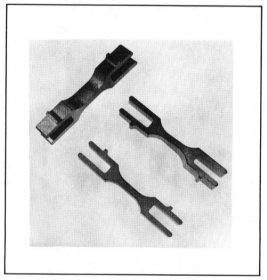

Photo courtesy of Imperial Clevite, Inc.

**Figure 9.4 - Connector link arm for
Pratt & Whitney F-100 engine**

Prealloyed Powders

Complex shapes are currently being produced principally by means of HIP or by the Colt-Crucible ceramic mold process (HIP-CCMP), a variation of the HIP process. CCMP is similar to the technology used for the investment casting process. Shaped wax patterns are made and a ceramic mold is prepared. After dewaxing and firing, the mold is evacuated, outgassed, and sealed. (This is similar to the way metal cans used in the standard HIP process are treated.) The sealed mold is then HIPed. HIP-CCMP is capable of producing more complex shapes than does HIP with standard technology.

Many of the P/M aerospace parts have been fabricated by means of the HIP-CCMP combination. The feasibility of manufacturing a small airframe component, the F-14A fuselage brace, shown in Figure 9.5(a), was established. The part has a high rib, deep pocket design which normally requires a great deal of machining. Another airframe part being produced by the same process is the F-18 engine mount support shown in Figure 9.5(b). The part is basically a sheet with stiffeners and involves extensive machining when it is made as a forging.

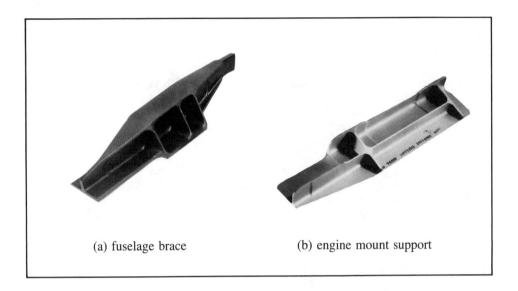

      (a) fuselage brace   (b) engine mount support

Figure 9.5 - Ti-6Al-4V P/M parts made by means of Colt-Crucible ceramic mold process

Recently, the Colt-Crucible ceramic mold technique has been advanced to the point where larger and more complex shapes can be made, including the F-107 radial compressor rotor for the Cruise missile engine, which is being produced to near-net shape. The ultimate goal is that no machine tool will touch the part; that is, it will be contoured to final dimensions by chemical milling only. The compressor rotor configuration is generic to many engines. Qualification of the part should lead to widespread application.

A PA Ti-6Al-4V powder was employed with standard HIP can technology to produce an airframe connecting arm. A total reduction of about 34% in production dollars was realized. HIP technology has been used to produce critical components for nonaerospace technology, too. Note the knee and hip prostheses shown in Figure 9.6.

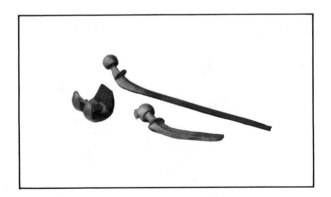

**Figure 9.6 - HIP prosthesis and knee
joint implant made from P/M Ti-6Al-4V**

APPLICATIONS OF Ti P/M

General

Although titanium P/M technology normally is associated with the aerospace industry, it has received limited use in the industrial and biomedical industries. It is used in the chemical industry for fasteners, fittings, and valve components. It currently is being considered for use for possible automotive applications.

PA powder achieves the necessary clean powder production standards, and it is the primary material used in aerospace P/M application efforts. BE powder has yet to achieve the necessary property levels owing to retained chlorides and porosity characteristics. (Porosity may, at times, result from the salt.) The lower production costs for BE powder titanium alloys, however, will ensure a significant advantage for them in less critical applications, even in the aerospace field.

Blended Elemental Powder Applications

Components made from BE powders are produced for aerospace and nonaerospace industries. Figure 9.7 illustrates typical nuts and fasteners produced for the electrochem-

Figure 9.7 - Typical cold-pressed and sintered Ti and Ti-6Al-4V nuts and parts produced from BE powder

ical industry from commercially pure titanium and Ti-6Al-4V alloy by cold pressing and sintering. Other blended elemental products manufactured for the chemical industry include valves, valve balls, and fittings made from commercially pure titanium and Ti-6Al-4V alloy.

Since the 1950s, extensive work has been conducted on the development of hot isostatically pressed titanium alloys for surgical implants. Most early use of titanium for prostheses entailed the unalloyed grades. Ti-6Al-4V has been investigated extensively for this purpose, because it retains the biocompatibility of the unalloyed metal with improved mechanical properties.

Partial and total joint replacements for the hip, knee, elbow, jaw, finger, and shoulder have been produced commercially from both unalloyed titanium and Ti-6Al-4V. Figure 9.6 shows a hip prosthesis and knee joint implant made from P/M Ti-6Al-4V; Figure 9.8 shows fasteners used for prosthetic fixation made from the same alloy.

Photo courtesy Imperial Clevite, Inc.

Figure 9.8 - Fasteners for external prosthetic fixation made of pressed and sintered Ti-6Al-4V

For partial joint replacements in which the implant articulates with the cartilage, titanium has been found to have the same minimal wear exhibited by stainless steel and cobalt-chromium-molybdenum alloys.

Prealloyed Powder Applications

The full density achieved by prealloying and the high level of mechanical properties and net shape capability of complex configurations made the PA powder P/M method suitable for aerospace components. In fact, the majority of aerospace prealloyed titanium P/M parts currently are produced by HIP-CCMP process, explained earlier. Recently, the process has advanced so that larger and more complex shapes can be made. These include the Cruise missile engine F-107 radial compressor rotor, which is produced so closely to net shape that only a final chemical mill should be required, thus eliminating expensive machining. The dimensional reproducibility of this method is demonstrated in Table 9.3. The generic nature of this configuration should lead to widespread P/M engine applications.

Table 9.3 Shape reproducibility of Ti-6Al-4V P/M impeller produced by the Colt-Crucible ceramic mold process (HIP-CCMP) (a)

Dimension	Median			Variation		Variation, %
		mm	in.	mm	in.	
Radius measured from centerline (OD)	88.9	3.500	±0.66	±0.026		±0.7
Radius measured from centerline (ID)	53.31	2.217	±0.63	±0.025		±1.1
Blade height on large impeller outer diameter	55.06	2.168	±0.38	±0.015		±0.7
(a) Values are based on 7 parts and 23 locations per part.						

The largest P/M part currently produced by HIP-CCMP is the F-14A Nacelle frame, which is 100 x 120 cm (40 x 48 in.). This component consists of a high rib and deep pocket design that currently requires significant machining during fabrication. The part is to be fabricated by electron beam welding of four hot isostatically pressed powder parts into an oval-shaped F-14A frame section, which increases the difficulty of producing this part. (Refer to Figure 9.9.)

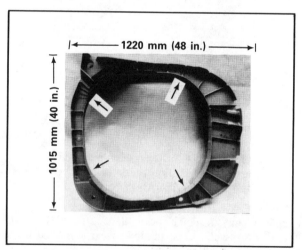

Photo courtesy of Colt Industries

Figure 9.9 - Largest Ti P/M component produced by HIP-CCMP

COST FACTORS

As discussed previously, cost reduction is the major reason, at this time, for using P/M processing instead of conventional alloy production. Because the bulk of the cost of titanium component fabrication is concentrated in forging and machining, selection of appropriate parts can be made only after consideration of these factors, as well as material selection. Forging and machining costs are directly proportional to the size and complexity of the part. Material costs are higher as the buy-to-fly ratio increases. (Buy-to-fly ratio indicates the number of kilograms or pounds of purchased material for each kilogram or pound of finished part.) Generally, P/M processing is attractive for large, complex parts with a high buy-to-fly ratio when fabricated by conventional methods. However, because the largest autoclave currently available is about 120 cm (48 in.) in diameter by 245 cm (96 in.) in height, an upper size limit exists, unless the process is modified (by subsequent welding, for example) for fabrication of large components. (See Figure 9.9.)

Table 9.4 lists the current forging weight, P/M product weight, final part weight, and anticipated potential cost savings for various parts produced by the prealloyed technique. These estimates suggest that cost savings realized by P/M processing compared to forged parts could range between 20 and 50%, depending on the size and complexity of the part and production quantity. Higher volume runs result in higher savings. An additional advantage, which assumes even greater importance with material shortage, is that lead-time can be reduced by 50% or more for P/M parts over equivalent forged parts.

Table 9.4 Typical titanium prealloyed P/M parts produced by Colt-Crucible ceramic mold process (HIP-CCMP)

Component	Part weight:						% save.
	Forged Billet		P/M part		Final part		
	kg	lb	kg	lb	kg	lb	
F-14 fuselage brace	2.8	6.2	1.1	2.5	0.77	1.7	50
F-18 engine mount support	7.7	17.0	2.5	5.5	0.5	1.1	20
F-18 arrestor hook support fitting	79.4	175.0	24.9	55.0	12.8	28.4	25
F-107 radial compressor impeller	14.5	32.0	2.8	6.2	1.6	3.6	40
F-14 Nacelle frame	142.8	315.0	82.1	181.0	24.1	53.2	50

Conventionally pressed and sintered P/M titanium represents a theoretical density of 94 to 96%. Further working or processing is required to achieve full density and to maximize properties. The use of high-density (but less than fully dense) titanium alloys as preforms may offer advantages here. A prototype compressor blade preform was cold isostatically pressed with blended elemental Ti-6Al-4V powder and sintered to approximately 95% density. The finished forged product was achieved with only one blow, resulting in minimal flash. (Refer to Figure 9.10.) Consequently, minimal scrap resulted. Conventional forging of this part would have required substantially more barstock and several sets of breakdown tooling. A substantial improvement in economics thus was realized with P/M processing.

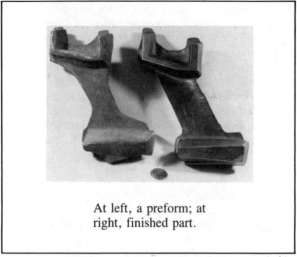

At left, a preform; at
right, finished part.

Photo courtesy of Imperial Clevite Technology Center

**Figure 9.10 - Cold isostatically pressed
compressor blade**

A P/M titanium alloy dome housing for the Sidewinder missile is in commercial production. Previously, this component was machined from a billet blank weighing over 2.2 kg (5 lb). Final shape was machined to close tolerances and a thin wall of 0.635 mm (0.025 in.). Alternatively, isostatic pressing and vacuum sintering of this component resulted in a preformed blank weighing only 0.56 kg (1.25 lb). It significantly reduced the amount of finish machining necessary. Additionally, because final machining was conducted with a free-machining material, close tolerances were easily achieved. The isostatic alloy pressed and sintered preform (left) and the finished machined component (right) are shown in Figure 9.11.

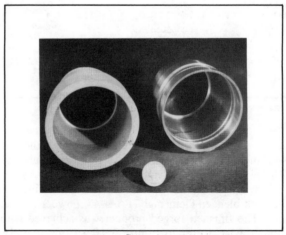

Photo courtesy of Dynamet Technology, Inc.

**Figure 9.11 - P/M Ti alloy dome
housing for Sidewinder missile**

In some cases, the cost of the P/M preform may exceed the cost of billet stock, but significant cost savings may be achieved through reduced machining costs. With the housing, however, the cost of preforming is only 60% of the cost of a wrought billet, with additional savings achieved in final machining.

A typical P/M titanium ordnance component is shown in Figure 9.12. The preform and finished machine part are shown. This non-rotating part is a lens housing for the Maverick missile. Powder metallurgy effectively reduced the buy-to-fly ratio from 15-to-1 to 3-to-1. Material savings were approximately 80%. Additional savings resulted from the superior machinability of P/M titanium, even though this part was machined to very close tolerances.

Photo courtesy of Imperial Clevite Technology Center

**Figure 9.12 - Preform and machined
lens housing of a P/M Ti alloy**

Commercially pure titanium fasteners produced by P/M fabricating methods are shown in Figure 9.13. Acceptance of these fasteners has been exceptional because of significant cost savings realized by P/M processing. This P/M fastener has captured a large segment of the titanium hex nut market and is likely to find use in ordnance applications.

Costs of titanium components are greatly influenced by the volume of the production run—as is common in all P/M applications. Add to this factor the costs of achieving critical properties by means of superclean powders and the even more costly compacting techniques (HIP, VHP). Powder costs may be volume driven, but volume is also cost driven. As more powder chemistries become available and volume creeps upward, overall costs should show a decrease.

For specialized applications in which rapid solidification rate powder is required, costs are likely to escalate, yet there will be a performance payback. A general rule of thumb for the use of titanium P/M would be to consider the technology from the very start

Photo courtesy of Imperial Clevite Technology Center

Figure 9.13 - Cp Ti hex nuts

when a wrought titanium alloy is typically applied. This rule assumes that commercially pure titanium or Ti-6Al-4V will provide more than ample mechanical properties, and excess-stock removal costs can be balanced against the higher initial costs of the titanium powder.

MECHANICAL PROPERTIES

Mechanical properties of titanium P/M parts are determined by the use of PA or BE powders along with the sophistication of the consolidation techniques employed. When high-temperature, long-time processes with slow cooling rates — HIP, for example — are used, microstructures "as-HIPed" are more coarse than desirable. When the microstructure is refined by heat treatment, working, combinations of both or other techniques, optimum properties result.

Tensile and creep-rupture properties also are moderately affected when less than full densification is achieved. Porosity and small inclusions do not affect static properties as much as microstructure does. Cyclic properties such as fatigue are critically dependent on defects in the P/M parts as well as on microstructure.

Tensile properties of several titanium alloys produced from BE powders compacted by cold pressing and sintering are shown in Figure 9.14. Strength levels of the pressed and sintered materials are comparable to wrought materials, but ductility generally is lower.

The fatigue properties of BE powder, pressed and sintered compacts, and hot isostatically pressed PA powder are compared in Figure 9.15. The fatigue properties of PA powder, when pressed to 100% density and with optimized microstructure, can be equivalent to wrought alloy product. Remnant salt and porosity in customary BE powder compacts result in lower fatigue capability. The application of cold isostatic pressing plus sinter plus HIP (CHIP) to BE powder and the reduction of salt content should increase fatigue strength.

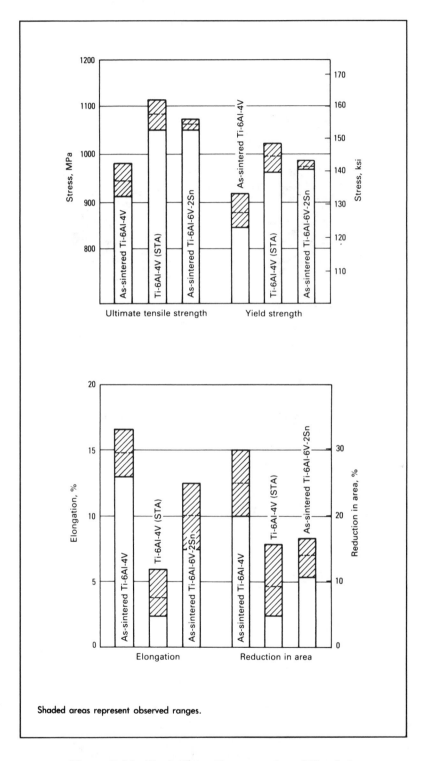

**Figure 9.14 - Typical tensile properties of blended
elemental P/M powders**

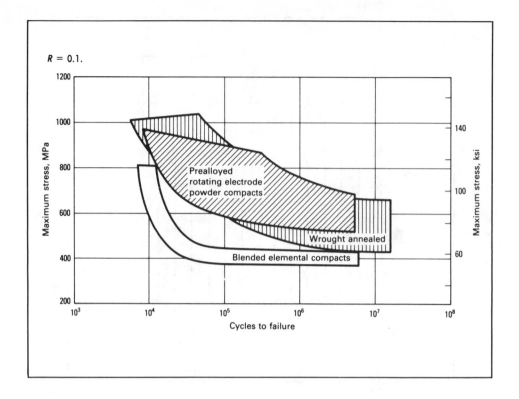

Figure 9.15 - Comparison of room-temperature smooth axial fatigue behavior of Ti-6Al-4V blended elemental and pre-alloyed P/M compacts with wrought annealed material

Chapter 10

Joining

GENERAL

Titanium, as a reactive material, requires special considerations both before and during joining activities. When proper precautions are taken and correct preparations are made, titanium and its alloys can be successfully joined for applications ranging from subzero levels to elevated temperatures. The specific techniques available for titanium joining discussed here are:

- Gas tungsten arc welding (GTAW or TIG)

- Gas metal arc welding (GMAW or MIG)

- Plasma arc welding (PAW)

- Electron beam welding (EBW)

- Laser beam welding (LBW)

- Resistance spot welding (SPOT) and seam welding (SEAM)

- Flash welding (FW)

- Diffusion bonding (DB)

Also available, but not discussed here, is ultrasonic welding (UW).

With respect to diffusion bonding, which is a specialized fabrication and welding operation, superplastic forming/diffusion bonding (SPFDB) has been used in the aerospace industry. Also, titanium can be brazed, but it is not adhesively bonded commercially at this time.

The fusion weldability of a specific titanium alloy is related to its grade or alloy composition and microstructure. Excess β phase restricts weldability; thus α, near-α, and α-β alloys with a low β contents are more readily fusion welded. Gaseous contamination is of great concern in fusion welds, but it is also a concern whenever temperatures exceed about 427°C (800°F). (Refer to Chapter 7, "Cleaning and Finishing.")

Open-air techniques can be used with fusion welding when the area to be joined is well shielded by an inert gas. By and large, however, atmospheric control by means of a "glove box," temporary bag, or chamber is preferred. Air must be excluded when solid state bonding and brazing methods are used.

The subsequent discussions in this chapter divide the subject of joining into three broad categories:

- Fusion welding

- Solid state welding

- Brazing

Throughout the discussions contained in this chapter, frequent references are made to titanium's various alloys and type: alpha, alpha-beta, beta, etc. Refer to Appendix J, "Common Designation Listing" or to Chapter 3, "Understanding the Metallurgy of Titanium."

ARC WELDING (A FUSION WELDING PROCESS)

General

Titanium and most titanium alloys can be welded by the following fusion welding techniques:

- Gas tungsten arc

- Gas metal arc

- Plasma arc

Each of these specific techniques is discussed below, but these introductory comments on arc welding apply in general ways to them.

Procedures and equipment are generally similar to those used for welding austenitic stainless steel or aluminum. Because titanium and titanium alloys are extremely reactive above 538°C (1000°F), however, additional precautions, ones exceeding those required during the welding of austenitic stainless steel or aluminum alloys, must be taken to shield the weld and hot root side of the joint from air. One method involves an inert, gas-filled, large Mylar bag in which the welding is performed.

Weldability

Unalloyed titanium and all-alpha titanium alloys are weldable. Although the alpha-beta alloy Ti-6Al-4V and other weakly beta-stabilized alloys are also weldable, strongly beta-stabilized alpha-beta alloys are embrittled by welding. Most beta alloys can be successfully welded, but because aged welds in beta alloys can be quite brittle, heat treatment to strengthen the weld by age hardening should be used with caution.

All grades of unalloyed titanium usually are welded in the annealed condition. Welding of cold-worked alloys anneals the heat-affected zone (HAZ) and negates the strength produced by cold working.

Alpha alloys and near-alpha alloys such as Ti-5Al-2.5Sn, Ti-6Al-2Sn-4Zr-2Mo, Ti-5Al-5Sn-2Zr-2Mo, Ti-6Al-2Cb-1Ta-1Mo, and Ti-8Al-1Mo-1V are always welded in the annealed condition.

Alpha-beta alloys of Ti-6Al-4V can be welded in the annealed condition or in the solution-treated and partially aged condition, with aging being completed during postweld stress relieving. In contrast to unalloyed titanium and the alpha alloys, which can be strengthened only by cold work, the alpha-beta and beta alloys can be strengthened by heat treatment.

The low weld ductility of most alpha-beta alloys is caused by phase transformation in the weld zone or in the HAZ. Alpha-beta alloys can be welded autogenously or with various filler metals. It is common to weld some of the lower-alloyed materials with matching filler metals. Where strength is not critical and more toughness is desired, unalloyed alpha titanium filler metals may be used. Filler metal of matching composition is used to weld the Ti-6Al-4V alloy. The extra-low-interstitial (ELI) grade is employed to improve ductility and toughness.

The use of filler metals that improve ductility may not prevent embrittlement of the HAZ in susceptible alloys. In addition, low-alloy welds can be embrittled by hydride precipitation. It should be noted, however, that with proper joint preparation, filler-metal storage, and shielding, hydride precipitation can be avoided.

Sheet thicknesses of 2.54 mm (0.100 in.) and thinner can be welded without filler metal additions by the fused-root welding technique. Filler metal may be added to repair unfused and sunken-weld metal areas. The lack of a joint line on the root face of the weld indicates 100% penetration.

Metastable beta alloys such as Ti-13V-11Cr-3Al, Ti-11.5Mo-6Zr-4.5Sn, Ti-8Mo-8V-2Fe-3Al, Ti-15V-3Cr-3Al-3Sn, and Ti-3Al-8V-6Cr-4Zr-4Mo are weldable in the annealed or solution heat-treated condition. In the as-welded condition, welds are low in strength but ductile. Beta alloy weldments are sometimes used in the as-welded condition. Welds in the Ti-13V-11Cr-3Al alloy embrittle more severely when age hardened. To obtain full strength, the metastable beta alloys are welded in the annealed condition; the weld is cold worked by peening or planishing, and the weldment is then solution treated and aged. This procedure also obtains adequate ductility in the weld.

Welding Processes

Gas tungsten arc welding (GTAW) is the most widely used process for joining titanium and titanium alloys, except when large thicknesses are involved. Square-groove butt joints can be welded without filler metal in base metal up to 2.54 mm (0.10 in.) thick. With thicker base metal, the joint should be grooved, and filler metal is required. Where possible, welding should be done in the flat position. Hot wire GTAW can be used for welding titanium alloys more than 6.35 mm (0.25 in.) thick.

Gas metal arc welding (GMAW) is employed to join titanium and titanium alloys more than 3.18 mm (1/8 in.) thick. It is applied using pulsed current or the spray mode. It is less costly than GTAW, especially when base-metal thickness is greater than 12.7 mm (0.5 in.).

Plasma arc welding (PAW) is also applicable to the welding of titanium and titanium alloys. It is faster than GTAW and can be used on thicker sections, such as one-pass welding of titanium alloy plate up to 12.7 mm (0.5 in.) thick, using square-groove butt joints and the keyhole technique.

Filler Metals

For welding titanium thicker than about 2.54 mm (0.10 in.) by the GTAW process, a filler metal must be used. For PAW, a filler metal may or may not be used for welding metal less than 12.7 mm (0.5 in.) thick.

Fourteen titanium and titanium alloy filler-metal (or electrode) classifications are given in AWS specification A5.16. (See Appendix D, "Filler Metals.") Five of these are essentially unalloyed titanium, and the remainder are titanium alloy filler metals. Maximums are set on carbon, oxygen, hydrogen, and nitrogen contents. Compositions for titanium and titanium alloy filler metals are given in Table 10.1.

Filler-metal composition usually is matched to the grade of titanium being welded. For improved joint ductility when welding the higher-strength grades of unalloyed titanium, filler metal of yield strength lower than that of the base metal occasionally is used. Because of the dilution that occurs during welding, the weld deposit acquires the required strength. Unalloyed filler metal is sometimes used to weld Ti-5Al-2.5Sn and Ti-6Al-4V for improved joint ductility. The use of unalloyed filler metals lowers the beta content of the weldment, thereby reducing the extent of the transformation which occurs and thus improving ductility.

Engineering approval, however, is recommended when employing pure filler metal to ensure that the weld meets strength requirements. Another option is filler metal containing lower interstitial content (oxygen, hydrogen, nitrogen, and carbon) or alloying contents that are lower than the base metal being used. The use of filler metals that improve ductility does not preclude embrittlement of the HAZ in susceptible alloys. In addition, low-alloy welds may enhance the possibility of hydrogen embrittlement.

The filler metal, as well as the base metal, should be clean at the time of welding. Wires of the size used for filler metals have a large surface-to-volume ratio. Therefore, if the wire surface is slightly contaminated, the weld may be severely contaminated. Some procedures require that the filler wire be cleaned immediately before use. An acetone-soaked, lint-free cloth serves to access surface contamination caused by the die lubricant used in the wire-drawing operation, in addition to cleaning the filler wire. Pickling in nitric/hydrofluoric acid solution is also used for cleaning.

Shielding Gases

When welding titanium and titanium alloys, only argon or helium, and occasionally a mixture of these two gases, are used for shielding. Since it is more readily available and less costly, argon is more widely used. Argon shielding gas was used in the examples given in this chapter.

Table 10.1 Chemical compositions of titanium and titanium alloy filler metals per AWS A5.16 (a)

AWS classification	Composition, %						
	C	O	H	N	Al	V	Sn
ERTi-1 (b)	0.03	0.10	0.005	0.012	-	-	-
ERTi-2	0.05	0.10	0.008	0.020	-	-	-
ERTi-3	0.05	0.10-0.15	0.008	0.020	-	-	-
ERTi-4	0.05	0.15-0.25	0.008	0.020	-	-	-
ERTi-0.2Pd	0.05	0.15	0.008	0.020	-	-	-
ERTi-3Al-2.5V	0.05	0.12	0.008	0.020	2.5-3.5	2.0-3.0	-
ERTi-3Al-2.5V-1 (b)	0.04	0.10	0.005	0.012	2.5-3.5	2.0-3.0	-
ERTi-5Al-2.5Sn	0.05	0.12	0.008	0.030	4.7-5.6	-	2.0-3.0
ERTi-5Al-2.5Sn-1 (b)	0.04	0.10	0.005	0.012	4.7-5.6	-	2.0-3.0
ERTi-6Al-2Nb-1Ta-1Mo	0.04	0.10	0.005	0.012	5.5-6.5	-	-
ERTi-6Al-4V	0.05	0.15	0.008	0.020	5.5-6.75	3.5-4.5	-
ERTi-6Al-4V-1 (b)	0.04	0.10	0.005	0.012	5.5-6.75	3.5-4.5	-
ERTi-8Al-1Mo-1V	0.05	0.12	0.008	0.03	7.35-8.35	0.75-1.25	-
ERTi-13V-11Cr-3Al	0.05	0.12	0.008	0.03	2.5-3.5	12.5-14.5	-

AWS classification	Composition, %						
	Cr	Fe	Mo	Nb	Ta	Pd	Ti
ERTi-1 (b)	-	0.10	-	-	-	-	rem
ERTi-2	-	0.20	-	-	-	-	rem
ERTi-3	-	0.20	-	-	-	-	rem
ERTi-4	-	0.30	-	-	-	-	rem
ERTi-0.2Pd	-	0.25	-	-	-	0.15-0.25	rem
ERTi-3Al-2.5V	-	0.25	-	-	-	-	rem
ERTi-3Al-2.5V-1 (b)	-	0.25	-	-	-	-	rem
ERTi-5Al-2.5Sn	-	0.40	-	-	-	-	rem
ERTi-5Al-2.5Sn-1 (b)	-	0.25	-	-	-	-	rem
ERTi-6Al-2Nb-1Ta-1Mo	-	0.15	0.5-1.5	1.5-2.5	0.5-1.5	-	rem
ERTi-6Al-4V	-	0.25	-	-	-	-	rem
ERTi-6Al-4V-1 (b)	-	0.15	-	-	-	-	rem
ERTi-8Al-1Mo-1V	-	0.25	0.75-1.25	-	-	-	rem
ERTi-13V-11Cr-3Al	10.0-12.0	0.25	-	-	-	-	rem

(a) Single values are maximum.
(b) Extra-low interstitials for welding similar base metals.

Because of high purity (99.985% min) and low moisture content, liquid argon is often preferred. The argon gas should have a dew point of -24°C (-75°F) or lower. The hose used for shielding gas should be clean, nonporous, and flexible. The recommended materials are Tygon or vinyl plastic. Because rubber hose absorbs air, it should not be used. Excessive gas flow rates that cause turbulence should be avoided, and flowmeters are usually employed for all gas shields. Pressure (psi) gages may be employed for trailing and backup shields.

The type of shielding gas used affects the characteristics of the arc. At a given welding current, the arc voltage is much greater with helium than with argon. Because the heat energy liberated in helium is about twice that in argon, higher welding speeds can be obtained, weld penetration is deeper, and thicker sections can be welded more rapidly using helium shielding. However, when using pure helium for welding, arc stability and weld-metal control are sacrificed.

Argon is used in the welding of thin and thick sections where the arc length can be altered without appreciably changing the heat input. Argon-helium mixtures are also employed, particularly 75% argon, which improves arc stability, and 25% helium, which increases penetration. The 75%Ar-25%He mixture is also frequently utilized as the shielding gas at the torch in automatic operations.

Helium is used in shielding of out-of-position welds.

Joint Preparation

If welding is done outside a controlled-atmosphere welding chamber, joints must be carefully designed so that both the top and the underside of the weld can be shielded. Dimensions of typical joints are given in Table 10.2. For welding titanium alloys, joint fit-up should be better than for welding other metals, because of the possibility of air entrapment in the joint. To prevent separation during welding, the joint should be clamped.

Table 10.2 Dimensions of typical joints for welding titanium and titanium alloys (a)

Base-metal thickness, in.	Root opening, in.	Groove angle,°	Weld bead width, in.
Square-groove butt joint:			
0.010-0.090	0	–	–
0.031-0.125	0 – 0.10t	–	–
Single-V-groove butt joint:			
0.062-0.125	0 – 0.10t	30 – 60	0.10 – 0.25t
0.090-0.125	(b)	90	–
0.125-0.250	0 – 0.10t	30 – 60	0.10 – 0.25t
Double-V-groove butt joint:			
0.250-0.500	0 – 0.20t	30 – 120	0.10 – 0.25t
Single-U-groove butt joint:			
0.250-0.750	0 – 0.10t	15 – 30	0.10 – 0.25t
Double-U-groove butt joint:			
0.750-1.500	0 – 0.10t	15 – 30	0.10 – 0.25t
Fillet weld:			
0.031-0.125	0 – 0.10t	0 – 45	0 – 0.25t
0.125-0.500	0 – 0.10t	30 – 45	0.10 – 0.25t

(a) t is base-metal thickness.
(b) Root face, 0.76 mm (0.030 in.)

Cleaning

To obtain a good weld, the joint and the surfaces of the workpiece at least 50 mm (2 in.) beyond the width of the gas trailing shield on each side of the weld groove must be meticulously cleaned. As shown in Figure 10.1, the cleaning procedure depends on whether the oxide layer in the joint area is light or heavy.

Titanium alloys may be descaled in molten salt baths or by abrasive blasting. Chlorinated solvents, such as trichlorethylene, should not be used for degreasing titanium alloys, because the chlorine residues cause intergranular attack in subsequent heating operations. Chemical cleaning may be performed by pickling for 1 to 20 min in solutions containing 20 to 47% nitric acid plus 2 to 4% hydrofluoric acid in water, or about a 10-to-1 ratio of nitric acid in hydrofluoric acid; bath temperature is 27° to 71°C (80° to 160°F). (Refer to Chapter 7, "Cleaning and Finishing," for additional information.)

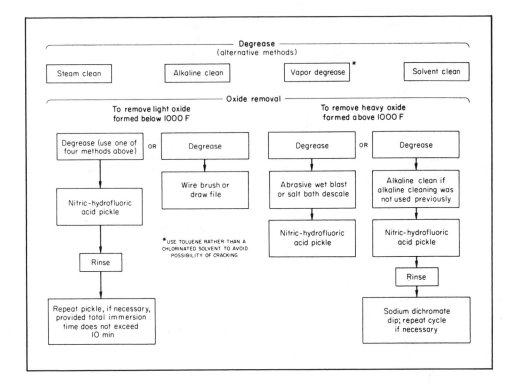

Figure 10.1 - Procedures flowchart for cleaning Ti alloys

Welding in Chambers

For successful arc welding of titanium and titanium alloys, complete shielding of the weld is necessary, because of the high reactivity of titanium to oxygen and nitrogen at welding temperatures. Excellent welds can be obtained in titanium and its alloys in a welding chamber, where welding is done in a protective gas atmosphere, thus giving adequate shielding. Welding in a chamber, however, is not always practical. For example, when manual welding within a metal chamber, the location of the glove ports and the presence of a chamber wall impose limitations on visibility, movement, and accessibility.

For large assemblies, welding in a chamber requires unloading of the chamber after each weldment is completed; this implies the loss or controlled removal of purging gas. The chamber must be repurged before welding the next assembly. Such procedures are time-consuming and expensive. Various types of modified chambers, such as clamp-on chambers, have been tried to remedy these problems.

Experimental GTA welding of titanium was first done in metal chambers that can be evacuated and then backfilled with argon or helium. Such chambers are equipped with glove ports so that the welder can handle the torch, separate filler metal (if used), and the weldment without admitting air to the chamber. Viewing ports enable the welder to see the welding operation. Although expensive to operate, especially for large weldments, metal chambers frequently are used in aerospace applications.

Generally, shielding gas is not supplied to the welding torch when welding titanium in a metal chamber. Excellent welds can be made if the chamber atmosphere is maintained properly. In some applications, however, where heavy or long welds are required, gas is supplied to the torch to improve shielding.

With flow-purged chambers, the atmosphere is often tested by welding a piece of scrap titanium before making the assembly weld. The color of the solidified weld metal is observed by gradually pulling the torch away from the molten pool. The weld-metal colors, in increasing order of contamination, are bright silver, light straw, dark straw, light blue, dark blue, gray-blue, gray, and white loose powder. A light straw color is generally considered acceptable for all but the most stringent requirements, although a bright silvery color like a newly minted dime is preferred.

To continuously monitor the inert atmosphere, a heated tungsten filament may be placed inside the chamber. Any discoloration or ignition of the filament indicates that the purity of the atmosphere has become degraded.

Rigid or collapsible chambers made of transparent plastic can be used when production runs are short, the assembly is large or complicated, and when manual welding is required. Rigid plastic chambers are flow-purged before welding is started with argon or helium in volumes equal to five to ten times the volume of the chamber. Collapsible plastic chambers are first collapsed and then flow-purged with argon or helium; they require less gas for purging than do rigid chambers. It is of interest that collapsible plastic chambers have been used on the titanium production line of a major gas turbine manufacturer for over 30 years.

Advantages of plastic chambers (either rigid or collapsible) are low cost and good visibility of the work. Because there is generally a greater probability of leakage occurring in a plastic chamber than in a metal chamber, the atmosphere must be checked frequently to ensure that it is of proper purity. In addition, torch shielding is usually employed to make certain that the weld zone is adequately protected.

Out-of-Chamber Welding

With proper tooling, joints in titanium can be adequately shielded for welding without using a chamber. However, both the weld and the heat-affected zone (HAZ) must be temporarily shielded during welding and until the temperature of the metal in the area of the weld is below 538°C (1000°F). If shielding is inadequate, the welds absorb oxygen and become brittle.

With the out-of-chamber method, the welding torch (or electrode holder) usually is equipped to supply a trailing shield that provides a diffuse, nonturbulent flow of gas to the solidifying weld. A typical trailing shield is shown in Figure 10.2. The length of the trailing shield must be adjusted to the speed of welding. Both straight and curvilinear welding can be shielded. In addition, the welding station must be shielded by curtains to prevent drafts. Most shields are designed and/or handcrafted for the particular weld.

GAS TUNGSTEN ARC WELDING (AN ARC WELDING TECHNIQUE)

Gas tungsten arc welding (GTAW), often called tungsten inert gas (TIG) welding, is an arc welding technique in which the heat is produced between a nonconsumable electrode

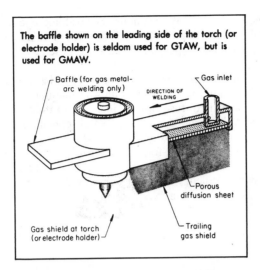

The baffle shown on the leading side of the torch (or electrode holder) is seldom used for GTAW, but is used for GMAW.

Figure 10.2 - Shielding arrangement for automatic welding of Ti alloys in air

and the work metal. The electrode, weld pool, arc, and adjacent heated areas of the workpiece are protected from atmospheric contamination by a gaseous shield.

Transformer-rectifiers are the preferred power supply for welding titanium, because the current can be controlled more closely than with motor-generator sets; slight variations in welding current may cause variations in penetration. Direct current electrode negative (DCEN) is always used for GTAW of titanium because deeper weld penetration and a narrower bead can be obtained than with direct current electrode positive (DCEP). Also, in manual welding, DCEN is easier to control.

The power supply should include accessories for arc initiation because of the danger of tungsten contamination of the weld if the arc is struck by torch starting. If welding is to be done in air, controls for extinguishing the arc without pulling the torch away from the workpiece are needed so that shielding-gas flow continues and the hot weld metal is not contaminated by air.

The conventional thoriated tungsten types of electrodes (EWTh-1 or EWTh-2) are used for GTAW of titanium. Electrode size is governed by the smallest diameter able to carry the welding current. To improve arc initiation and control the spread of the arc, the electrode should be ground to a point. The electrode may extend one and a one-half times the size of the diameter beyond the end of the nozzle.

To ensure a diffuse, nonturbulent flow of shielding gas, nozzles of torches for welding titanium are larger than those used for welding other metals. With a 1/32-in.-diam electrode, a 9/16-in.-ID nozzle is ordinarily used. With a 1/16-in.-diam electrode, a 3/4-in.-ID nozzle is used. (For metric equivalents, consult the authorized equipment dealer or titanium supplier.) Phenolic or other plastic nozzles should not be used to avoid the danger of contaminating the weld with carbon.

Because titanium has low thermal conductivity, the area ahead of the arc does not become heated above 538°C (1000°F). Therefore, leading shields are seldom required when welding is done by the GTAW technique. For welding operations where a trailing shield is not adaptable, the nozzle of the torch is fitted with a concentric outer shroud through which a supplementary supply of shielding as is fed.

Shielding of the underside of a weld is provided by slotted backing bars, usually copper, through which a diffuse flow of argon or helium is maintained. Gas channels in the clamping fixtures also provide diffuse flow of inert gas to the weld area. These fixtures are placed close to the weld to avoid the danger of air contamination.

Copper fixtures are usually employed for GTAW. Although other metals are used, they should be nonmagnetic; arc blow tends to occur with magnetic metal fixtures. Metal fixtures are sometimes water cooled, but this method introduces the possibility of moisture from the air condensing on the fixtures.

Procedures for GTAW

Generally, procedures for GTAW of titanium alloys are similar to those used for austenitic stainless steel.

Preheating is not required for titanium alloys. Although cracking may occur in titanium alloy weldments, it is most often related to contamination and cannot be prevented by preheating. Also, maintenance of a specific interpass temperature is not necessary. Preheating from 82° to 93°C (180° to 200°F) may be used to eliminate surface moisture. Rather than an open-flame torch, the usual method is a heat lamp, hot air gun or infrared heater.

Tack welding is used to pre-position parts cr subassemblies for final welding operations. Elaborate fixturing often can be eliminated when tack welds are used to their full advantage. Various tack welding procedures can be used, but in any procedure good cleaning practices and adequate shielding must be provided to prevent contamination of the welds. Contamination or cracks developed in tack welds can be transferred to the finish weld. One procedure is to tack weld in such as way that the finish weld never crosses over a previous tack weld. To accomplish this, sufficient filler metal is used in tack welding to completely fill the joint at a particular location. The finish weld beads may then be blended into the ends of the tack welds.

Arc length for welding without filler metal (as with stainless steel and the nickel-base alloys) should have a maximum size about equal to the electrode diameter. With longer arc length, there is danger of turbulence, which may draw air into the weld pool. In addition, increasing the arc length produces wider weld beads. When filler metal is used, the maximum arc length should be about one and a one-half times the electrode diameter, depending on the thickness of the base metal.

Suggested welding conditions or schedules for GTAW of sheet are given in Table 10.3. In welding titanium alloys, the best heat input to use is a temperature just above the minimum required to produce the weld. If heat input is greater, the possibility that the weld will become contaminated, distorted, or embrittled increases.

Avoiding porosity in welds is an important consideration in welding titanium alloys. If the joint and filler wire are properly cleaned and the tooling does not chill the weld too rapidly, porosity can be reduced or eliminated by using a slower welding speed. This retards weld solidification and allows entrained gases to escape.

Table 10.3 Suggested welding procedure schedule for GTAW of titanium (a)

Material thickness (b), in.	Tungsten electrode diameter, in.	Filler rod diameter, in.	Nozzle size ID, in.	Shielding gas flow, ft^3/h	Welding current (c), A	Number of passes	Travel speed (d), in./min
Square-groove and fillet welds:							
0.024	1/16	--	3/8	18	20-35	1	6
0.063	1/16	--	5/8	18	85-140	1	6
0.093	3/32	1/16	5/8	25	170-215	1	8
0.125	3/32	1/16	5/8	25	190-235	1	8
0.188	3/32	1/8	5/8	25	220-280	2	8
V-groove and fillet welds:							
0.25	1/8	1/8	5/8	30	275-320	2	8
0.375	1/8	1/8	3/4	35	300-350	2	6
0.50	1/8	5/32	3/4	40	325-425	3	6

(a) Note: Tungsten used for the electrode; first choice 2% thoriated EWTh2, second choice 1% thoriated EWTh1. Use filler metal one or two grades lower than the base metal. Adequate gas shielding is essential not only for the arc, but for heated material also. Backing gas is recommended at all times. A trailing gas shield is also recommended. Argon is preferred. For higher heat input, or thicker material, use argon-helium mixture. Without backing or chill bar, decrease current 20%.
(b) Or fillet size.
(c) Direct current electrode negative.
(d) Per pass

GAS METAL ARC WELDING (AN ARC WELDING TECHNIQUE)

Gas metal arc welding (GMAW), which often is called metal inert gas (MIG) welding, is an arc welding technique in which the heat for welding is generated by an arc between a consumable electrode and the work metal. The electrode, a bare, solid wire that is continuously fed to the weld area, becomes the filler metal as it is consumed. The electrode, weld pool, arc, and adjacent areas of the base metal are protected from atmospheric contamination by a gaseous shield provided by a stream of gas, or mixture of gases, fed through the welding gun. The gas shield must provide full protection, because even a small amount of entrained air can contaminate the weld deposit.

Gas metal arc welding overcomes the restriction of using an electrode of limited length and overcomes the inability to weld in various positions. Gas metal arc welding is widely used in semiautomatic, machine and automatic modes.

Gas metal arc welding normally is used for welding titanium and titanium alloys 3.18 mm (1/8 in.) or more in thickness. This process is frequently used for welding 12.7 mm (1/2 in.) plate.

Metal transfer through the arc in GMAW can lead to difficulty in meeting stringent aerospace quality requirements. For example, weld spatter is often associated with inferior weld quality and arc instability. It is a potential cause of weld contamination and defect formation. Some users of titanium alloys prefer GTAW over GMAW (even for joining thick plate) because with the gas tungsten arc process more uniform and predictable transverse shrinkage is obtained.

Typical conditions for GMAW of Ti-6A-4V and Ti-5Al-2.5Sn plate are given in Tables 10.4 and 10.5. Electrode wires for GMAW are available in several grades of unalloyed titanium and titanium alloys that match the composition of the base metal. Joint preparation is the same as in Table 10.3.

Shielding for out-of-chamber GMAW is provided by inert gas being fed through the nozzle of the electrode holder, through the backing bar or plate, and as a trailing shield, much as in GTAW. (See Figure 10.2.) The electrode holder is basically the same as the GMAW of steel. To avoid contamination and porosity in GMAW, a leading shield is necessary, as well as a trailing shield and a suitable baffle added on the leading edge of the electrode holder. (Again, see Figure 10.2.) A leading shield prevents oxidation of spatter before it is melted in the weld metal.

PLASMA ARC WELDING (AN ARC WELDING TECHNIQUE)

Plasma arc welding (PAW) is an arc welding technique in which heat is produced by a constricted arc between an electrode and a workpiece (transferred arc), or between a nonconsumable tungsten electrode and a constricting orifice (nontransferred arc). Shield-

Table 10.4 Typical conditions for manual and automatic GMAW of Ti-6Al-4V plate (a)

Plate thickness, in.	Current (DCEP), A	Voltage, V	Welding speed, in./min	Torch:	Argon flow rate, ft³h Trailing	Backing
Manual welding:						
0.625	310	38	--	36	(b)	(b)
2.00 (b)	310	38	--	36	(b)	(b)
Automatic welding:						
0.625 (c)	360	45	15	50	60	6
2.00 (c)	325	33	25	(d)	(d)	(d)

(a) Using a 0.062-in.-diam electrode.
(b) Not reported.
(c) Multiple passes.
(d) Argon chamber.

Table 10.5 Typical conditions for GMAW of 1/2-in. thick Ti-5Al-2.5Sn plate

Electrode wire	1/16-in.-diam ERTi-5Al-2.5Sn
Wire-feed rate	300 in./min
Current	300 A (DCEP)
Voltage	30 V
Nozzle	1-in. ID
Backing bar	Copper, with 1/16-in.-deep by 1/4-in.-wide groove
Shielding gas flow:	
At torch	Argon, at 50 ft³/h
Trailing	Argon, at 50 ft³/h
Backing	Helium, at 20 ft³/h
Welding speed	20 in./min

ing generally is obtained from the hot, ionized gas issuing from the orifice of the constricting nozzle, which may be supplemented by an auxiliary source of shielding gas. Shielding gas may be an inert gas or a mixture of gases. Plasma arc welding is closely related to gas tungsten arc welding (GTAW).

The joining of titanium alloys is one of the major applications of PAW. Because titanium has a lower density, keyhole welds can be made through thicker titanium square butt joints than for steel. As with GTAW, PAW requires backing gas and a trailing gas shield to prevent atmospheric contamination of the weld and adjacent weld metal. Typical operating conditions for PAW of titanium alloys are given in Table 10.6.

Table 10.6 Typical operating conditions for PAW of titanium alloys (a)

Thickness, in.	Travel speed, in./min	Current (DCEN), A	Arc voltage, V	Gas	Gas flow, ft³/h (b)	(c)	Joint type
0.125	20	185	21	Argon	8	60	Square butt
0.187	13	175	25	Argon	18	60	Square butt
0.390	10	225	38	75He-25A	32	60	Square butt
0.500	10	270	36	50He-50A	27	60	Square butt
0.600	7	250	39	50He-50A	30	60	V-groove (d)

(a) Backing gas and trailing shield required; keyhole technique used with orifice-to-work distance of 3/16 in.
(b) Orifice gas
(c) Shielding gas
(d) 30° included angle; 3/8 in. root face.

STRESS RELIEVING OF FUSION WELDS

Most titanium weldments are stress relieved after welding to prevent weld cracking and susceptibility to stress-corrosion cracking in service. Stress relief also improves fatigue strength. An assembly subjected to a substantial amount of welding and severe fixturing restraint may require intermediate stress relieving of the partially welded structure, which should be done in an inert atmosphere; otherwise, the unwelded joints may have to be recleaned before being welded.

With unalloyed titanium and alpha titanium alloys, time and temperature should be controlled to prevent grain growth. Stress-relieving times and temperatures for titanium alloys are noted in Chapter 5, "Heat Treatment."

Stress relieving of some beta alloys such as Ti-13V-11Cr-3Al weldments may cause aging and subsequent embrittlement of the weld and HAZ and, therefore, is not recommended. Resolution heat treatment (re-annealing) may be used to relieve stresses if the welded assembly is amenable to such treatment.

All titanium surfaces should be free of dirt, fingerprints, grease, and residues before stress relieving. Contaminated surface metal must be removed from the entire weldment by machining or descaling and pickling to remove 0.025 to 0.051 mm (0.001 to 0.002 in.) per surface. (See Chapter 7, "Cleaning and Finishing," for additional information.)

REPAIR WELDING

Repair welds should follow the established specification requirements for the original welds and be made prior to final heat treatment. Manual or automatic GTAW is generally used for repairing butt and fillet welds. Repairs can also employ a combination of welding processes such as GTAW and the initial welding process (GMAW or PAW).

Repair welds always must be carefully executed, and all traces of liquid-penetrant inspection material must be removed. Generally, inspection is performed on both faces of the repair weld and several inches beyond the repaired area.

SUCCESS AND FAILURE IN ARC WELDING

Commercially pure titanium and most titanium alloys can be arc welded satisfactorily by a wide range of welding processes. Unalloyed titanium and all the α titanium alloys, being substantially single-phase materials, may be welded with little effect on the microstructure. For this reason, the mechanical properties of a correctly welded joint of these alloys are equal to those of the base metal and have good ductility.

The two-phase microstructures of the α-β titanium alloys respond to thermal treatment. Consequently they can be altered by welding. The result is an extreme brittleness in some alloys that renders them nonweldable. The most commonly used α-β titanium alloys that respond well to welding include Ti-6Al-4V, Ti-3Al-2.5Sn, and Ti-6Al-6V-2Sn.

Most β-titanium alloys can be successfully welded. These, however, require particular care if heat treatment is to be employed to strengthen the welds. Aging of certain β-titanium alloy welds, such as Ti-3Al-13V-11Cr alloy, renders the welds susceptible to embrittlement.

The factors that cause failure of titanium or titanium alloy welds may be metallurgical, mechanical, or a combination of both. Basically, the metallurgical origin is the presence of a phase or combination of phases that possess a very limited capacity to tolerate strain within a critical temperature range. The critical phase or phase mixture will be influenced by the composition of the alloy; the thermal and mechanical processing the alloy has received; the welding conditions used; and the weld shape, which affects the temperature distribution in the adjacent metal. The extent to which either the metallurgical structure or the imposed stresses contribute to a failure can vary considerably.

Mechanical origins include the occurrence of thermal and restraint stresses that, as affected by the presence of weld defects and within this critical temperature range, produce strain in excess of the tolerance of the microstructure.

The majority of weld failures in titanium alloys, however, arise from residual stresses in the welded joint from normal weld shrinkage or from weldment restraint. If these residual stresses are multiaxial and exist in areas of limited ductility because of contamination by carbon, oxygen, hydrogen, or nitrogen, the result may be cracking immediately after welding, or later under applied service loads. Abuse in service is another common source of mechanical origins for failures of weldments made of titanium alloys.

ELECTRON BEAM WELDING (A FUSION WELDING PROCESS)

General

Electron beam welding (EBW) is a high-energy density fusion welding process that is accomplished by bombarding the joint to be welded with an intense (strongly focused) beam of electrons that have been accelerated up to velocities 0.3 to 0.7 times the speed of light at 25 to 200 kV, respectively. The instantaneous conversion of the kinetic energy of these electrons into thermal energy as they impact and penetrate into the workpiece on which they are impinging causes the weld-seam interface surfaces to melt and produces the weld-joint coalescence desired.

Originally, EBW generally was performed only under high-vacuum (1×10^{-4} torr, or lower) conditions. Because an ambient vacuum environment was required to generate the beam, welding the part within the same clean atmosphere was considered beneficial. However, as the demand for greater part production increased, it was found that the weld chamber vacuum level need not be as high as that needed for the gun region; ultimately, the need for any type of vacuum surrounding the workpiece was totally eliminated for some applications. Currently, three distinct modes of EBW are employed:

- High-vacuum (EBW-HV), where the workpiece is in the ambient pressure ranging from 10^{-6} to 10^{-3} torr

- Medium-vacuum (EBW-MV), where the workpiece may be in a "soft" or "partial" vacuum ranging from 10^{-3} to 25 torr

- Nonvacuum (EBW-NV), which is also referred to as atmospheric EBW, where the workpiece is at atmospheric pressure in air or protective gas

In all EBW applications, the electron beam gun region is maintained at a pressure of 10^{-4} torr or lower.

One of the prime advantages of EBW is the ability to make welds that are deeper and narrower than arc welds, with a total heat input that is much lower than that required in arc welding. This ability to achieve a high depth-to-width ratio for the weld eliminates the need for multiple-pass welds, as is required in arc welding. The lower heat input results in a narrow workpiece heat-affected zone (HAZ) and noticeably less thermal effects on the workpiece.

Equipment costs for EBW generally are higher than those for conventional welding processes. However, when compared to other types of high-energy density welding (such as LBW), production costs are not as high. The cost of joint preparation and tooling is more than that encountered in arc welding processes, because the relatively small electron beam spot size that is used requires precise joint gap and position.

Welding of Ti Alloys

All of the commercial alloys of titanium that can be joined by arc welding can also be joined by EBW. Their relative weldability and response to heat cycling in EBW are generally the same as in arc welding.

The vacuum environment of EBW prevents exposure to the atmospheric contaminants that cause embrittlement of titanium alloys, whereas arc welding processes must use elaborate and costly shielding methods to accomplish this. Cost studies show that direct labor costs for EBW of titanium sections more than 25.4 mm (1 in.) thick are less than for arc welding, provided a suitably large vacuum chamber is available.

Filler metal is not ordinarily used, and the work is not preheated. Tack welding, contrary to experience in GTAW, presents no difficulties in EBW. For optimum results, welding is done in a high vacuum, but medium-vacuum welding is satisfactory for many applications. Nonvacuum welding is not a preferred technique.

Ti-6Al-4V, the alloy most frequently used in assemblies to be welded, can be electron beam welded in either the annealed or solution-treated-and-aged condition. For weldments that will be used at elevated temperatures, a preferred process sequence is anneal, weld, solution treat, and age. For other service conditions, a process sequence of solution treat, age, and weld gives almost the same strength properties and only slightly lower fracture toughness.

LASER BEAM WELDING (A FUSION WELDING PROCESS)

General

Laser beam welding (LBW) is a fusion welding process that produces coalescence of materials with the heat obtained from the application of a concentrated coherent light beam impinging upon the surfaces to be welded. The word laser is an acronym for "light amplification by stimulated emission of radiation." The laser can be considered, for metal joining applications, as a unique source of thermal energy, precisely controllable in intensity and position. For welding, the laser beam must be focused to a small spot size to produce a high-power density. This controlled power density melts the metal and, in the case of deep penetration welds, vaporizes some of it. When solidification occurs, a fusion zone, or weld joint, results. The laser beam, which consists of a stream of photons, can be focused and directed by optical elements (mirrors or lenses). The laser beam can be transmitted through the air for appreciable distances without serious power attenuation or degradation.

Welding of Ti Alloys

The EBW technique has been used more frequently than LBW for the welding of Ti-6Al-4V, an alloy widely used in the aerospace industries for its high strength-to-weight ratio. However, the deep penetration of EBW can be obtained only up to a short distance under nonvacuum conditions. For optimum efficiency, EBW is carried out in an evacuated chamber. In contrast, CO_2 laser beams can be transmitted for appreciable distances through the atmosphere without serious attenuation or optical degradation. Thus, the laser offers an easily maneuvered, chemically clean, high-intensity, atmospheric welding process that produces deep-penetration welds (aspect ratio greater than 1-to-1) with a narrow HAZ and subsequent low distortion.

The application of the laser technique to a metal such as titanium alloy, which requires extreme cleanliness for attainment of sound welds, is of great interest to the aerospace and chemical industries. More generally, laser techniques are of interest from the standpoint of welding a chemically sensitive metal with a complex, temperature-dependent structure. The importance and the need for better joining methods for titanium and its alloys resulted in several investigations of LBW techniques over various power ranges.

The relationship between LBW parameters and the metallurgical and mechanical properties of laser welded Ti-6Al-4V and commercially pure titanium has been reported. Welding speeds in excess of 15.24 m/min (50 ft/min) were obtained for 1.02 mm-thick (0.04 in.) Ti-6Al-4V using 4.7 kW of laser power.

X-ray radiographs of successful laser butt welds of Ti-6Al-4V and commercially pure titanium showed no cracks, porosity, or inclusions. Low porosity in a laser welded titanium alloy was also observed, and radiographically sound welds were produced. Undercutting was not prominent.

A comparative study of electron beam, laser beam, and plasma arc welds in Ti-6Al-4V alloy has been conducted. Radiographically sound welds were produced by all three techniques. The electron beam welds were produced by all three techniques. The electron beam welds were quite narrow and exhibited a somewhat nonuniform radiographic appearance due to lower surface weld spatter, whereas the arc welds were considerably broader, but also quite uniform in density. Laser welds were narrower than arc welds and were comparable to, but more uniform than, electron beam welds. Following stress relieving for two hours at 538°C (1000°F), the welds produced by all three techniques had tensile strengths equivalent to or exceeding those of the base metal.

RESISTANCE SPOT WELDING (A FUSION WELDING PROCESS)

General

Resistance welding, which is another fusion welding process, occurs when heat is generated by resistance to electrical current at two surfaces in contact with each other. When heat is generated, the metal melts in the vicinity of the current flow. Pressure keeps the faces together. When the current is interrupted, a solidified weld nugget is formed. The nugget is contained within the metal being joined and does not reach an external surface.

In cases when the electrode is applied discretely at separate locations, the technique is called spot welding.

When continuous cylindrical electrode wheels are used, and a sheet metal to be welded passes continuously through the electrodes which pulse current at regular intervals, a series of overlapping spots occurs, and a seam weld is generated. The thickness of metal which can be joined by resistance welding is a function of the base alloy together with the actual chemistry. Spot welded lap joints have been made up to a total thickness in excess of 12.7 mm (0.5 in.) of steel; however resistance welding of thick stock is generally uneconomical.

Welding of Ti Alloys

Titanium alloy sheet may be resistance welded. The thermal conductivity of titanium alloys compares favorably with the conductivity of stainless steel. The processing techniques for titanium alloys are similar to those used to weld stainless. It is claimed that spot and seam welding do not require inert gas shielding because electrode pressure excludes air and, also, because there is a very short weld duty cycle. Inevitably there is some oxygen pickup, however, even when fully cleaned surfaces are resistance welded. Properties of resistance welded joints made in air are not as good as those made with a shielding gas.

During spot welding operations, an oxide film can form around the weld area of the touching surfaces of the sheet. With respect to seam welding, it is claimed that the speed of continuous welding and the pressure prevent film formation ahead of the weld. If a nugget dissolves an oxide film, the weld's ductility is reduced.

FLASH WELDING (A SOLID STATE PROCESS)

General

Flash welding commonly is used to join sections of metals and alloys in production quantities. It is a resistance/forge welding process in which the items to be welded are securely clamped to electric current-carrying dies, heated by the electric current, and then upset.

Early machines were manually operated. To make a weld, the operator energized the transformer and brought the pieces to be joined into light contact so that localized melting would occur at a few spots. Molten metal was violently expelled, and heat flowed back into the workpiece from the molten layers on the surfaces. When a sufficiently high temperature had been reached some distance from the abutting surfaces, the operator rapidly moved his lever to extinguish the arcing, pushed out the molten and overheated metal on the surfaces in the form of flash, and completed the weld by forging the two pieces together while interrupting the current, thus making a solid-state weld. Although the process has remained essentially the same since its inception, major advances include automated control, increased knowledge of the interactions of process variables along with their influences on the metal being welded, and improved weld quality.

Welding of Ti Alloys

An inert gas shield is needed to prevent embrittlement of titanium alloys. Appropriate flash schedules or welding schedules must be developed for each alloy to be welded. While not the major technique for weld processing, at least with respect to aerospace alloys, flash welding has been used for alloys such as CP Ti, Ti-6Al-4V, Ti-8Al-1Mo-1V, Ti-5Al-2.5Sn, Ti-6Al-2Sn-4Zr-2Mo, and Ti-6Al-2Sn-4Zr-6Mo.

DIFFUSION BONDING (A SOLID STATE PROCESS)

General

Diffusion bonding (DB), at times called diffusion welding, is a solid state welding process in which the surfaces are placed in proximity under a moderate pressure at an elevated temperature. Coalescence occurs across the interface. Because diffusion bonding requires heat, pressure, and a vacuum, inert gas or a reducing atmosphere, equipment is frequently custom built by the user.

Surface cleanliness is essential to diffusion bonding. Thus prior surface deformation — by scratch brushing, for example — may be beneficial. Cleanliness must be maintained up to and including the application of heat and pressure. In many instances no intermediate layer is required to effect a satisfactory diffusion bond. In other cases, intermediate layers of foil (or of surface activation) may be necessary to develop a sound bond.

Recrystallization may occur across the bond line, but it is not necessary for achieving a full-strength joint. There is no gross deformation of the parts being joined by diffusion bonding. Stop-off may be used with this technique to prevent a specific portion of the bond line from being welded.

Under actual shop conditions, surface contaminants are invariably present, and, depending on the materials being welded, mechanisms must exist for dispersion of contaminants away from, or into, localized areas on the faying surface. Metals that have a high solubility for interstitial contaminants, such as oxygen, can easily accommodate removal of these contaminants from the faying surfaces by assimilation into the base metal by volume diffusion. Thus, the surface is decontaminated during welding by diffusion, and short-range interatomic forces can operate. Metals such as titanium fall into this class and are among the easiest to diffusion weld.

Diffusion bonding is usually performed at a welding temperature equal to, or greater than, one-half the melting temperature of the material being welded. However, the choice of welding temperature is strongly influenced by the time required for surface contaminants to diffuse away, the tendency to weld above or below a phase transformation, and the amount of load available at the faying surfaces.

Welding of Ti Alloys

More diffusion welding has been conducted on titanium and its alloys than any other material. Titanium diffusion welding is readily adaptable to production applications because:

- Titanium and its alloys have a high solubility for interstitial oxygen and can assimilate contaminant films into the bulk material during welding.

- Conventional fusion welds in titanium are not as strong as the base metal, whereas titanium diffusion welds possess properties equivalent to the base metal.

- Most titanium structures or components are high-technology items utilized in aerospace applications, and extra costs for diffusion welding can be justified based on improved performance.

An example of diffusion bonding involves the production of Ti-6Al-4V alloy turbofan aircraft engine discs with hollow hubs. Figures 10.3, 10.4, and 10.5 show a schematic illustration of the disc profile, including the location of the diffusion welds, the welding assembly, and the hot press used for welding. Fabrication of this disc by diffusion welding resulted in weld properties equivalent to the base metal and a weight savings of 47.25 kg (105 lb) or about 30%. This was done without sacrificing stiffness or durability.

Other titanium alloys that have been successfully diffusion welded include: Ti-8Al-1Mo-1V, Ti-13V-11Cr-3Al, Ti-8Mo-8V-3Al-2Fe, Ti-6Al-6V-2Sn, Ti-6Al-2Sn-4Zr-6Mo, and Ti-11.5Mo-6Zr-4.5Sn.

As many as 66 different titanium components have been fabricated successfully by diffusion welding for use in the United States Air Force's B-1 bomber. Very large parts have been fabricated in this manner. Figure 10.6 shows a wing carry-through section that contains 533 individual parts that were diffusion welded together. A 61-m diam

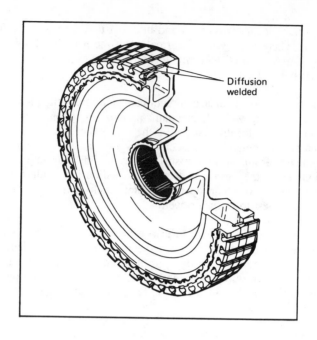

**Figure 10.3 - Diffusion welds located in hollow
fan hub from a turbofan engine**

(200 ft) titanium alloy chamber for the zero gradient synchrotron at Argonne National
Laboratory also has been produced by diffusion welding.

Figure 10.4 - Diffusion-welded hollow-rim fan disc and split tracer ring

SPFDB

Superplastic forming of titanium plate and sheet was developed as a reduced-cost
method for processing of material. A combination of superplastic forming and diffusion

Photo courtesy of Pratt & Whitney Aircraft Group

Figure 10.5 - Vacuum hot press (300 ton) used for diffusion-welding of turbofan components

bonding (SPFDB) has been used on titanium to produce complex structures. An example of SPFDB is the bonding of three Ti-6Al-4V sheets and, in the same, single operation, their expansion to form an integrally stiffened structure.

An illustration of this technique is shown in Figure 10.7 in a variety of specific applications. Figure 10.7(a) shows the method described in the preceding paragraph. The application shown in Figure 10.7(c) requires that three cleaned and treated alloy sheets be created. Two are treated with stop-off at the indicated points. A third sheet is inserted between them. The edges are welded to exclude air, but a pipe is attached for later pressurization. Initial bonding occurs under pressure at temperature, and then internal pressure, through the pipe, causes superplastic flow of the bottom two sheets to fill the die and create the configuration shown.

A number of aerospace applications are using this unique process.

Photo courtesy of Rockwell International Corp.

Figure 10.6 - Wing carry-through fabricated by diffusion welding 533 individual details

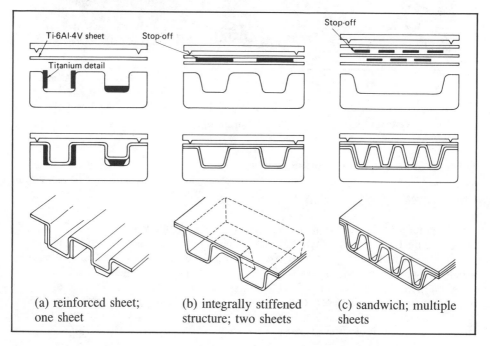

(a) reinforced sheet; one sheet

(b) integrally stiffened structure; two sheets

(c) sandwich; multiple sheets

Figure 10.7 - Basic shapes of superplastically formed/diffusion-welded structures

BRAZING

General

Brazing, the third broad category of joining, is the bonding of metal that occurs when a high-temperature filler is allowed to melt and flow into a joint by capillary action before the filler solidifies. Braze joints are metal-metal bonds. They can be very strong, depending on the filler metal and base alloys.

Brazing takes place above 538°C (1000°F). It usually, but not always, is used to join two, or more, dissimilar metals into an assembly. While torch brazing is used, furnace brazing is more common for most applications, since an atmosphere can be used, if necessary. Also process conditions produce more consistent, and usually stronger, joints.

For reactive metals, brazing processes should be used that do not allow joint surfaces to come in contact with air during heating. Induction and furnace brazing in inert gas or vacuum atmospheres can be used successfully, but torch brazing is difficult and requires special precautions and techniques. Induction brazing of small, symmetrical parts is very effective, because its speed minimizes reaction between braze filler metal and base metal. Furnace brazing is favored for large parts, since uniformity of temperature throughout the heating and cooling cycle can be readily controlled.

Brazing Ti Alloys

Titanium assemblies frequently are brazed in high-vacuum, cold-wall furnaces. Ideally, brazing should be done in a vacuum at a pressure of 10^{-5} to 10^{-3} torr or in dry inert-gas atmosphere if vacuum brazing is not possible. Argon, helium, and vacuum atmospheres are satisfactory for brazing titanium. For torch brazing, special fluxes must be used on the titanium. Fluxes for titanium are primarily mixtures of fluorides and chlorides of the alkali metals, sodium, potassium, and lithium.

Vacuum and inert-gas atmospheres protect titanium during furnace and induction-brazing operations. A vacuum of 10^{-3} torr, or more, is required to braze titanium. However, it is necessary to have a dew point of -21°C (-70°F), or less, is necessary to prevent discoloration of the titanium in a brazing temperature range of 760° to 927°C (1400° to 1700°F).

Material properties of alpha alloys are not affected greatly by brazing, and assemblies are not heat treatable. The mechanical properties of alpha-beta titanium alloys may be altered by heat treatment and variations in microstructure. Wrought alpha-beta titanium alloys generally are fabricated to obtain a fine-grain, equiaxed, duplex microstructure to produce maximum ductility. It is desirable to maintain this microstructure by requiring that the brazing temperature not exceed the beta-phase transformation temperature (beta transus), which varies from 899° to 1038°C (1650° to 1900°F). Alpha-beta alloy assemblies are used in annealed, solution-treated, and aged conditions. Several beta base metals are available commercially. In the annealed condition, these metals are unaffected by brazing; however, if heat treated, the brazing temperature may have an effect on the beta alloy's properties. Ductility in base metal is obtained by brazing at the solution treatment temperature, while the ductility of the base metal decreases as braze temperature increases.

Braze filler metals initially used for brazing titanium and its alloys were silver with additions of lithium, copper, aluminum, or tin. Most of these braze filler metals were used in low-temperature applications 538° to 593°C (1000° to 1100°F). Commercial braze filler metals, including silver-palladium, titanium-nickel, titanium-nickel-copper, and titanium-zirconium-beryllium are now available. These can be used in the 871° to 927°C (1600° and 1700°F) range. Higher strengths and improved resistance to crevice-type corrosion are desirable characteristics that current braze filler metals enjoy. For joining applications requiring a high degree of corrosion resistance, the 48Ti-48Zr-4Be and 43Ti-43Zr-12Ni-2Be braze filler metals were developed on an experimental basis by Argonne National Laboratory. A silver-palladium-gallium braze filler metal (Ag-9Pd-9Ga), which flows at 899° to 913°C (1650° to 1675°F), is another excellent filler metal with which to fill large gaps.

In general, aluminum-silicon filler metals are unsuitable for brazing aluminum to uncoated titanium because of the formation of brittle intermetallic compounds. Titanium can be hot-dip coated with aluminum, however, after which it can be brazed to aluminum with the usual aluminum filler metals. During the SST supersonic transport program, methods to braze titanium honeycomb sandwich assemblies with aluminum braze filler metal were developed. Aircraft structures up to 7 m (23 ft) in length were brazed successfully using 3003 brazing foils. Use of aluminum brazing filler metal (3003) provided satisfactory strength up to about 316°C (600°F). If temperatures of 538° to 593°C (1000° to 1100°F) are required, high-strength, corrosion-resistant titanium-zirconium-beryllium or titanium-zirconium-nickel-beryllium braze filler metals should be used.

A typical titanium honeycomb assembly used for aircraft structures is shown in Figure 10.8.

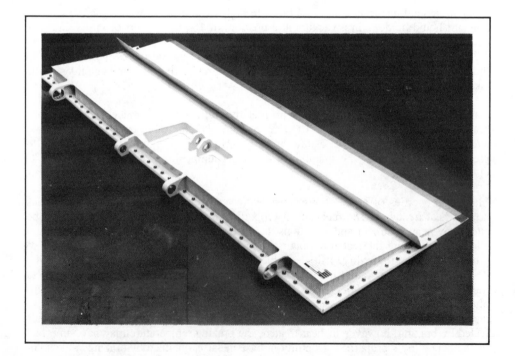

Figure 10.8 - Brazed titanium honeycomb sandwich aerospace assembly

Selection of filler metals and brazing cycles that are compatible with the heat treatment required for alpha-beta and beta-titanium base metals may present some difficulty. Ideally, brazing should be conducted from 38° to 66°C (100° to 150°F) below the beta transus. The ductility of alpha-beta base metals may be impaired if this temperature is exceeded. The beta transus can be exceeded when beta-titanium base metals are brazed; however, if the brazing temperature is too high, base-metal ductility after heat treatment may be impaired and considerable interaction between the filler metal and base metal may occur.

The tensile properties of heat-treatable titanium alloys also may be adversely affected by brazing, unless the assembly can be heat treated after brazing. For example, alpha-beta titanium alloys must be solution treated, quenched, and aged to develop optimum properties. It is difficult to select a filler metal that is suitable for brazing and solution treating in a single operation. Similarly, it is not always possible to quench a brazed assembly at the desired cooling rate. Certain configurations, such as honeycomb sandwich structures, cannot be quenched rapidly without distortion. Brazing at the aging temperature is impractical, because few filler metals melt and flow at these temperatures.

The possibility of galvanic corrosion must be considered when filler metals are selected for brazing titanium-base metals. While titanium is an active metal, its activity tends to decrease in an oxidizing environment, because its surface undergoes anodic polarization similar to that of aluminum. Thus, in many environments, titanium becomes more chemically inactive than most structural alloys. The corrosion resistance of titanium is generally not affected by contact with structural steels, but other metals, such as copper, corrode rapidly in contact with titanium under oxidizing conditions. Thus, filler metals must be chosen carefully to avoid preferential corrosion of the brazed joint.

When titanium is brazed, precautions must be taken to ensure that the brazing retort or chamber is free of contaminants from previous brazing operations. Mechanical properties of titanium may deteriorate because of gaseous contamination from the brazing furnace. Also, the choice of materials to be used in fixtures must be carefully considered. Nickel or materials containing high amounts of nickel generally should be avoided; nickel and titanium form a low-melting eutectic at about 942°C (1728°F) 28.4 wt% Ni. If titanium workpieces come in contact with fixtures or a retort made of a nickel-base alloy, the parts may fuse together if the brazing temperature is in excess of 942°C (1728°F). If a fixture material, such as stainless steel, which may contain a high nickel content is used, it should be oxide coated. In most applications, coated graphite or carbon steel fixture materials are used.

Brazing Applications

Titanium and its alloys are used as brazed assemblies for aerospace hardware and jet engine inlet vanes, hydraulic tubing or fittings, and honeycomb sandwich panels. For example, anti-icing airfoils have been made with wrap-around skins and corrugated core members, which are brazed in vacuum with 70Ti-15Cu-15Ni braze filler metal.

Titanium plumbing systems providing additional weight savings are an outgrowth from aircraft and aerospace brazed stainless steel high-pressure fluid line joints and fittings. Application of hinged induction heater tools and protective argon atmosphere permit rapid, reliable brazes to be made in open lines. Close-out joints or dead-end lines present difficulties in ensuring adequate backside shielding and should be avoided.

In one example, a wide-body jetliner utilized over 250 brazed joints involving titanium tubing and Ti-6Al-4V fittings. The braze filler metal was 90Ag-10Pd. The SST supersonic transport design relied heavily on high-efficiency honeycomb structures. Metallurgically bonded titanium sandwich panels were considered for fuselage, wing, and control surfaces. The honeycomb core was 0.254 mm (0.010 in.) thick, and aluminum alloy 3003 braze filler metal was successfully used with ribbon heated-ceramic tooling and argon atmosphere at rarefied pressure to produce aluminum brazed titanium alloy honeycomb sandwich panels. Engine cowls were made successfully in production for other aircraft systems.

Successful joining of dissimilar metal combinations has been achieved with titanium and stainless steel and copper. A brazed transition section located between a titanium tank and stainless steel feed lines was evaluated during the course of the space program. Titanium alloy Ti-6Al-4V was vacuum induction brazed to type 304L stainless steel with Au-18Ni braze filler metal. The presence of a brittle intermetallic compound and an indication of cracking led to extended joint evaluations. It was concluded that the brazed joint could sustain loads in excess of the yield strength of the stainless steel. The successful performance of this joint has been attributed to the rigid control of all the brazing process variables. The formation of brittle intermetallic compounds was minimized because the joint was brazed rapidly and holding time at temperature was kept to a minimum.

A braze filler metal, Pd-9.0Ag-4.2Si, was developed and successfully brazed steel and titanium in vacuum at 738°C (1360°F). Excellent flow properties were exhibited by this filler metal. Titanium alloy Ti-3Al-1.5Mn has been brazed to Cu-0.8Cr alloy using silver-base braze filler metals. Three braze filler metals were evaluated during this program: Ag-28Cu, Ag-40Cu-35Zn, and Ag-27Cu-5Sn.

A process for vacuum brazing copper-plated titanium has also been developed. Copper was electroplated on the surface of titanium alloy Ti-3Al-1.5Mn after the surface had been etched in a sulfuric acid solution. The etching of the titanium surface removed hydrogen or hydrogen compounds, thereby creating an effective surface for the copper plate. Joints were made between the titanium alloy and commercially pure copper, as well as between Ti-3Al-1.5Mn and stainless steel or nickel-base alloy using Ag-27Cu-5Sn braze filler metal.

Chapter 11

Relationships of Properties and Processes

STRUCTURE AND PROPERTIES

Titanium and its alloys have many desirable properties, as has been extensively discussed in preceding chapters. Pure titanium can be strengthened appreciably by alloying and heat treatment, yet the alloys still retain low density levels. Consequently, their mechanical properties are attractive, particularly with respect to strength-to-density ratios. Figure 11.1 indicates this ratio for pure titanium, steel, and high-strength aluminum- and magnesium-base alloys. Table 11.1 compares typical properties for pure titanium and two titanium alloys with those of ultra-high-strength steel. The table shows that titanium alloys are remarkably strong on this basis.

Although physical properties are unaffected by processing, the mechanical properties and microstructures of titanium alloys are directly affected. The kinetics of the titanium and titanium alloy beta phase transformations after heating, cooling, and aging strongly influence microstructure and, therefore, mechanical properties. Elastic properties are affected by chemistry and texture, yet they are not particularly affected by heat treatment with respect to the development of microstructure.

The microstructures of titanium can be quite complex, and therefore they may easily be misunderstood. In this chapter, the microstructural description aims at a quick understanding of the effects of processing and heat treatment on properties. Fine points of detail are generally avoided.

Structure Defined

Structure may be defined as the macro- and microappearance of the units of a polished and etched cross section of metal that are visible up to and including a 10 000x magnification. Here the instrument is assumed to be an electron microscope. The basic features that are visible are:

- Flow lines (macroscopic)   • Grain shape (microscopic)

- Grains (microscopic)   • Secondary (intermetallic) phases (microscopic)

- Grain size (microscopic)

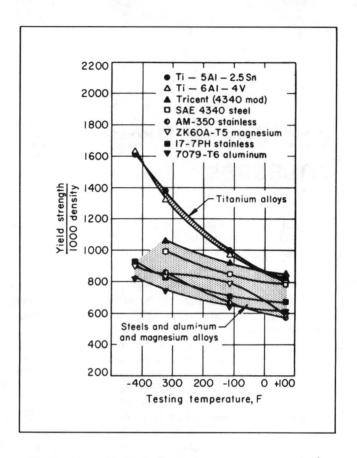

Figure 11.1 - Yield strength-to-density ratio (lb$_f$/in.^{2}) as a function of temperature for Ti alloys compared with that of several other alloys

The following information centers on grains and secondary phases. It also covers two other structural items which require chemical and X-ray techniques for proper identification. These are:

- Phase type (alpha, beta)   • Texture (orientation) of grains

Table 11.1 Comparison of typical strength-to-weight ratios at 20°C

Metal	Specific gravity	Tensile strength, lb/in.^{2}	Tensile strength + specific gravity
CP	4.5	58 000	13 000
Ti-6Al-4V	4.4	130 000	29 000
Ti-4Al-3Mo-1V	4.5	200 000	45 000
Ultra-high-strength steel (4340)	7.9	287 000	36 000

With this simplified list, it is possible to begin a consideration of the relationships of properties and processing.

CP Ti Property Development

As with any single-phase alloy, the microstructure of commercially pure (CP) titanium depends upon whether or not it has been cold worked and on the specific type of annealing employed. In addition, upon cooling from the β region, which begins at 882°C (1620°F), the structure depends upon the cooling process, because the process directly affects the β-to-α transformation along with the final α grain size and shape.

The equiaxed microstructure of α titanium after annealing at 800°C (1472°F) in the α region is shown in Figure 11.2(a). Here the grain size—and hence properties—can be varied only by cold working and annealing. Properties typical of this microstructure are shown in the figure.

Annealing in the β region at 1000°C (1832°F), and then cooling rapidly by means of water quenching to 25°C (77°F) produces the typical structure shown in Figure 11.2(b). Even rapid cooling does not suppress the β-to-α transformation; the structure is entirely transformed to α. Note that the α grain boundaries are serrated and quite irregular. Properties typical of this structure are shown in the figure. This structure is stronger than the equiaxed structure developed by annealing only in the α region. For both treatments, the titanium is still quite ductile.

Cooling slowly—e.g., 20 h to 25°C (77°F)—produces the structure in Figure 11.2(c). The structure is again completely α, but the grain boundaries are less irregular than those produced upon cooling rapidly. This structure is somewhat weaker than that produced upon cooling rapidly, but it is still stronger than the equiaxed structure shown in Figure 11.2(a).

Properties of these pure titaniums are primarily a function of grain size and shape: fineness of structure is more desirable from the point of view of tensile strength. These are wrought, annealed materials, but cast CP titaniums look similar.

A more significant influence on mechanical behavior of CP titanium is brought about by hydrogen, nitrogen, carbon, and oxygen which dissolve interstitially in titanium and have a potent effect on mechanical properties.

Hydrogen in solution has little effect on the mechanical properties. Upon precipitation of the hydride, however, the ductility suffers, as shown in Figure 11.3.

The solubility of hydrogen in α Ti at 300°C (572°F) is about 8 at.% (about 0.15 wt %, or about 1000 ppm by weight). Twenty ppm corresponds to about 0.1 at.%. The data in Figure 11.4 show that this low a concentration has little effect on the impact strength, independent of the heat treatment. This correlates with the data in Figure 11.3. However, note in Figure 11.4 that 0.5% hydrogen (about 100 ppm) causes measurable embrittlement. Slow cooling from the α region—e.g. 400°C (752°F) allows sufficient hydride to precipitate in order to reduce the impact energy.

Rapid cooling, by water quenching, from 400°C (752°F) suppresses the precipitation, thereby retaining the high impact energy. However, aging at room temperature, even for

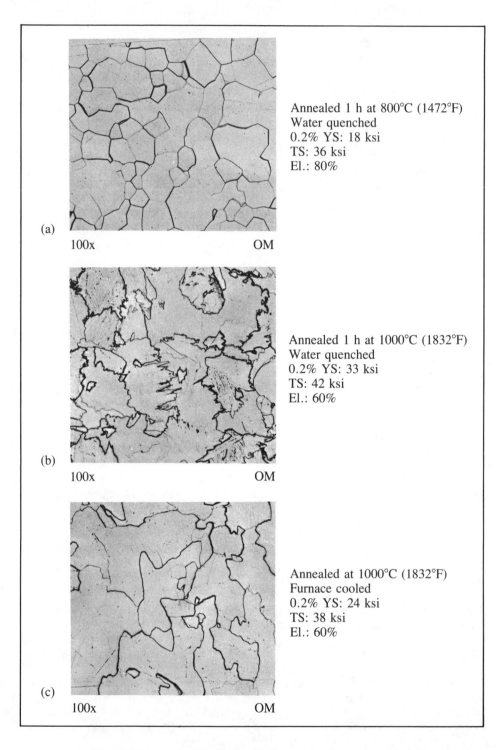

(a)

Annealed 1 h at 800°C (1472°F)
Water quenched
0.2% YS: 18 ksi
TS: 36 ksi
El.: 80%

100x   OM

(b)

Annealed 1 h at 1000°C (1832°F)
Water quenched
0.2% YS: 33 ksi
TS: 42 ksi
El.: 60%

100x   OM

(c)

Annealed at 1000°C (1832°F)
Furnace cooled
0.2% YS: 24 ksi
TS: 38 ksi
El.: 60%

100x   OM

**Figure 11.2 - Microstructure and mechanical properties (at 25°C/77°F)
of pure titanium; typical of those developed upon annealing in the α region,
and upon annealing in the β region, and then cooling to the α region**

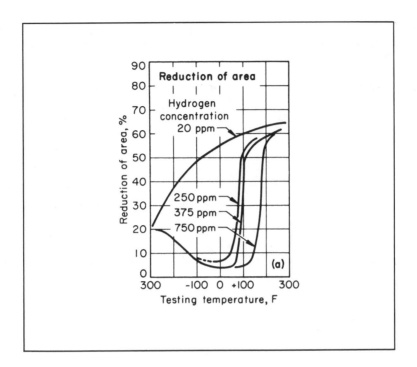

Figure 11.3 - Effect of hydrogen in α-titanium on reduction in area in a tension test; shows embrittling effect of hydrogen

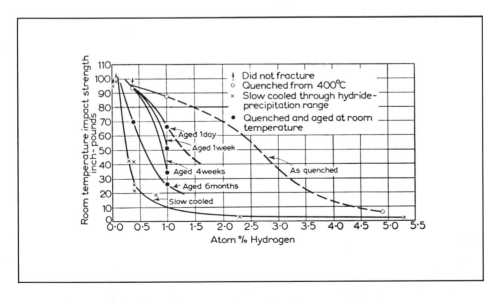

Figure 11.4 - Effects of hydrogen content and heat treatment on the impact energy of α-titanium

a few days, allows sufficient hydride to precipitate, thus lowering the impact energy. Therefore, even though the hydride precipitation may be controlled by heat treatment, aging at 25°C (77°F) results in sufficient precipitate formation and coarsening to embrittle the alloy. The only practical approach to control the hydrogen problem is to maintain a low concentration. As the result commercially pure titanium will usually have a maximum allowable hydrogen content, about 0.015 wt% (about 100 ppm). For example, for the grades of commercially pure titanium shown in Table 11.2, the level is about 0.01%.

Oxygen and nitrogen have a potent effect on strength, as shown in Figure 11.5. As the amount of O and N increases, the toughness decreases until the material eventually becomes quite brittle. Referring to the data given in Table 11.2, note that the allowed O content is higher than the allowed N content. This is consistent with the relative effects shown in Figure 11.5. This same figure indicates why the grades of titanium with the higher allowable O and N contents have the higher strengths. (Reference Table 11.2.) Note, though, that even the grade with the highest interstitial content has good ductility. However, due to the oxygen and nitrogen, embrittlement occurs at a concentration considerably below the solubility limit.

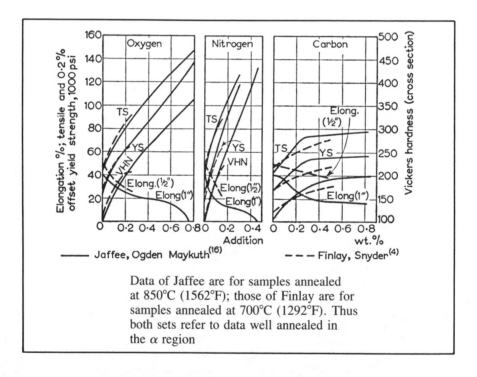

Figure 11.5 - Effects of hydrogen, nitrogen and carbon content on the tensile mechanical properties and hardness of α-titanium

In a titanium which contains O and N, the α formed from β has a much more distinctive Widmanstätten structure than a titanium essentially free of these elements. (See Figure 11.6.) This effect, however, may have little direct bearing on the mechanical properties.

The addition of carbon, up to about 0.3%, strengthens titanium greatly and reduces ductility somewhat. (See Figure 11.5.)

Table 11.2 ASTM maximum interstitial content and minimum mechanical properties

Item	Grade 1	Grade 2	Grade 3	Grade 4
Maximum interstitial content allowed, wt %:				
Nitrogen	0.03	0.03	0.05	0.05
Carbon	0.10	0.10	0.10	0.10
Oxygen	0.18	0.25	0.35	0.40
Hydrogen	0.0125(b) 0.0100(c)	0.0125(b) 0.0100(c)	0.0125(b) 0.0100(c)	0.0125(b) 0.0100(c)
Maximum tensile strength, ksi	35	50	65	80
Minimum yield strength, ksi	25	40	55	70
Elongation, %	24	20	18	15
Reduction in area, %	30	30	30	25

(a) Of four Grades of commercially pure alpha titanium for bars and billets (ASTM specification B 348-78)
(b) Bars only
(c) Billets only

Although the preceding discussion deals with interstitial effects in CP titanium, the concepts can be carried over to the commercially available titanium alloys in which the oxygen and nitrogen levels play a role in defining alloy strength and, in particular, ductility. The ELI (extra-low interstitial) titaniums implicitly acknowledged this effect on ductility with respect to critical applications, and so they enhanced the alloys' usefulness by keeping interstitials at a very low level.

Hydrogen is always kept at a low level to avoid embrittlement, yet there still remains concern about the most reasonable level to specify in both CP and alloyed titanium.

COMMERCIAL (WROUGHT) ALPHA AND ALPHA-BETA ALLOYS

The following discussions center on the effects of working and annealing of commercial alpha and alpha-beta alloys; on their aging and stabilizing; and on longitudinal and transverse properties.

Working, Annealing Effects

The α alloys have α as their common phase at low temperature, below 800°C (1472°F). Thus the properties in general cannot be altered by heat treatment. The only strengthening mechanisms are cold work; cold work and annealing (to control the α grain size); and solute additions for solid solution strengthening. Alpha formers such as Al and Sn can increase the strength of titanium by solid solution strengthening. Both aluminum and tin have a significant effect on titanium's mechanical properties: there is an increase in strength of approximately 55 MPa (8 ksi) for each 1% of Al, and 28 MPa (4 ksi) for

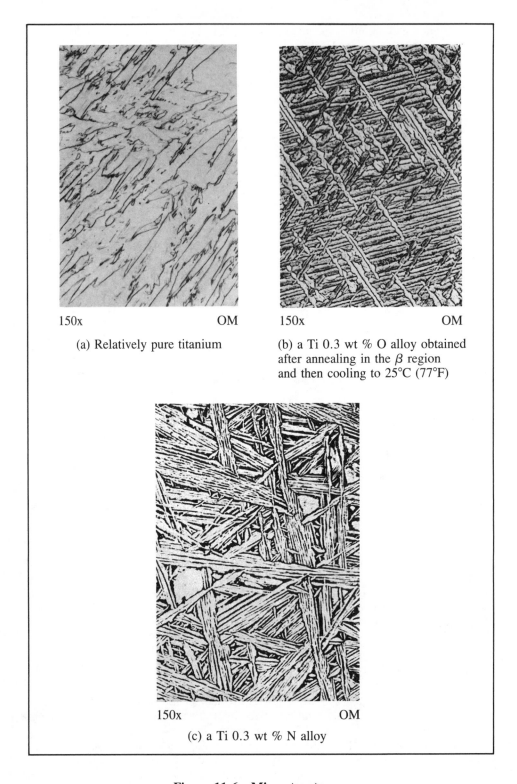

150x   OM

(a) Relatively pure titanium

150x   OM

(b) a Ti 0.3 wt % O alloy obtained after annealing in the β region and then cooling to 25°C (77°F)

150x   OM

(c) a Ti 0.3 wt % N alloy

Figure 11.6 - Microstructures

each 1% of Sn. (However, for Zr the effect is only about 3.5 MPa (0.5 ksi) for each 1% addition. Thus Zr is not used to strengthen α titanium alloys.)

In sum, Al and Sn are added to titanium to promote α stabilization (over β) and to increase strength of titanium alloys. The effectiveness of Sn as a strengthener soon begins to level off, whereas it does not with Al. Still, a practical upper limit to the aluminum content is about 7%. Above this value the alloy is difficult to hot work, and, at low temperature, embrittlement may occur.

To put the behavior of alpha-beta alloys into perspective, alpha-favoring elements can be added to aid solid solution hardening. The addition of beta-favoring elements:

- Permits solution heat treatment at lower temperatures

- May solid-solution harden the alloy still further

- May retard alpha formation so that beta transforms to martensite, or it is retained to transform later into alpha upon reheating (stabilizing, aging) to temperatures from 427° to about 816°C (800° to 1500°F).

The relative amounts of primary alpha, present at high-temperature solution anneal, of retained beta and of martensitic alpha are a function of chemistry and prior thermal treatment. However, they are also a function of mechanical processing history. One of the main purposes of wrought processing of titanium alloys, as indicated in Chapter 4, is to control microstructure and, thus, the properties themselves. Tensile strength, fatigue strength, and toughness as well as creep resistance all may be better in forgings than in other wrought — or even cast or powder — forms. Proper working of titanium alpha-beta alloys enables the microstructure at working temperature to be more homogenous, while any gross microstructural anomalies, as shown in Figures 11.7 and 11.8, are removed or prevented from forming.

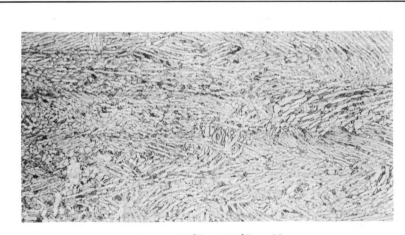

Rolled at 955°C (1750°F); 100x

Figure 11.7 - Distorted Widmanstätten alpha remaining as result of limited working in the $\alpha + \beta$ field

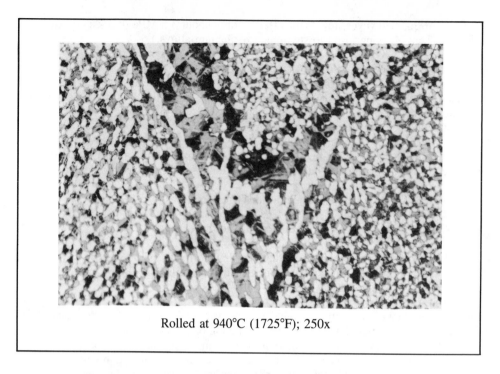

Rolled at 940°C (1725°F); 250x

Figure 11.8 - Grain boundary alpha remnants not broken up after forging due to improper cooling from the β region

The effects of prior processing on microstructure are quite varied. Extensive hot working in the alpha + beta field is required to produce the proper microstructure typified by Figure 11.9. If hot working of the alpha + beta phases has been limited, microstructures

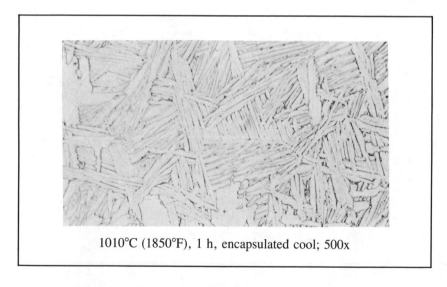

1010°C (1850°F), 1 h, encapsulated cool; 500x

Figure 11.9 - Typical microstructure of Ti-6Al-4V alloy; effect of solution temperature very close to β transus

such as that shown in Figure 11.7 will occur. As may be observed, the Widmanstätten platelets are quite distorted but are not yet broken up. This condition is not particularly detrimental to fracture toughness, although it may affect fatigue crack propagation.

When alloy Ti-6Al-4V is processed improperly after heating into the beta field, alpha phase can form preferentially along the prior beta grains. Extensive hot work is required to break up such structures. An example of grain boundary alpha not completely broken up is shown in Figure 11.8. Because cracks tend to propagate in, or near, interfaces, this type of structure can provide loci for crack initiation and propagation and thereby lead to premature failure.

Microstructural control is effected by using proper combinations of hot work and heat treatment. Heat treatment alone does not suffice to convert the Widmanstätten structure to an equiaxed form; therefore, heat treatment alone is not used unless a transformed structure is desired. Grain refinement cannot be obtained by heat treatment, and, after one β to $\alpha + \beta$ sequence has been accomplished, additional $\alpha + \beta$ to β to $\alpha + \beta$ cycles have no effect on the basic crystallographic texture, although the structure may be coarsened as a consequence of beta grain growth.

In actual material components, the structure of titanium alpha-beta type alloys is controlled not only by how much working is done and by how close to, or above, the beta transus the alloy is processed, but also by the section size of the component. This last factor can drastically change cooling rates. Thick sections may behave as if furnace cooled, while thinner ones may display, at room temperature, a microstructure more characteristic of a rapid cooling.

Rapid cooling, in general, promotes finer structure and better properties; thus thick sections possess lower tensile strengths than thinner ones. The effects of variations in cooling rates can be seen in Table 11.3 for alloy Ti-6Al-4V. Here furnace cooling produces lower strengths than water quenching.

Table 11.3 Typical tensile mechanical properties and microstructure at 25°C of a Ti-6Al-4V alloy for different thermal treatments

Thermal treatment	Yield strength: MPa	ksi	Tensile strength: MPa	ksi	Elongation at fracture, %	Reduction in area, %	Microstructure at 25°C (77°F), approx. vol % phases
955°C (1751°F) Furnace cooled	834	121	937	136	19	46	90% alpha; 10% beta
955°C (1751°F) Water quenched	951	138	1117	162	17	60	50% primary alpha; 60% alpha prime + alpha double prime + retained beta
900°C (1652°F) Furnace cooled	855	124	965	140	17	43	90% alpha; 10% beta
900°C(1652°F) Water quenched	923	134	1117	162	15	54	60% primary alpha; 40% alpha prime + alpha double prime + retained beta

Table 11.3 also indicates a characteristic of higher-temperature thermal treatment. Here the temperature that is just below beta transus produces slightly better yield strengths than a treatment lower in the alpha-beta temperature range.

When titanium alloys are heat treated high in the alpha-beta range, and then cooled, the resulting structure, because of the presence of globular (equiaxed) primary alpha in the transformed beta (plate like) matrix is called equiaxed. When a 100% transformed beta structure is achieved by cooling from above the beta transus, the structure may be called acicular, or needlelike. Table 11.4 lists the strength advantages and disadvantages of both structural types. Table 11.5 clearly indicates that higher yield strength is favored by equiaxed structures. Table 11.5 also indicates that better toughness is characteristic of transformed, or acicular, structures.

Table 11.4 Relative advantages of equiaxed and acicular microstructures

Equiaxed:

* Higher ductility and formability
* Higher threshold stress for hot-salt stress corrosion
* Higher strength (for equivalent heat treatment)
* Better hydrogen tolerance
* Better low-cycle fatigue (initiation) properties

Acicular:

* Superior creep properties
* Higher fracture-toughness values

Generally speaking, alpha-beta alloys would be annealed just below the beta transus to produce a maximum of transformed acicular beta with approximately 10% of equiaxed alpha present. Some titanium alloys—for example, Ti-6Al-2Sn-4Zr-2Mo—are given beta heat treatments to enhance high-temperature creep resistance. (Castings and powder products may be given a beta anneal, too, in order to break up the structure, although not necessarily for optimizing creep strength.)

Table 11.5 Typical fracture-toughness values of high-strength titanium alloys

Alloy	Alpha Morphology	Yield strength		Fracture toughness K_{Ic}	
		MPa	ksi	MPa·m$^{1/2}$	ksi·in.$^{1/2}$
Ti-6Al-4V	Equiaxed	910	130	44-66	40-60
	Transformed	875	125	88-110	80-100
Ti-6Al-6V-2Sn	Equiaxed	1085	155	33-55	30-50
	Transformed	980	140	55-77	50-70
Ti-6Al-2Sn-4Zr-6Mo	Equiaxed	1155	165	22-23	20-30
	Transformed	1120	160	33-55	30-50

Aging, Stability

One of the least understood concepts in the behavior or alpha-beta titanium alloys is that of aging. With few exceptions titanium alloys do not age in the classical sense: that is, a secondary, strong intermetallic compound appears and strengthens the matrix by its dispersion. A dispersion is produced, on aging of alpha-beta alloys, but it is thought to be beta dispersed in the alpha or martensitic alpha prime. Beta is not materially different from alpha phase with respect to strength; however the effectiveness of strengthening in titanium alloys appears to center in the number and fineness of alpha-beta phase boundaries. Annealing and rapid cooling, which maximize alpha-beta boundaries for a fixed primary alpha content, along with aging, which may promote additional boundary structure, can significantly increase alloy strength. (Refer to Table 11.6.)

Table 11.6 Effect of aging on the tensile mechanical properties at 25°C of a Ti-6Al-4V alloy

	Yield strength: ksi	Tensile strength: ksi	Elonga- tion at fracture, %	Reduc- tion in area, %
955°C (1751°F) Water quenched	138	162	17	62
955°C (1751°F) Water quenched	155	172	17	56
900°C (1652°F) Water quenched	134	162	15	54
900°C (1652°F) Water quenched	147	162	15	48

It is interesting to note that the strength for Ti-6Al-4V alloy may increase by 10 to 15%, while ductility remains unchanged, although there may be a slight decrease after aging treatment. The key to aging response is the ability to cool all section locations to produce a fine martensitic structure before aging.

Aging of titanium alpha-beta alloys may be used to stabilize the alloy during service exposure against additional transformation of unstable martensite and retained beta. While true aging tends to occur at temperatures in the approximate range of 538° to 593°C (1000° to 1100°F), stabilization may occur at that point, or up to 38°C (100°F) higher.

Longitudinal vs. Transverse Properties

Another influencing factor on titanium alloy strengthening is the effect of testing direction. The texture and mechanical working effects on directionality of structure can be significant, especially in bar, plate, or sheet mill products. Table 11.7 shows the effect of test direction on properties of textured Ti-6Al-2Sn-4Zr-6Mo plate. Substantial differences are obtained with test direction.

PROPERTY STRUCTURE: OVERVIEW

The following paragraphs contain a number of general comments concerning titanium's property and structure as related to the alpha-beta and near-alpha (superalpha) alloys.

Table 11.7 Effect of test direction on mechanical properties of textured Ti-6Al-2Sn-4Zr-6Mo plate

Test direc-tion (a)	Tensile strength MPa	Yield strength MPa	Elonga-tion, %	Reduction in area, %	Elastic modulus, GPa	K_{Ic} MPa·m$^{1/2}$	K_{Ic} ksi·in.$^{1/2}$	K_{Ic} specimen orien-tation (b)
L	1027	952	11.5	18.0	107	75	68	L-T
T	1358	1200	11.3	13.5	134	91	83	L-T
S	938	924	6.5	26.0	104	49	45	S-T

(a) High basal pole intensitites reported in the transverse direction, 90° from normal, and also intensity nodes in positions
(b) 45° from the longitudinal (rolling) direction and about 40° from the plate normal. L=longitudinal; T=transverse

The major topics are:

- General comments

- Tensile, creep-rupture properties

- Fatigue, fracture toughness properties

- Beta Ti alloys

- Cast Ti alloy properties

- P/M Ti alloy properties

General Comments

The mechanical properties of titanium alloys depend on alloy chemistry, microstructure, and metallographic texture (through its influence on elastic and plastic anisotropy). The influence of these factors on strength, toughness, and resistance to environmental effects on crack propagation will be discussed further in the following paragraphs.

The all-purpose, alpha-beta Ti-6Al-4V alloy derives its annealed strength from several sources, the principal source being substitutional and interstitial alloying of elements in solid solution in both alpha and beta phases. Aluminum is the most important substitutional solid solution strengthener. Its effect on strength is linear. Other less important sources of strengthening are:

- Interstitial solid solution strengthening

- Grain size effects

- Second phase (beta) effects

- Ordering in alpha

- Age hardening

- Effects of crystallographic texture

Aluminum in Ti-6Al-4V gives rise to some tendency toward ordering in the alpha phase, the ordered product being Ti_3Al, alpha-2. Ordering in the alpha phase contributes perhaps 15 to 35 MPa (2 to 5 ksi) to the strength of standard Ti-6Al-4V and contributes less than this to the strength of the ELI grade. Ordering also appears to degrade toughness. The effect of crystallographic texture is to introduce directionality into the strength equation. Relative to the hexagonal axis in alpha, strength (and modulus) is high in the parallel direction and low in the normal direction. Because metalworking operations tend to produce preferred crystallographic orientations in alpha grains, strength becomes an anisotropic quantity in most product forms. This feature can be minimized by proper processing and is rarely of direct concern. In some instances, it can be an advantage.

Because the beta phase present in alloy Ti-6Al-4V can be manipulated in amount and composition by heat treatment, the alloy is responsive to heat treatment. The β- to- $\alpha + \beta$ reaction at low temperature leads to increased strength. The key is to quench from high in the $\alpha + \beta$ field and then age at a lower temperature. A typical strengthening heat treatment consists of heating for 1 h at 955°C (1750°F) and water quenching, followed by heating for 4 h at 540°C (1000°F) and air cooling. Response is limited in a practical sense, however, by two factors:

- The small amount of beta in Ti-6Al-4V

- Section size

The first factor puts an intrinsic ceiling on the increased strengthening response available — about 280 MPa (40 ksi) in thin-gage material. The second factor relates to depth of hardening, because Ti-6Al-4V is not effectively hardenable in sections greater than 25 mm (1 in.) in thickness. The Ti-6Al-4V alloy is, therefore, most commonly used in the annealed condition.

Two alloys that fall in the high-strength alpha-beta class are Ti-6Al-6V-2Sn, which is used in airframes, and Ti-6Al-2Sn-4Zr-6Mo, which is used in jet engines. Alloy Ti-6Al-2Sn-4Zr-6Mo may more often be classified as a superalpha alloy. Both of the latter alloys are stronger and more readily heat treated than Ti-6Al-4V. These features arise from the increased solid solution strengthening afforded by tin and zirconium, which have relatively small effects on the transformation temperature, and from the increased amounts of beta phase that result from the larger vanadium and molybdenum additions. (Both vanadium and molybdenum are beta stabilizers.) The Ti-6Al-6V-2Sn alloy contains the beta stabilizers copper and iron in combined amounts up to 1.4 wt% for enhanced strength and response to aging. Alloy Ti-6Al-2Sn-4Zr-6Mo also is useful at the moderately elevated temperatures from 425° to 480°C (800° to 900°F). This alloy combines high tensile strength with good creep resistance. The alpha phase tends to order more readily in these alloys than in alloy Ti-6Al-4V. Moreover, the transformed alpha platelets in Ti-6Al-2Sn-4Zr-6Mo tend to be narrower than those in Ti-6Al-4V, and formation of packets of parallel platelets is less likely. For both Ti-6Al-6V-2Sn and Ti-6Al-2Sn-4Zr-6Mo, martensite does not form in ordinary situations. Alpha is the dominant phase in these alloys but to a lesser extent than in Ti-6Al-4V. The metallurgy of these alloys is otherwise very similar to that of Ti-6Al-4V.

Alloys Ti-6Al-2Sn-4Zr-2Mo and Ti-8Al-1Mo-1V are set in the near-alpha (superalpha) class. They are used primarily in jet engine applications and are useful at temperatures above the normal range for Ti-6Al-4V. Alloy Ti-6Al-2Sn-4Zr-2Mo may be modified with silicon additions of up to 0.1%, and, when beta annealed (i.e., annealed by heating above the transformation temperature), the modified alloy provides the highest creep strength and temperature capability of all commercial titanium alloys currently produced in the United States. The Ti-8Al-1Mo-1V alloy has the highest modulus and the lowest density of any commercial titanium alloy. Each of these alloys tends to order in the alpha phase more readily than does Ti-6Al-4V. Martensite forms more readily in either of these alloys than in Ti-6Al-4V.

Generally speaking, these alloys contain less beta phase than Ti-6Al-4V. Age hardening treatments are thus not very effective and are, moreover, deleterious to creep resistance. These alloys therefore are usually employed as solution annealed and stabilized. Solution annealing may be done at a temperature some 35°C (63°F) below the transformation temperature, and stabilization is commonly produced by heating for 8 h at about 590°C (1100°F).

At high temperatures, strain aging arising from aluminum, silicon, and tin, and perhaps from oxygen and zirconium, is thought to contribute to the creep resistance of these materials.

The alpha phase dominates the properties of these alloys to a greater extent than it does in Ti-6Al-4V. The metallurgy of the near-alpha alloys is otherwise similar to that of Ti-6Al-4V.

Tensile, Creep-Rupture Properties

Typical minimum property guarantees for titanium alloy mill products are listed in Table 11.8. The effects of temperature on strength for the same alloys are shown in Table 11.9. Data for unalloyed titanium is included in this table to illustrate that the alloys not only have higher room temperature strengths but also retain much larger fractions of that strength at elevated temperatures.

Table 11.8 Typical mill-guaranteed room temperature tensile properties for selected titanium alloys

Alloy	Ultimate strength MPa	ksi	Yield strength MPa	ksi	Ductility Elongation, %	Reduction in area, %
Ti-6Al-4V	895	130	825	120	10	20
Ti-6Al-6V-2Sn	1065	155	995	145	10	20
Ti-6Al-2Sn-4Zr-6Mo	1030	150	965	140	10	20
Ti-6Al-2Sn-4Zr-2Mo	895	130	825	120	10	25
Ti-8Al-1Mo-1V	895	130	825	120	10	20

In terms of the principal heat treatments used for titanium, beta annealing decreases strength by 35 to 100 MPa (5 to 15 ksi) depending on prior grain size, average crystallographic texture and testing direction. Solution treating and aging can be used to enhance strength at the expense of fracture toughness in alloys containing sufficient beta stabilizer (that is, 4 wt %, or more).

Typical tensile strengths and 0.1% creep strengths as functions of temperature of some prominent alloys are shown in Figure 11.10 and 11.11, respectively.

Table 11.9 Fraction of room-temperature strength retained at elevated temperature for several titanium alloys (a)

Temperature		Unalloy-ed Ti		Ti-6Al-4V		Ti-6Al-6V-2Sn		Ti-6Al-2Sn-4Zr-6Mo		Ti-6Al-2Sn-4Zr-2Mo		Ti-8Al-1Mo-1V	
°C	°F	TS	YS	TS	YS	TS	YS	TS	YS	TS	YS	TS	YS
93	200	0.80	0.75	0.90	0.87	0.91	0.89	0.90	0.89	0.93	0.90	0.93	0.92
204	400	0.57	0.45	0.78	0.70	0.81	0.74	0.80	0.80	0.83	0.76	0.84	0.79
316	600	0.45	0.31	0.71	0.62	0.76	0.69	0.74	0.75	0.77	0.70	0.76	0.67
427	800	0.36	0.25	0.66	0.58	0.70	0.63	0.69	0.71	0.72	0.65	0.67	0.60
482	900	0.33	0.22	0.60	0.53	-	-	0.66	0.69	0.69	0.62	0.61	0.55
538	1000	0.30	0.20	0.51	0.44	-	-	0.61	0.66	0.66	0.60	0.53	0.46

(a) Short-time tensile test with less than 1 h at temperature prior to test.

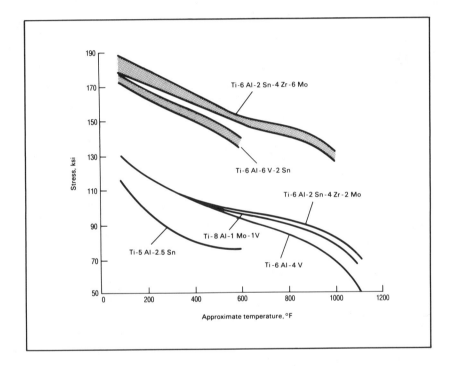

Figure 11.10 - Comparison of typical tensile strengths for various Ti alloys

Fatigue, Fracture Toughness Properties

Fatigue and fracture toughness tend to be considered as one area, but fatigue tests of titanium alloys have tended to be reported in recent years in terms of crack-propagation results; that is, da/dN vs. delta-K_{Ic} (growth rate vs. incremental fracture toughness change). While da/dN vs. delta-K_{Ic} curves have their uses, it is highly desirable to describe the isothermal, cyclic fatigue behavior of titanium alloys in some graphic manner. Limited studies have shown, as expected, that notch properties in fatigue are significantly different from smooth properties. They also show that the stress ratio (R value,

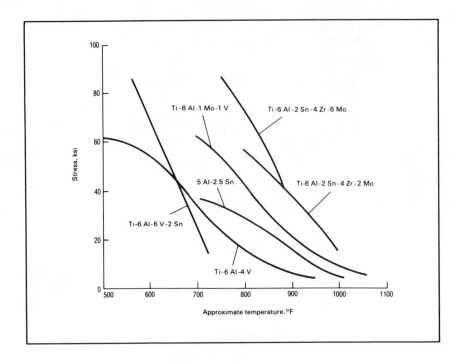

Figure 11.11 - Comparison of typical 150-h, 0.1% creep strengths for various Ti alloys

min/max stress ratio) is vitally important. Static properties such as tensile and creep rupture are thus fairly easy to depict by contrast to dynamic properties of fatigue and fracture toughness. Factors affecting ductility in static properties act to reduce cyclic ductility and, also, life.

Some limiting aspects to the depiction of titanium fatigue are:

- The very restricted data available

- The need to describe both low-cycle fatigue (LCF: less than about 10^5 cycles to failure)

- High-cycle fatigue (HCF: usually 10^6 to 10^8 cycles to failure)

The high inherent fatigue strength of titanium alloys and the probability that fatigue will initiate at stress concentrators in components make crack propagation a concern regarding these materials. For LCF and HCF there is a great value in high fracture toughness and low crack-propagation rates.

Titanium fatigue (and toughness), even more than static properties, may be structure dependent. Results of an LCF study on Ti-6Al-4V are shown in Figure 11.12. In the figure, time to the first crack (at a fixed strain) varies with microstructure. Note that time to crack initiation is optimized with a structure having high amounts of transformed beta, yet still having about 10% of primary alpha. (However, the crack-propagation resistance of the beta-processed structure still exceeds that of alpha-processed material.)

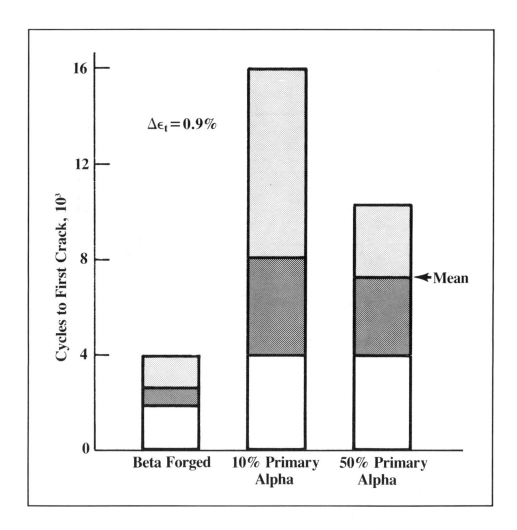

Figure 11.12 - LCF life of Ti-6Al-4V alloy with different structures: beta forged (100% transformed beta); 10% primary alpha (bal. transformed beta); 50% primary alpha

Other interesting aspects of titanium alloys' fatigue lines are in their response to R value. (That is, preload; R = -1 indicates similar tension and compression stresses.) Also, the alloys' notch concentrations and surface conditions play a part. The beneficial aspects of peening and glass bead blasting on Ti-6Al-4V LCF at 21°C (70°F) are shown in Figure 11.13. The effects of K_t and crack propagation on LCF life on preloaded Ti-6Al-4V at 204°C (400°F) are shown in Figure 11.14.

Low cycle fatigue is very difficult to quantify owing to the wide range of variables and to the limited amount of published data. In general, available data are for load-controlled, not strain-controlled, data. Barstock defers from forgings. Component full-scale test results on forgings agree with test results on specimens machined from forgings. The effect of temperature is barely reported; the stress capability of Ti-6Al-4V

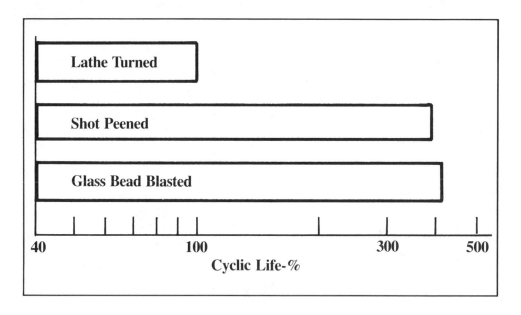

Figure 11.13 - Effects of surface condition on LCF life of Ti-6Al-4V at 21°C (70°F)

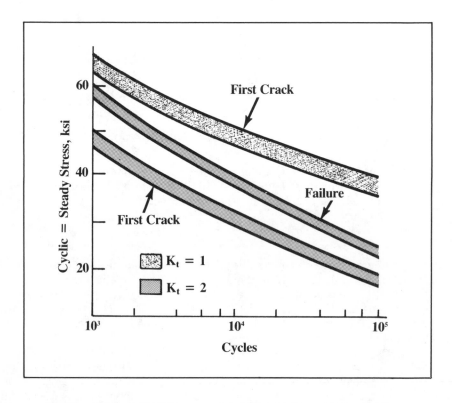

Figure 11.14 - LCF life strength of Ti-6Al-4V at 204°C (400°F); shows effects on life of K_t and of crack propagation to failure

at R = 0 (cyclic = steady, preloaded stress) was about 11% less at 204°C (400°F) than
it was at 21°C (70°F).

In high-cycle fatigue, depending on the alloy, the fatigue endurance limit tends to be relatively flat with temperatures out to 316°C (600°F), or above, as shown in Figure 11.15.
This figure also illustrates the benefits of titanium alloys over steels in fatigue strength
capability. Surface condition affects HCF, as illustrated in Figure 11.16. This figure
also indicates, in a broad way, the range of data scatter that can be found in a given titanium alloy at a single K_t and temperature. Applications of titanium alloys to fatigue-
limited components should include verification of the fatigue strength under expected
service conditions.

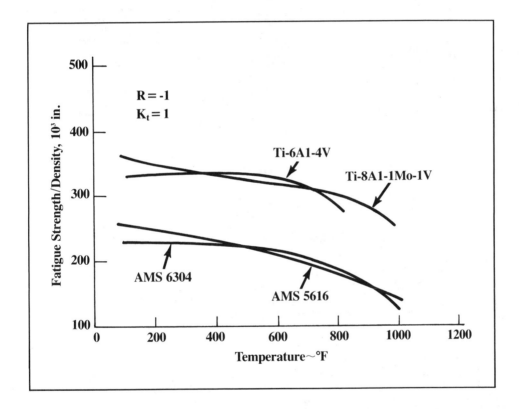

**Figure 11.15 - HCF (5×10^7 cycles) strength; density of
alloys shows limited temperature dependence and benefits
of titanium over steels**

One further observation about titanium alloys is that fatigue data reported in the literature often may be on material with favorable surface residual stress induced by turning,
milling, etc. Fully stress-relieved or chem-milled surfaces probably have fatigue
strengths below the reported alloy capabilities, because the latter have been biased
upward by the favorable — that is compressive — stresses.

Fracture toughness can be varied within a nominal titanium alloy by as much as a multiple of two or three. This may be accomplished by manipulating alloy chemistry,

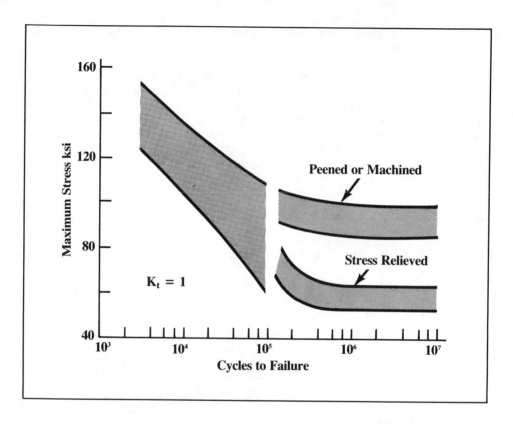

Figure 11.16 - Fatigue scatter band for Ti-6Al-4V at 21°C (70°F)

microstructure, and texture. Some trade-offs of other desired properties may be necessary to achieve high-fracture toughness. Plane-strain fracture toughness, K_{Ic}, is of special interest because the critical crack size at which unstable growth can occur is proportional to $(K_{Ic})^2$. Strength is often achieved in titanium alloys at the expense of K_{Ic}.

There are significant differences among titanium alloys in fracture toughness, but there also is appreciable overlap in their properties. Table 11.5 gives examples of typical plane-strain fracture toughness ranges for alpha-beta titanium alloys. From these data it is apparent that the basic alloy chemistry affects the relationship between strength and toughness. From Table 11.5 it also is evident that transformed microstructures may greatly enhance toughness while only slightly reducing strength.

Within the permissible range of chemistry for a specific titanium alloy and grade, oxygen is the most important variable insofar as its effect on toughness is concerned. In essence, if high fracture toughness is required, oxygen must be kept low, other things being equal. Reducing nitrogen, as in Ti-6Al-4V-ELI, is also indicated, but the effect is not as strong as it is with oxygen.

Improvements in K_{Ic} can be obtained by providing either of two basic types of microstructures:

• Transformed structures, or structures transformed as much as possible, because such structures provide tortuous crack paths

- Equiaxed structures composed mainly of regrowth alpha that have both low dislocation-defect densities and low concentrations of nitrogen and oxygen (the so-called "recrystallization annealed" structures)

Transformed structures appear to be tough primarily because fractures in such structures must proceed along tortuous, many-faceted crack paths.

According to some work, the K_{Ic} is proportional to the fraction of transformed structure, from the beta, in the alloy. The subject is a complex one without clear-cut, empirical rules. Furthermore, the enhancement of fracture toughness at one stage of an operation — for example, a forging billet — does not necessarily carry over to a forged part. Because welds in alloy Ti-6Al-4V contain transformed products, one would expect such welds to be relatively high in toughness. This is, in fact, the case, as shown in Table 11.10.

Table 11.10   Fracture toughness of alloy Ti-6Al-4V (0.11 wt % O) in welds and heat affected zones

Post stress relief	Weld		Heat-affected zone K_{Ic}		Base metal (a)	
	$MPa{\cdot}m^{1/2}$	$ksi{\cdot}in.^{1/2}$	$MPa{\cdot}m^{1/2}$	$ksi{\cdot}in.^{1/2}$	$MPa{\cdot}m^{1/2}$	$ksi{\cdot}in.^{1/2}$
2 h at 590°C (1100°F), AC	87(b)	79(b)	81(c)	74(c)	92	84
1 h at 650°C (1200°F), AC	85(d)	77(d)	77(d)(e)	70(d)(e)	92	84
1 h at 760°C (1400°F), AC	-	-	76(d)	69(d)	92	84

(a) Recrystallization anneal at 0.11 wt % O.
(b) Based on data from 2 samples.
(c) Based on data from 20 samples.
(d) Based on data from 1 sample.
(e) Annealed for 2 h at 650°C (1200°F), air cool (AC)

In addition to welding, many other factors such as environment, cooling rates in large sections, hydrogen content, etc. may affect K_{Ic}.

Just as K_{Ic} is important in calculating loads that a structural member can carry in the presence of a flaw, so also it is important in many cases to know that the remaining fatigue life is in the presence of a fatigue crack or other sharp crack. In a very general way, fatigue-crack propagation (FCP) behavior in titanium parallels fracture toughness; that is, for a given alloy, those conditions giving highest toughness tend also to give, under fatigue loading, lowest cyclic growth rates.

In FCP measurements, there can be a significant amount of scatter in the data. The example shown in Figure 11.17 is one of the more extreme cases encountered and arises for the mill annealed condition where uncertainties of microstructure, texture, and strength may exist. Part of the scatter is due to test reproducibility. There also may be a point-to-point material variability. The latter is due to minor processing and material inhomogeneities. Variations in chemistry, microstructure, and texture effects within a given lot may, in some cases, be additive even under controlled conditions. There are, of course, lot-to-lot differences. For design purposes, users are well advised to use statistical data derived from information in digital form from several lots.

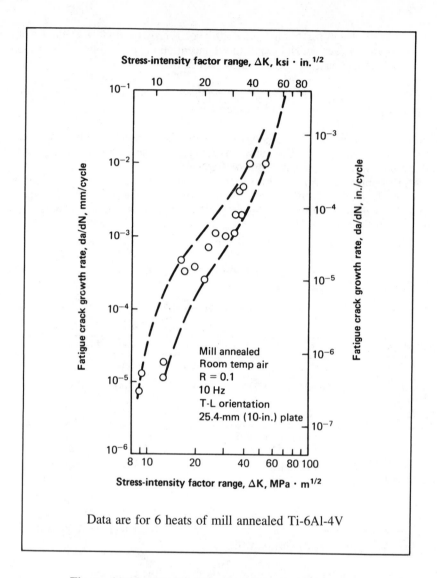

**Figure 11.17 - Illustration of scatter that can occur
in fatigue crack propagation measurements**

Titanium alloys have different FCP characteristics just as they have different K_{Ic} characteristics. Selected data indicate that fatigue cracks propagate more rapidly in Ti-6Al-2Sn-4Zr-6Mo than in Ti-8Al-1Mo-1V or Ti-6Al-2Sn-4Zr-2Mo under the same test conditions. This may be a simple effect of strength. However, the relative amounts of beta phase may lead to intrinsically different fatigue-crack propagation characteristics. The Ti-6Al-2Sn-4Zr-6Mo alloy is also more easily textured. In addition, different phases, such as orthorhombic alpha double prime martensite, may exist and could effect FCP problems after aging.

Microstructure variations also produce general parallels between K_{Ic} and FCP. As for K_{Ic} FCP is favorably influenced by transformation microstructure and also by application of recrystallization-anneal-type thermal cycles. Microstructure in a given lot of Ti-6Al-4V can affect FCP by a factor of more than ten and can affect delta-K by 5 to 30 MPa · m$^{1/2}$ (4 to 27 ksi · in.$^{1/2}$), depending on where these parameters are measured on the da/dN curve. Generally speaking, beta-annealed microstructures have the lowest fatigue-crack growth rates, whereas mill annealed microstructures yield the highest growth rates. A typical example of such behavior is shown in Figure 11.18.

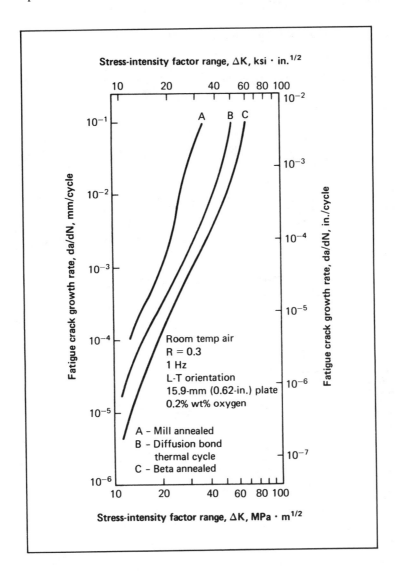

Figure 11.18 - Effect of heat treatment of fatigue crack growth rate of Ti-6Al-4V

Because of the reactive nature of titanium, it should come as no surprise that environment affects FCP rates in titanium just as it affects fracture toughness. The only surprise in the available data is that the chemical environment does not have a larger effect than it does. In general, only the more severe environments (such as a 3.5% NaCl solution) affect FCP rates by an order of magnitude, or more. Gaseous atmospheres also may play a role in affecting FCP rates. All chemical or gaseous environmental effects undoubtedly are sensitive to some degree both microscopically and chemically.

BETA TITANIUM ALLOYS

A large data base has not been developed for beta titanium alloys. Current efforts are concentrated on two areas:

- Ti-10V-2Fe-3Al, as a forging alloy

- Ti-15V-3Al-3Sn-3Cr, as a sheet alloy

The Ti-10V-2Fe-3Al alloy has been cast, HIPed (hot isostatic pressed), and powder HIPed processed. Other beta alloys have been evaluated, but only Ti-13V-11Cr-3Al has a significant published data base.

It has been claimed that forgings of Ti-10V-2Fe-3Al have achieved strengths at greater section sizes than with Ti-6Al-4V alloy. HIP castings of Ti-10V-2Fe-3Al also may offer improved strength and fatigue resistance over their Ti-6Al-1V casting counterparts. Significant work is required for both cast and HIP P/M techniques in order to develop a good data base on Ti-10V-2Fe-3Al.

Typical property data for Ti-10V-2Fe-3Al and Ti-15V-3Al-3Sn-3Cr are indicated in Table 11.11 and 11.12, respectively.

Table 11.11 Typical data for Ti-10V-2Fe-3Al (a)

Beta transus temperature: 795°C (1465°F)
Modulus of elasticity: 14.0×10^6 psi as solution treated (ST)
    15.5×10^6 psi as ST and aged
Density: 0.168 lb/in^3, 4.65 g/cm^3
Tensile properties of forgings:

Condition	UTS		YS		El, %	RA, %	K_{Ic}
	MPa	ksi	MPa	ksi			ksi in.
STA: 1 h 2525-2570°C (1385-1410°F), WQ + 8 hr 1652-1742°C (900-950°F), AC	1241-1379	180-200	1158-1269	168-184	4-12	10-30	42-56
STOA: 1 hr 2462°C (1350°F), AC + 8 hr 1967-2012°C (1075 -1100°F), AC	965-1034	140-150	896-965	130-140	20	45	93
BAOA: 1 hr 2732°C (1500°F), AC + 8 hr 2102°C (1150°F), AC	1000	145	930	135	17	46	100

(a) Maximum oxygen 0.13%

Fatigue properties for Ti-10V-2Fe-3Al are shown in Figure 11.19. Fatigue crack propagation rates are shown in Figure 11.20. In both cases comparisons are made with Ti-6Al-4V.

Table 11.12 Typical data for Ti-15V-3Cr-3Al-3Sn (a)

Beta transus temperature: 760°C (1400°F)
Modulus of elasticity: 12 x 10⁶ psi, annealed (ANN)
15 x 10⁶ psi, aged
Density: 0.170 lb/in³, 4.71 g/cm³
Tensile properties:

Product	Condition	UTS MPA	UTS ksi	YS MPa	YS ksi	El,%
Strip, 0.035-0.07 in.	ANN 4-30 min 2642°C (1450°F), AC	786	114	772	112	21
	ANN + 14 hr 1742°C (950°F)	1310	190	1207	175	7
	ANN + 14 hr 1832°C (1000°F)	1103	160	1000	145	12

(a) Maximum oxygen 0.13%

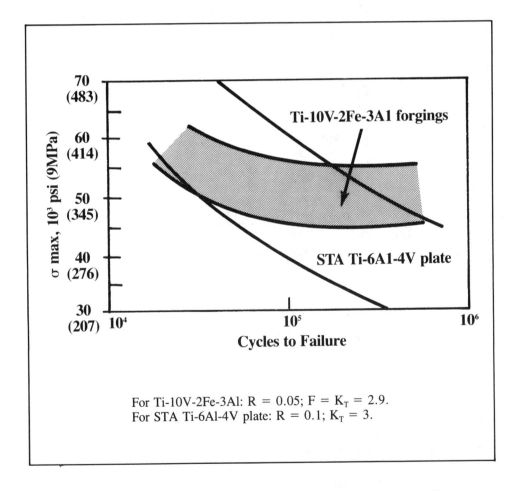

Figure 11.19 - Comparison of notched fatigue curves for Ti two alloys

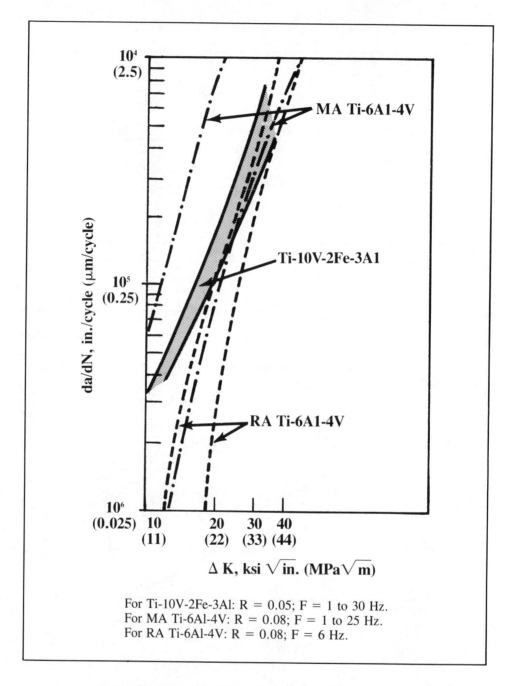

Figure 11.20 - Comparison of fatigue crack growth rates

CAST Ti ALLOY PROPERTIES

As indicated earlier, cast titanium alloys are generally alpha-beta alloys. They are equal, or nearly equal, in strength to wrought alloys of the same compositions. Typical room-temperature tensile properties of several cast titanium alloys are shown in Table 11.13. Although a number of titanium alloys have been studied in cast form, virtually all existing data have been generated from alloy Ti-6Al-4V. Thus this alloy is the basis for the following discussion.

Table 11.13 Typical room-temperature tensile properties of several cast titanium alloys

Alloy	Condition	Tensile strength MPa	ksi	Yield strength MPa	ksi	Elongation % (b)	Reduction in area %
Commercially pure titanium	As-cast or annealed	550	80	450	65	17	32
Ti-6Al-4V	As-cast or annealed	1035	150	890	129	10	19
Ti-6Al-2Sn-4Zr-2Mo	Duplex annealed	1035	150	895	130	8	16
Ti-5Al-2.5Sn-ELI	Annealed	805	117	745	108	11	-

(a) In 50 mm or 2 in.

A distribution of room-temperature tensile properties for Ti-6Al-4V castings from a particular producer is shown in Figure 11.21. Elevated-temperature tensile properties for

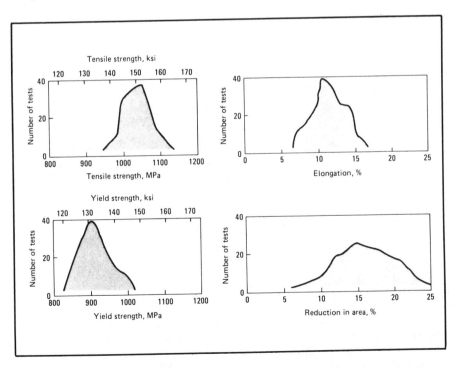

Figure 11.21 - Typical distribution of tensile properties for as-cast Ti-6Al-4V

the same alloy are shown in Figure 11.22. Tensile properties for cast and wrought alloy Ti-6Al-2Sn-4Zr-2Mo are indicated in Figure 11.23. The plane-strain fracture-toughness values for Ti-6Al-4V castings are compared with values for Ti-6Al-4V plate and, also, with other wrought titanium alloys in Figure 11.24.

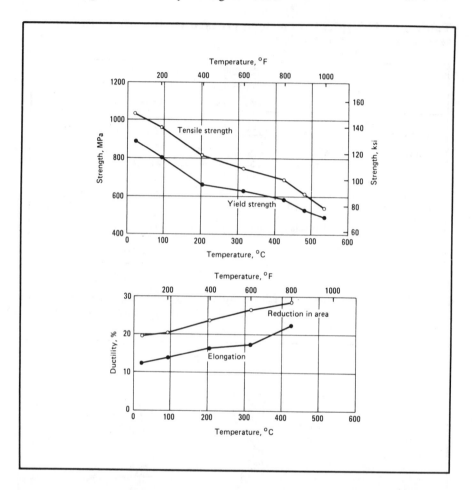

Figure 11.22 - Typical elevated-temperature properties of cast Ti-6Al-4V

Although properties of Ti-6Al-4V alloy castings generally meet the properties of beta-annealed, forged-ingot metallurgy (wrought) products, the forged material has superior high-cycle fatigue (HCF) properties. Forged products typically are processed in the alpha-beta phase field, yielding a refined alpha + beta microstructure with good fatigue resistance. By contrast, castings cool slowly from the beta phase field, producing a coarse microstructure; this is aggravated by additional coarsening during hot isostatic pressing (HIP).

Generally, an improvement in fatigue resistance is gained by HIPing of cast material. In addition, results of research suggest that substantial improvement in resistance to fatigue-crack propagation can be obtained by beta heat treating and over-aging of cast alloys.

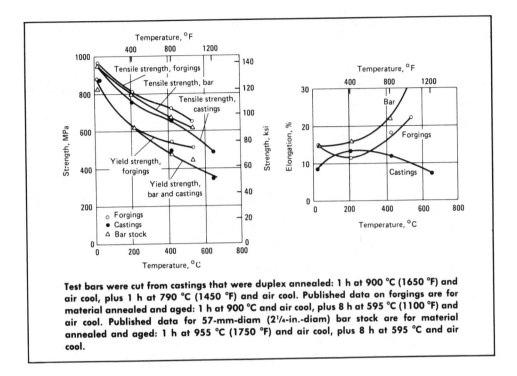

Test bars were cut from castings that were duplex annealed: 1 h at 900 °C (1650 °F) and air cool, plus 1 h at 790 °C (1450 °F) and air cool. Published data on forgings are for material annealed and aged: 1 h at 900 °C and air cool, plus 8 h at 595 °C (1100 °F) and air cool. Published data for 57-mm-diam (2¼-in.-diam) bar stock are for material annealed and aged: 1 h at 955 °C (1750 °F) and air cool, plus 8 h at 595 °C and air cool.

Figure 11.23 - Comparison of short-time tensile properties for wrought and cast forms of Ti-6Al-2Sn-4Zr-2Mo

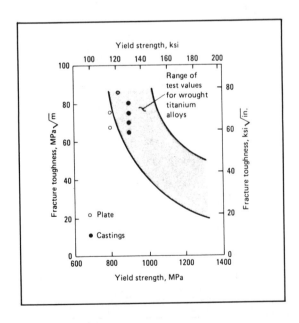

Figure 11.24 - Fracture toughness of Ti-6Al-4V castings compared to Ti-6Al-4V plate and to other Ti alloys

Actual crack-growth rates will be influenced by casting quality and postcasting heat treatment, including HIP.

Research has been conducted aimed at improving the tensile and fatigue strength of HIPed Ti-6Al-4V investment castings by post-HIP heat treating. Baseline microstructure of cast plus HIP parts consists of colonies of alpha plates and grain boundary alpha phase. The baseline data are for HIP at 900°C (1650°F) and 105 MPa (15 ksi). Both of the microstructural features noted above reduce fatigue crack-initiation resistance. Post-HIP heat treating is intended to break up this structure. Heat treatments included beta solution treatment with various rates of cooling from the solution temperature. This was followed by aging and solution treatment in the alpha-beta region and, finally, by similar aging.

Tensile and fatigue strength is improved significantly above the cast-plus-HIP levels by all heat treatments. However this is at the expense of ductility. Water quenching and gas fan cooling from a beta solution temperature of 1025°C (1880°F) produce the best combination of strength and ductility with tensile strengths of 1035 MPa (150 ksi), or better, and ductilities of 7%, or better.

Fine martensite produced by a beta quench is transformed into a fine "basketweave" lenticular alpha structure, called "broken up structure" (BUS). This may be contrasted with an alpha colony arrangement in a cast-plus-HIP product. Although grain boundary alpha still exists, it is not continuous like that in the cast-plus-HIP condition. While grain boundary alpha may be a source for fatigue-crack initiation, fatigue life is improved due to the absence of continuous alpha-beta interfaces. (See Figure 11.25.)

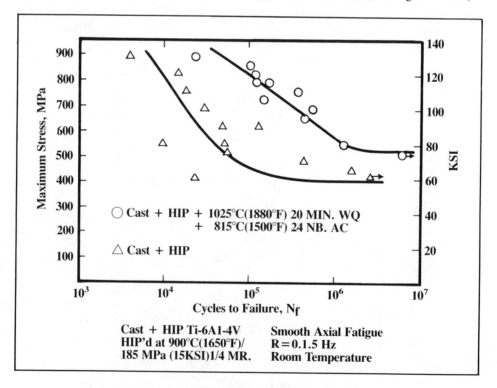

Figure 11.25 - Fatigue strength improvement of Ti-6Al-4V rammed graphite, HIPed cast parts by the "broken up" structure (BUS) treatment

The data presented in Figure 11.19 and 11.20 represent material cast in rammed graphite molds. The investment casting technique, rather than rammed graphite mold, was applied to a helicopter damper bracket. It was reported that variability of fatigue properties in HIP-cast alloy Ti-6Al-4V was decreased. This particular product was given a beta solution annealing at 1038°C (1900°F) for 2 to 3 h with an oil quench. It was aged at 732°C (1350°F) for 2.5 to 3 h after HIPing. Full-scale fatigue tests on the brackets indicated that the cast-plus-HIP-plus-heat treatment resulted in a product equal to forged alloy.

The general fatigue capability of cast-plus-HIP Ti-6Al-4V is shown in Figures 11.26 and 11.27. With optimum casting techniques, HIP and a post-HIP heat treatment, cast-plus-HIP strengths in simple parts should be comparable to those in wrought material. Of continuing concern, however, is the ability to verify for all cast-plus-HIP titanium properties that test bar properties are representative of component values. (At times this is a problem with wrought ingot metallurgy material, too.)

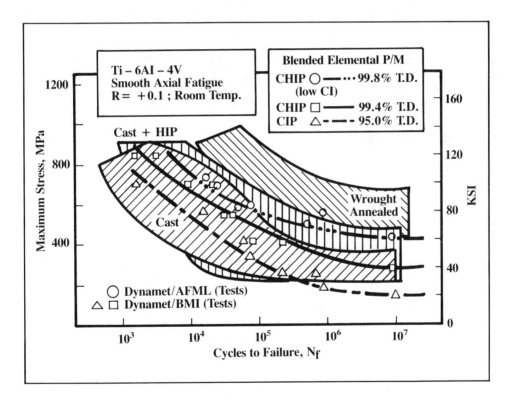

Figure 11.26 - Room temperature smooth axial fatigue data comparing Ti-6Al-4V cast and cast plus HIP material with annealing ingot metallurgy material

P/M Ti ALLOY PROPERTIES

Despite the extensive work on powder processing techniques, only limited property data exist for titanium alloys. Selection of the powder source — blended elemental, BE, or prealloyed, PA — and of the processing technique(s) — CIP + sinter, CIP + sinter +

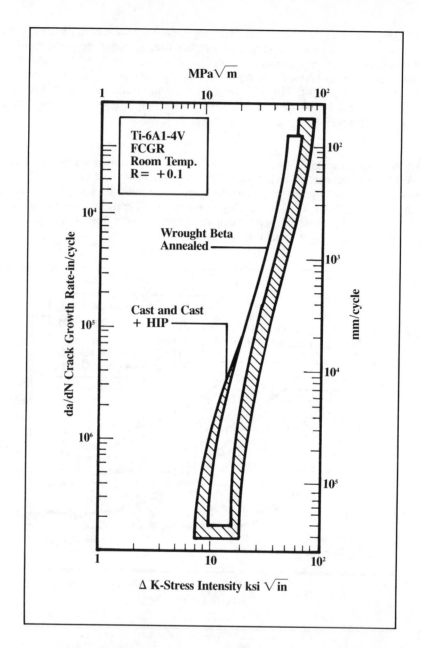

Figure 11.27 - Fatigue crack growth rate scatterband data comparing Ti-6Al-4V cast and cast plus HIP material with beta-annealed ingot metallurgy material

HIP, etc. — and post-processing heat treatment all greatly influence mechanical property capability.

Mechanical properties of P/M structural materials depend on the composition, density, and heat treatability of the material, as well as on processing and design considerations.

Final density, which is expressed in g/cm^3 as well as percent of theoretical density has the greatest effect on properties of P/M materials. (Theoretical density is the ratio of the density of a P/M material to that of its wrought counterpart.) Powder metallurgy parts with theoretical densities of less than 75% are considered to be low density; those above 90% are high density; and those in between are classified as medium density. Generally, structural parts have densities ranging from 80% to above 95%. However, fatigue-critical and aircraft quality parts aim for 100% density. The mechanical properties of powder metallurgy CP titanium and some alloys, along with some wrought properties, are listed in Table 11.14.

Note that in some instances Table 11.14 includes the oxygen content. One of the more important considerations in the manufacture of a titanium powder metallurgy product is control of oxygen content, because oxygen has the same undesirable effects on proper-ties of P/M parts as it has on those of wrought products. Powders must be handled very carefully, because they have a very high affinity for oxygen. Table 11.15, which shows room-temperature tensile properties of compacts made from alloy powders of varying oxygen content, illustrates the importance of low oxygen content in obtaining satisfac-tory ductility in hot-pressed powder compacts.

Currently, the most available prealloyed powder (PA) source appears to be the rotating electrode process. HIP produces the best consolidation. PA powders plus HIPing plus post-HIP heat treatment optimizes properties. Tensile tests conducted on hot isostatically pressed Ti-6Al-4V compacted from samples of rotating-electrode-processed powder indicate excellent properties, as shown in Table 11.16. Subsequent improvements in the process, which include plasma torch melting, have provided plasma rotating-electrode-processed Ti-6Al-4V P/M compacts which are reported to have fatigue properties that are comparable, or superior to, those for cast and wrought materials.

Fatigue properties of powder products are compared with those of wrought titanium in Figure 11.28. Note that the improvements in blended elemental powder (BE), especially regarding reduced chlorides, have led to claims of dramatically increased fatigue capa-bility of titanium BE compacts. Thus Figure 11.28 should be considered only as a guide. In the future BE compacts may be quite favorable with respect both to costs and to achieved property levels.

Beta alloy Ti-10V-2Fe-3Al also has been studied in the P/M environment. P/M tech-niques such as HIP plus heat treatment, or HIP preform plus forge plus heat treatment, may offer property advantages.

Alpha-beta alloys can be HIP preformed and forged.

For a given alloy composition, densification and structural control are the prime factors influencing cyclic properties. Fatigue-fracture life can be comparable to wrought product. Moreover, fatigue-crack-propagation (FCP) characteristics can be comparable to those for wrought (ingot metallurgy: I/M) materials having the same chemistry and structure. (Refer to Figure 11.29.) In addition, the fracture toughness of compacts made of blended elemental and plasma rotating-electrode powder are frequently the equiva-lent, or superior, to I/M mill annealed forgings.

POWDER vs. CAST vs. WROUGHT Ti ALLOY

Typical room-temperature properties of P/M, cast, and wrought titanium products from a variety of alloys are compared in Table 11.17. A comparison of the fatigue properties of the same three material forms for Ti-6Al-4V, only, is illustrated in Figure 11.30.

Table 11.14 Mechanical properties of P/M and wrought Ti and Ti alloys

Alloy	Processing	Density, %	Ultimate tensile strength, MPa (ksi)	Yield strength, MPa (ksi)	Elongation, %	Reduction in area, %	Elastic modulus, GPa (10^6 psi)	Fatigue limit, notched MPa (ksi)	Fracture toughness, MPa $\sqrt{m}$ (ksi $\sqrt{in.}$)
Wrought commercial purity titanium grade II	...	100	345 (50)	344 (50)	5	35	103 (14.9)	...	...
	...	95.5	414 (60)	324 (47)	15	14	103 (15)	...	...
Sponge commercial-purity P/M titanium(a)	...	94	427 (62)	338 (49)	15	23	...	...	...
	Forged	100	455 (66)	365 (53)	23	30	...	...	...
Wrought Ti-6Al-4V (AMS 4298)	...	100	896 (130)	827 (120)	10	20	114 (16.5)	427 (62)	55(e)
P/M Ti-6Al-4V	Blended elemental alloy, cold pressed	95.5	876 (127)	786 (114)	8	14	117 (17)	193 (28)	50(e) 45(e) 40(e)
	...	98+	919 (133)	839 (121.6)	10.9	19.0	...	262 (38)	56(e)
	Blended elemental alloy, forged preforms or vacuum hot pressed	99 min	937 (136)	862 (125)	12-18	15-40	116 (16.8)	414 (60)	51(e) 61(e) 56(e)
	Hot isostatically pressed prealloy	100	947 (137.4)	868 (125.9)	18.8	43.2	117 (17)	414 (60)	...
	Solution treated and aged	99	1103 (160)	1013 (147)	4.9	7.6	...	(60)	...
	Rapid omnidirectional compacted(c)	100	1014 (147)	944 (137)	18.4	40.9	...	...	...

(a) 0.12% oxygen. (b) 0.2% oxygen. (c) Consolidated at 811 MPa (58.8 tsi), 0.5-s dwell in low-carbon steel fluid dies. Preheat temperature was 940 °C (1725 °F), held at temperature $3/4$ h. Powder was vacuum filled into fluid dies following cold static outgassing for 24 h. (d) K_t = 3. (e) K_c. (f) K_{Ic}

(Continued)

Table 11.14 (Continued)

Alloy	Processing	Density, %	Ultimate tensile strength, MPa (ksi)	Yield strength MPa (ksi)	Elongation, %	Reduction in area, %	Elastic modulus, GPa (10⁶ psi)	Fatigue limit, notched MPa (ksi)	Fracture toughness, MPa √m (ksi √in.)
Plasma rotating electrode processed Ti-6Al-6V-2Sn	Hot isostatically pressed	100	1053 (152.7)	1008 (146.3)	18	36.5	110 (16)	448 (65)	...
Plasma rotating electrode processed Ti-6Al-4V	Hot isostatically pressed	100	951 (138)	910 (132)	15	39	...	414(d) (60)(d)	83(f) (76)(f)
P/M Ti-6Al-4V(a)	...	94	827 (120)	738 (107)	5	8	...	...	...
	Forged	100	920 (133.5)	841 (122)	11.5	25	...	...	...
P/M Ti-6Al-4V(b)	Hot isostatically pressed	100	917 (133)	827 (120)	13	26	...	...	...
P/M Ti-6Al-6V-2Sn	...	99	1067 (155)	977 (142)	10	14	...	...	...

(a) 0.12% oxygen. (b) 0.2% oxygen. (c) Consolidated at 811 MPa (58.8 tsi), 0.5-s dwell in low-carbon steel fluid dies. Preheat temperature was 940 °C (1725 °F), held at temperature ³/₄ h. Powder was vacuum filled into fluid dies following cold static outgassing for 24 h. (d) $K_t = 3$. (e) K_c. (f) K_{Ic}

Table 11.15 Typical room-temperature tensile properties of titanium P/M compacts as influenced by oxygen content (a)

Powder manufacturing process	Oxygen content, ppm	Tensile strength		Yield strength		Elongation, %
		MPa	ksi	MPa	ksi	
Ti-6Al-4V:						
Mechanical attrition	1750	1000	145	940	136	1.5
Rotating electrode	900	1000	145	925	134	7.5
Hydride-to-hydride	1570	1025	149	970	141	2.0
Ti-5Al-2.5Sn:						
Rotating electrode	980	905	131	905	131	4.0
Gas attrition	3530	895	130	–	–	–

(a) All data for compacts hot pressed to 1380 MPa (100 tons/in.^{2}) at 1010°C (1850°F).

Head-to-head comparisons of any alloy are difficult to find, and so property comparisons such as these tend to involve different heats, processing sources, test sources, etc. Under ideal conditions the mechanical properties of cast or prealloyed P/M material of Ti-6Al-4V can be close to those of wrought ingot metallurgy product. The effects of heat treatment, including cast HIP or powder HIP cycle, along with cooling rates within a given component, markedly affect properties. When comparable microstructures are produced, properties of fully densified castings or P/M material should be identical to wrought material. (Usually only beta-annealed structures can be comparably produced in all three product forms.) Note, however, that the alpha-beta processed condition is more frequently used for alloys such as Ti-6Al-4V. Explicit comparisons of cast or P/M products with the best alpha-beta processed wrought alloys are not readily available.

Table 11.16 Typical tensile properties of hot isostatically pressed Ti-6Al-4V rotating electrode processed powder (a)

Orientation	Tensile strength		Yield strength		Elongation % (b)	Reduction in area %
	MPa	ksi	MPa	ksi		
L	938.4	136.1	850.8	123.4	20.0	37.0
	936.3	135.8	868.1	125.9	18.0	37.4
					18.0	40.2
T	950.8	137.9	863.3	125.2	18.0	35.6
	936.3	135.8	848.8	123.1	23.0	42.2
					20.0	39.1
S	932.9	135.3	843.3	122.3	10 (b)	25 (b)
	941.9	136.6	848.8	123.1		
AMS 4928-H	896.4 (b)	130 (a)	827.4 (a)	120 (b)		

(a) Consolidated material made by hot isostatic pressure at 950°C (1750°F) for 10 h at 100 MPa (15 ksi). Vacuum annealed for 10 h at 700°C (1300°F). Hydrogen after vacuum annealing equals 0.0057%.
(b) Minimum.

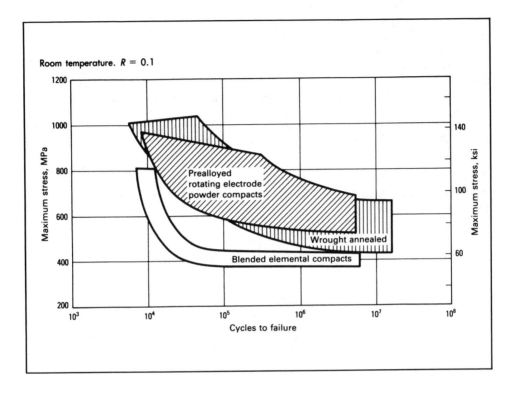

Room temperature. $R = 0.1$

Figure 11.28 - Comparison of smooth axial fatigue behavior of Ti-6Al-4V blended elemental and prealloyed P/M compacts with wrought annealed material

Alpha alloys such as Ti-5Al-2.5Sn have had little evaluation in cast or P/M form since the wrought alloy properties do not lend themselves to most new aerospace component designs. The Ti-10V-2Fe-3Al alloy has been evaluated in all three product forms with somewhat indeterminate results. No property data base is readily available for the beta alloys in cast or P/M form, although the wrought alloy data base is being expanded.

SUBZERO APPLICATIONS

The phrase *subzero applications* may be misleading here since the following discussion centers on titanium's tensile properties and fatigue strength when subzero temperatures exist. However, from this data designers may gain a fundamental understanding of the limits and benefits of the commercially available titanium alloys.

General

All structural metals undergo changes in properties when cooled from room temperature to temperatures below 0°C (32°F) in the "subzero" range. The greatest changes in properties occur when the metal is cooled to very low temperatures near the boiling points of liquid hydrogen and liquid helium.

Many of the available titanium alloys have been evaluated at subzero temperatures, but service experience at such temperatures has been gained only for Ti-5Al-2.5Sn and Ti-6Al-4V alloys. These alloys have very high strength-to-weight ratios at cryogenic temperatures and have been the preferred alloys for special applications at temperatures from -196° to -269°C (-320° to -452°F). Among these applications are spherical pressure vessels that are part of the propulsion and reaction-control systems for the Atlas and Centaur rockets and the Apollo and Saturn launch boosters. These pressure vessels were fabricated by welding together hemispherical forgings that had been machined to the desired thickness. The Ti-5Al-2.5Sn alloy also has been used for fuel-pump impellers for pumping liquid hydrogen.

Commercially pure titanium may be used for tubing and other small-scale cryogenic applications that involve only low stresses in service.

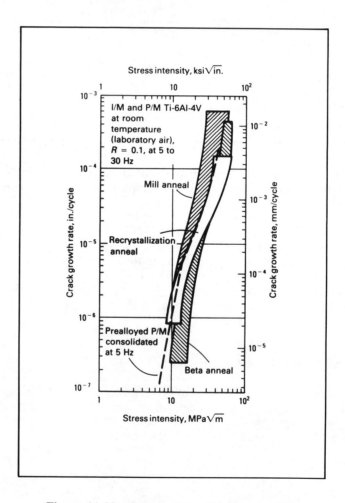

Figure 11.29 - Comparison of fatigue crack propagation rate of Ti-6Al-4V P/M compact with P/M material heat treated to various conditions

Table 11.17 Comparison of typical room temperature properties of wrought, cast, and P/M titanium products

Product and condition	Tensile strength MPa	ksi	Yield strength MPa	ksi	Elonga-tion, %	Reduction in area, %	Impact strength (a) J	ft·lb
Unalloyed Ti								
Wrought bar, annealed	550	80	480	70	18	33	35	26
Cast bar, as cast	635	92	510	74	20	31	26	19
P/M compact, an-nealed(b).	480	70	370	54	18	22	· · ·	· · ·
Ti-5Al-2.5Sn-ELI								
Wrought bar, annealed	815	118	710	103	19	34	· · ·	· · ·
Cast bar, as cast	795	115	725	105	10	17	· · ·	· · ·
P/M compact, annealed and forged(c)	795	115	715	104	16	27	· · ·	· · ·
Ti-6Al-4V								
Wrought bar, annealed	1000	145	925	134	16	34	22	16
Cast bar								
As cast	1025	149	880	128	12	19	19	14
Annealed	1015	147	890	129	10	16	· · ·	· · ·
Solution treated and aged(d)	1180	171	1085	157	6	11	· · ·	· · ·
P/M compact								
Annealed(b)	825-855	120-124	740-785	107-114	5-8	8-14	· · ·	· · ·
Annealed and forged(c).	925	134	840	122	12	27	· · ·	· · ·
Solution treated and aged(d)	965	140	895	130	4	6	· · ·	· · ·
Ti-6Al-6V-2Sn								
Wrought bar, annealed	1125	163	1055	153	16	38	20	15
Cast bar, as cast	1105	160	965	140	6	11	14	10
P/M compact, an-nealed(b)	965	140	840	122	5	5	· · ·	· · ·

(a) Charpy, at −40 °C (−40 °F), (b) About 94% dense. (c) Almost 100% dense. (d) Aging treatment not specified.

The Ti-5Al-2.5Sn alloy usually is used in the mill-annealed condition and has a 100% alpha microstructure. The Ti-6Al-4V alloy may be used in the annealed condition or in the solution-treated and aged condition. For maximum toughness in cryogenic applications, the annealed condition usually is preferred. The Ti-6Al-4V alloy is an alpha-beta alloy that has significantly higher yield and ultimate tensile strengths than the all-alpha alloy.

Impurities such as iron and the interstitials oxygen, carbon, nitrogen, and hydrogen tend to reduce the toughness of these alloys at both room and subzero temperatures. For maximum toughness, extra-low-interstitial (ELI) grades are specified for critical applications. Note that the iron and oxygen contents of the ELI grades are substantially lower than those of the normal interstitial (NI) grades. Iron is a strong stabilizer of the beta phase. The NI grades are suitable for service to -195°C (-320°F); for temperatures below -195°C, ELI grades generally are specified. For ELI grades, reduced creep strength at room temperature must be considered in design for pressure vessel service. In Ti-5Al-2.5Sn, stress rupture occurs at stresses below the yield strength.

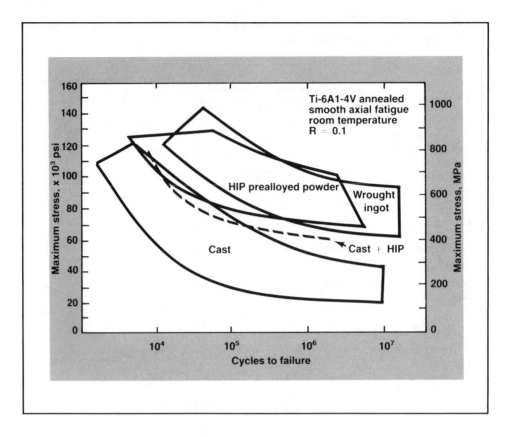

Figure 11.30 - Fatigue properties of Ti-6Al-4V wrought, cast, and prealloyed HIP P/M material at room temperature

There are two precautions that should be emphasized in considering titanium and titanium alloys for service at cryogenic temperatures:

- They must not be used for transfer or storage of liquid oxygen

- Titanium must not be used where it will be exposed to air while below the temperature at which oxygen will condense on its surfaces.

Any abrasion or impact of titanium that is in contact with liquid oxygen causes ignition. (Pressure vessels in contact with liquid oxygen in the Apollo launch vehicles were produced from Inconel 718 rather than from Ti-6Al-4V alloy to avoid this problem.)

Tensile Properties

Typical tensile properties of titanium and of titanium alloys Ti-5Al-2.5Sn and Ti-6Al-4V at room temperature and at subzero temperatures are presented in Table 11.18. Marked increases in yield and tensile strengths are evident for commercial titanium and for titanium alloys as test temperature is reduced from room temperature to -253°C

Table 11.18 Typical tensile properties of Ti and Ti alloys from room temperature to subzero temperatures

Temperature °C	°F	Tensile strength MPa	ksi	Yield strength MPa	ksi	Elongation, %	Reduction in area, %	Notch tensile strength(a) MPa	ksi	Young's modulus GPa	10⁶ psi

Ti-75A sheet, annealed, longitudinal orientation

24	75	580	84.3	465	67.6	25	...	785	114	...	...
−78	−108	750	109	615	89.2	25	...	...	...	...	...
−196	−320	1050	152	940	136	18	...	1100	159	...	...
−253	−423	1280	186	1190	173	8	...	875	127	...	...

Ti-75A sheet, annealed, transverse orientation

24	75	585	85.1	475	69.0	25	...	800	116	...	...
−78	−108	760	110	645	93.4	20	...	905	131	...	...
−196	−320	1060	153	965	140	14	...	1120	163	...	...
−253	−423	1340	194	1260	182	7	...	880	128	...	...

Ti-5Al-2.5Sn sheet, nominal interstitial annealed, longitudinal orientation

24	75	850	123	795	115	16	...	1130	164	105	15.4
−78	−108	1080	156	1020	148	13	...	1310	190	115	16.6
−196	−320	1370	199	1300	188	14	...	1630	236	120	17.7
−253	−423	1700	246	1590	231	7	...	1430	208	130	18.5

Ti-5Al-2.5Sn sheet, nominal interstitial annealed, transverse orientation

24	75	895	130	860	125	14	...	1170	170	...	...
−78	−108	1050	152	1020	148	12	...	1250	181	...	...
−196	−320	1430	208	1370	198	12	...	1630	236	...	...
−253	−423	1670	242	1610	234	6	...	1290	187	...	...
−268	−450	1590	231	...	...	1.5	...	...	...	...	...

(a) K_t = 6.3 for all three sheet forms; K_t = 5 to 8 for Ti-6Al-4V (ELI) forgings. (b) Recrystallization annealing treatment: 930 °C (1700 °F) 4 h, FC to 760 °C (1400 °F) in 3 h, cooled to 480 °C (900 °F) in 3/4 h, AC.

(Continued)

Table 11.18 (Continued)

Temperature °C	°F	Tensile strength MPa	ksi	Yield strength MPa	ksi	Elongation, %	Reduction in area, %	Notch tensile strength(a) MPa	ksi	Young's modulus GPa	10⁶ psi
Ti-5Al-2.5Sn (ELI) sheet, annealed, longitudinal orientation											
24	75	800	116	740	107	16	...	1060	154	115	16.4
−78	−108	960	139	880	128	14	...	1190	173	125	18.0
−196	−320	1300	188	1210	175	16	...	1560	226	130	18.6
−253	−423	1570	228	1450	210	10	...	1670	242	130	19.2
Ti-5Al-2.5Sn (ELI) sheet, annealed, transverse orientation											
24	75	805	117	760	110	14	...	1100	159	110	16.0
−78	−108	950	138	895	130	12	...	1260	182	125	18.1
−196	−320	1300	188	1230	179	14	...	1570	228	130	18.9
−253	−423	1570	228	1480	214	8	...	1530	222	140	20.1
Ti-5Al-2.5Sn (ELI) sheet/weldment, annealed, EB weld											
24	75	815	118	785	114	...	...	...	...	...	...
−196	−320	1300	189	1210	176	...	...	...	...	...	...
−253	−423	1510	219	1380	200	...	...	...	...	...	...
Ti-5Al-25Sn (ELI) plate, annealed, longitudinal orientation											
24	75	765	111	705	102	33	43	...	...	...	...
−253	−423	1430	208	1390	202	17	32	...	...	...	...
Ti-5Al-25Sn (ELI) forgings, as forged, tangential orientation											
24	75	835	121	760	110	15	36	...	...	...	...
−78	−108	980	142	905	131	12	31	...	...	...	...
−196	−320	1260	182	1100	159	15	30	...	...	...	...
−253	−423	1420	206	1260	182	13	22	...	...	...	...

(a) K_t = 6.3 for all three sheet forms; K_t = 5 to 8 for Ti-6Al-4V (ELI) forgings. (b) Recrystallization annealing treatment: 930 °C (1700 °F) 4 h, FC to 760 °C (1400 °F) in 3 h, cooled to 480 °C (900 °F) in 3/4 h, AC.

(Continued)

Table 11.18 (Continued)

Temperature °C	°F	Tensile strength MPa	ksi	Yield strength MPa	ksi	Elongation, %	Reduction in area, %	Notch tensile strength(a) MPa	ksi	Young's modulus GPa	10⁶ psi
Ti-6Al-4V (ELI) sheet, annealed, longitudinal orientation											
24	75	960	139	890	129	12	...	1120	162	110	16.2
−78	−108	1160	168	1100	160	9	...	1220	177	115	16.6
−196	−320	1500	217	1420	206	10	...	1460	211	120	17.5
−253	−423	1770	256	1700	246	4	...	1500	217	130	18.6
Ti-6Al-4V (ELI) sheet, annealed, transverse orientation											
24	75	960	139	895	130	12	...	1130	164	110	16.0
−78	−108	1170	169	1100	160	12	...	1260	183	115	16.5
−196	−320	1500	218	1460	212	11	...	1440	209	125	18.2
−253	−423	1750	254	1700	246	4	...	1550	225	130	19.2
Ti-6Al-4V (ELI) plate, annealed, longitudinal orientation											
24	75	890	129	840	122	15	37	...	...	...	...
−253	−423	1640	238	1600	232	...	8	...	...	...	...
Ti-6Al-4V (ELI) forgings, as forged, longitudinal orientation											
24	75	970	141	915	133	14	40	1330	193	...	...
−78	−108	1160	168	1120	163	13	31	1560	226	...	...
−196	−320	1570	227	1480	214	11	31	1900	276	...	...
−253	−423	1650	239	1570	227	11	24	1820	264	...	...
Ti-6Al-4V (ELI) forging, recrystallization annealed(b)											
24	75	890	129	825	120	14	41	...	...	110	16.1
−196	−320	1430	207	1370	198	10	16	...	...	120	17.5

(a) K_t = 6.3 for all three sheet forms; K_t = 5 to 8 for Ti-6Al-4V (ELI) forgings. (b) Recrystallization annealing treatment: 930 °C (1700 °F) 4 h, FC to 760 °C (1400 °F) in 3 h, cooled to 480 °C (900 °F) in 3/4 h, AC.

(-423°F). In the cryogenic temperature range, these alloys have the highest strength-to-weight ratios of all fusion-weldable alloys that retain nearly the same strength in the weld metal as in the base metal. Yield and tensile strengths of an electron-beam weldment of Ti-5Al-2.5Sn-ELI sheet also are presented in Table 11.18.

The notch strengths given in Table 11.18 indicate that these two alloys retain sufficient notch toughness for use in temperatures as low as -253°C (-423°F). However, the tensile data do not show any substantial improvement in ductility or notch toughness for the ELI grade of Ti-5Al-2.5Sn sheet over the normal interstitial grade—except at very low temperatures. The recrystallization annealing treatment used for the Ti-6Al-4V-ELI forging was developed as a means of improving fracture toughness in large forgings and thick plate.

Values of Young's modulus for titanium alloys increase substantially as test temperature is decreased, as shown in Table 11.18 and by the data obtained ultrasonically and plotted in Figure 11.31. Values of Poisson's ratio for the two alloys in Figure 11.31 are plotted in Figure 11.32.

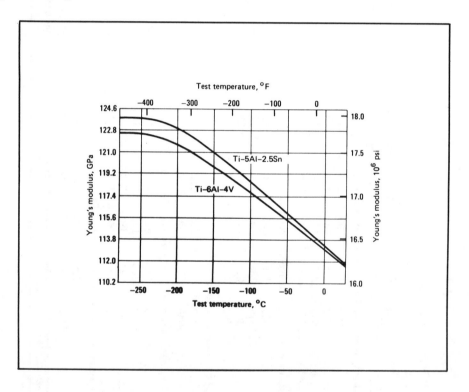

Figure 11.31 - Young's modulus for Ti-5Al-2.5Sn and Ti-6Al-4V alloys as determined ultrasonically

Fracture Toughness

Available data on plane-strain fracture toughness (K_{Ic}) at subzero temperatures for alloys Ti-5Al-2.5Sn and Ti-6Al-4V are presented in Table 11.19 along with corresponding

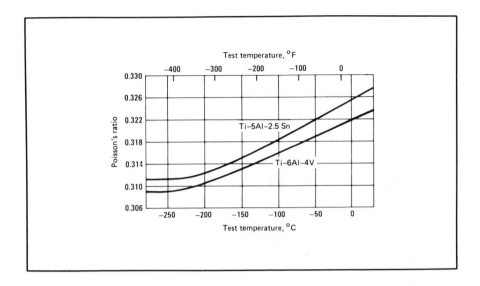

Figure 11.32 - Poisson's ratios for Ti-5Al-2.5Sn and Ti-6Al-4V alloys as determined ultrasonically

data for weldments. These data indicate that there is a modest reduction in fracture toughness as test temperature is reduced from room temperature to subzero temperatures. However, the ELI grades have better toughness than the corresponding normal interstitial grades at subzero temperatures. The limited data for electron-beam weldments indicate that at -196°C (-320°F) there is a slight reduction in toughness in both fusion and heat-affected zones when compared to the base metal in Ti-6Al-4V-ELI weldments.

FCP Rates

Data on fatigue-crack propagation (FCP) rates for Ti-5Al-2.5Sn and Ti-6Al-4V alloys are plotted in Figure 11.33. The corresponding data for determining the da/dN data from the equation da/dN = C (delta-K)^n are presented in Table 11.20. These data indicate that the exposure temperature has no effect on the fatigue-crack-growth rates for Ti-5Al-2.5Sn and Ti-6Al-4V. However, over part of the delta-K range, the fatigue-crack-growth rates for Ti-6Al-4V-ELI are higher at cryogenic temperatures than at room temperature at the same delta-K values.

Fatigue Strength

Values of fatigue strength at 10^6 cycles for titanium alloy base metal and weldments at room temperature and at subzero temperatures, based on results of axial and flexural fatigue tests, are presented in Table 11.21. For the unnotched specimens of parent metal, fatigue strength increased substantially when the test temperature was reduced from room temperature to -196°C (-320°F). When the test temperature was reduced to -253°C (-423°F), the fatigue strengths for some series of alloys were lower than at -196°C (-320°F). Fatigue strengths were much lower in the notched specimens than in

Table 11.19 Fracture toughness of two Ti alloys and weldments

Alloy and condition(a)	Form	Room temperature yield strength MPa	ksi	Specimen design	Orientation	Fracture toughness, K_{Ic} 24 °C(75 °F) MPa√m	ksi√in.	−196 °C(−320 °F) MPa√m	ksi√in.	−253 °C(−423 °F) MPa√m	ksi√in.	−269 °C(−452 °F) MPa√m	ksi√in.
Ti-5Al-2.5Sn(NI), annealed	Plate	876	127	CT	L-T	71.8	65.4	53.4	48.6	...	...	...	...
		876	127	Bend	L-T	...	...	...	...	51.4	46.8	...	...
		876	127	Bend	L-S	...	...	...	...	50.2	45.7	...	...
	Bar	871	126	CT	T-S	77.2	70.3	42.1	38.3	...	...	42.0	38.2
Ti-5Al-2.5Sn(ELI), annealed	Plate	703	102	CT	L-T	...	...	111	101	...	...	...	...
		703	102	Bend	L-T	...	...	...	...	89.6	81.5	...	...
Ti-5Al-2.5Sn(ELI), as forged	Forging	760	110	CT	R-L	...	...	...	...	79.4	72.3	...	...
					R-C	...	...	...	...	58.5	53.2	...	...
Ti-5Al-2.5Sn(ELI)	Forging(b)	779	113	CT	...	...	...	...	...	54.4 to 75.3	49.5 to 68.5	...	...
Ti-6Al-4V (NI), annealed	Bar	942	136	CT	T-L	47.4	43.2	38.8	35.3	...	...	38.5	35.1
Ti-6Al-4V (ELI), as forged	Forging	830	120	CT	T-L	...	...	61.0	55.5	...	...	54.1	49.2
Ti-6Al-4V (ELI), RA	Forging	830	120	CT	M-L(c)	...	...	62.8	57.2	...	...	...	...
					M-R(c)	...	...	62.0	56.4	...	...	...	...
Ti-6Al-4V (ELI), RA, electron beam welded, SR	Forging	830	120	CT	M-R(c)	...	...	61.1(d)	55.6(d)	...	...	...	...
	Weldment	...	...	...	M-L(c)	...	...	56.9(d)	51.8(d)	...	...	...	...
					M-R(c)	...	...	57.1(e)	52.0(e)	...	...	...	...
					M-R(c)	...	...	51.0(f)	46.4(f)	...	...	...	...

(a) SR = stress relieved: 540 °C (1000 °F) 50 h, AC. FC = furnace cool. AC = air cool. NI = normal interstitial content. ELI = extra low interstitial content. RA = recrystallization annealed: 930 °C (1700 °F) 4 h, FC to 810 °C (1400 °F) in 3 h, cooled to 480 °C (900 °F) in 3/4 h, AC. (b) Range for 18 tests. (c) M-L and M-R are specific orientations in a spherical forging. (d) Fusion zone. (e) Heat affected zone. (f) Heat affected zone boundary.

the corresponding unnotched specimens. At about -196° and -253°C (-320° and -423°F), the welded specimens had lower fatigue strengths than the base metal specimens.

Therefore, in designing welded structures of titanium alloys that will be subjected to fatigue loading at subzero temperatures, the weld areas usually should be thicker than the remaining areas. Hemispheres for spherical pressure vessels are machined so that the butting sections for the equatorial welds are thicker than the remaining sections, excluding inlet and discharge ports.

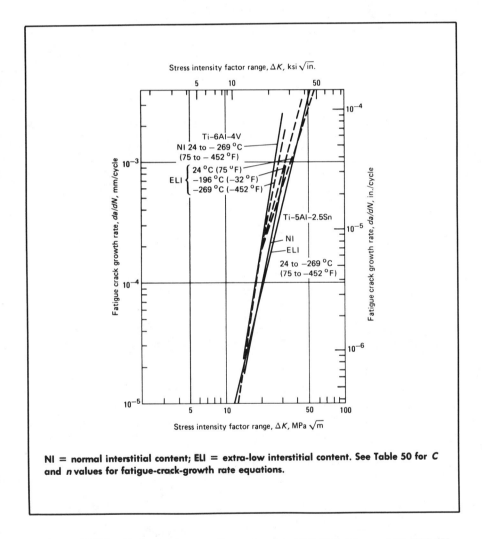

NI = normal interstitial content; ELI = extra-low interstitial content. See Table 50 for C and n values for fatigue-crack-growth rate equations.

Figure 11.33 - Fatigue propagation rates for Ti-5Al-2.5Sn and Ti-6Al-4V

Table 11.20 Fatigue crack propagation rate (da/dN) data for two Ti alloys (a)

Alloy and condition (b)	Orientation	Test temperature °C	Test temperature °F	C da/dN:mm/cycle ΔK:MPa√m	da/dN:in./cycle ΔK:ksi√in.	n	Estimated range for ΔK MPa√m	ksi√in.
Ti-5Al-2.5Sn (NI), annealed bar	T-S	24, −196, −269	75, −320, −452	5.1×10^{-11}	3.2×10^{-12}	4.8	14 to 30	13 to 27
Ti-5Al-2.5Sn (LI), annealed bar	T-L	24, −196, −269	75, −320, −452	4.9×10^{-10}	2.8×10^{-11}	4.0	10 to 60	9 to 54
Ti-6Al-4V (NI), annealed bar	T-L	24, −196, −269	75, −320, −452	3.1×10^{-12}	2.2×10^{-13}	6.0	14 to 30	13 to 27
Ti-6Al-4V (ELI), recrystallization annealed bar	T-L	24, −196, −269	75, −320, −452	1.9×10^{-13}	1.4×10^{-14}	7.0	10 to 20	9 to 18
Ti-6Al-4V		24, −196	75, −320	3.0×10^{-8}	1.6×10^{-9}	3.0	20 to 40	18 to 36

(a) Stress ratio: R = 0.1, at 20 to 28 Hz; compact specimens. (b) NI = normal interstitial, LI = low interstitial, ELI = extra low interstitial.

Table 11.21 Results of fatigue life tests on two titanium alloys

Alloy and condition	Stressing mode	Stress ratio, R	K_t	Fatigue strengths at 10^6 cycles 24 °C(75 °F) MPa	ksi	−196 °C(−320 °F) MPa	ksi	−253 °C(−423 °F) MPa	ksi
Ti-5Al-2.5Sn (ELI) sheet, annealed	Axial	0.01	1	495	72	815	118	760	110
			3.5	220	32	205	30	160	23
Ti-5Al-2.5Sn (ELI) sheet(a)	Axial	0.01	1	485	70	565	82	425	62
Ti-5Al-2.5Sn (ELI) bar, annealed(b)	Axial	0	1	760	110	985	143	925	134
Ti-6Al-4V (ELI) sheet(c)	Axial	0.01	1	505	73	675	98	895	130
			3.5	285	41	295	43	275	40
Ti-6Al-4V (ELI) sheet(a)	Axial	0.01	1	600	87	595	86	560	81
Ti-6Al-4V sheet, annealed	Flex	−1.0	1	345	50	550	80	530	77
			3.1	170	25	185	27	255	37

(a) Gas tungsten arc welded, base metal filler. (b) Cyclic frequency, 28 Hz. (c) STA: 900 °C (1650 °F) 5 min, WQ; 540 °C (1000 °F) 4 h, AC.

Chapter 12

Corrosion Resistance

GENERAL

Titanium and its alloys are used chiefly for their desirable mechanical properties, among which the most notable is their high strength-to-weight ratio. Somewhat more important for certain applications is titanium's outstanding resistance to corrosion. This metal in its various grades and alloys long has been used in a wide variety of aerospace, marine, and chemical applications because it has been found to be largely immune to corrosion-related failure in most environments.

Unalloyed titanium is highly resistant to the corrosion normally associated with many natural environments, including seawater, body fluids, and fruit and vegetable juices. Titanium exposed continuously to seawater for about 18 years has undergone only superficial discoloration. Wet chlorine, molten sulfur, many organic compounds (including acids and chlorinated compounds), and most oxidizing acids have essentially no effect on this metal. Titanium is used extensively for handling salt solutions (including chlorides, hypochlorides, sulfates, and sulfides), wet chlorine gas, and nitric acid solutions.

On the other hand, hot, concentrated, low-pH chloride salts (such as boiling 30% $AlCl_3$ and boiling 70% $CaCl_2$) corrode titanium. Warm or concentrated solutions of HCl, H_2SO_4, H_3PO_4 and oxalic acid also are damaging. In general, all acidic solutions that are reducing in nature corrode titanium unless they contain inhibitors. Strong oxidizers, including anhydrous red fuming nitric acid and 90% H_2O_2, also cause attack. Ionizable fluoride compounds, such as NaF and HF, activate the surface and can cause rapid corrosion; dry chlorine gas is especially harmful.

Titanium has limited oxidation resistance in air at temperatures above about 650°C (1200°F) and interacts with oxygen to dissolve it interstitially at temperatures as low as about 427°C (800°F). Chlorides or hydroxides deposited on its surface can accelerate oxidation. Exposure to liquid or gaseous oxygen, nitrogen tetroxide, or red fuming nitric acid can cause titanium to react violently under impact loading.

Titanium as been used to contain liquid or supercritical hydrogen at cryogenic temperatures, but above -100°C (-150°F) hydrogen may be absorbed and go on to diffuse into an alloy; it thus may severely embrittle the titanium. The potential for embrittlement is increased where hydrogen flow rates are high or where coatings on the titanium become damaged.

In unalloyed titanium and many titanium alloys, weld zones are just as resistant to corrosion as the base metal. Other fabrication processes (such as bending or machining) also appear to have no influence on basic corrosion resistance.

Commercially pure (CP) titanium is the preferred material of construction for much of the equipment built to handle industrial brines. It is used for pumps, piping, thermowells, heat exchangers, crystallizers, evaporators, condensers and many other items that are subject to the corrosive action of these brines.

Titanium alloys are generally less resistant to corrosion than commercially pure titanium.

PASSIVATION AND INHIBITION

Although titanium is chemically reactive, the thin oxide film that forms on titanium surfaces in most corrodents is relatively impervious and therefore quite protective. When titanium is not corrosion resistant, it is because the film is not fully protective. Reducing conditions, very powerful oxidizing environments, and the presence of fluoride ions diminish the protective nature of the oxide film, but its stability and integrity can be improved substantially by adding inhibitors to the environment.

Most acidic solutions (except those containing soluble fluorides) can be inhibited by the presence of even small amounts of oxidizing agents and heavy-metal ions. Thus, titanium can be used in certain industrial process solutions (including hydrochloric and sulfuric acids) that otherwise would be corrosive. Nitric and chromic acids, and dissolved salts of iron, nickel, copper, and chromium, are especially effective inhibitors. Attack by red fuming nitric acid and chlorine gas can be inhibited by small amounts of water.

Metallic ions and oxygen from the air are apparently absorbed into the titanium's surface, whereas strong oxidizing conditions (such as nitric acid, air at moderately high temperatures, and anodic treatments) promote resistance through growth of the oxide film. Dissolved oxygen is an important inhibitor in hot or mildly reducing chlorine solutions, but if its supply is restricted, as in deep crevices, corrosion may be accelerated.

GALVANIC CORROSION

Coupling of titanium to dissimilar metals usually does not accelerate corrosion of the titanium except in reducing environments, where titanium does not become passivated. Under reducing conditions, it has a galvanic potential similar to that of aluminum and undergoes accelerated corrosion when coupled to more noble metals.

One version of the galvanic series in seawater is indicated in Table 12.1. In this environment, titanium exhibits a potential of about -0.1 V versus a saturated calomel reference cell, which places it high on the passive (noble) end of the series.

In most environments, titanium is the cathodic member of any galvanic couple. It may accelerate corrosion of the other member, but in most instances it is itself unaffected. If the surface area of the titanium exposed to the environment is small in relation to the exposed surface area of the other metal, the effect of the titanium on the corrosion rate of the other metal is negligible; if, however, the exposed area of titanium greatly exceeds that of the other metal, severe corrosion of the other metal may result.

Table 12.1   Galvanic series in seawater

Series	Metal
Cathodic end (noble metals) ↓	Platinum
	Gold
	Silver
	Titanium
	Cr-Ni stainless steels, passive
	Straight Cr stainless steels, passive
	Ni-Cu alloys (Monels)
	Ni-Cr-Fe alloys (Inconels), passive
	Nickel, passive
	Silver solder
	Tin bronzes
	Copper nickels
	Silicon bronzes
	Copper
	Red and yellow brasses
	Aluminum bronzes
	Ni-Cr-Fe alloys (Inconels), active
	Nickel, active
	High-Zn yellow brasses (>30% Zn)
	Manganese bronzes
	Tin
	Lead
	Cr-Ni stainless steels, active
	Cast iron
	Wrought iron
	Low-carbon steel
	2xxx and 7xxx aluminum alloys
	Cadmium
	Alclad aluminum alloys
	6xxx aluminum alloys
↑	Galvanized steel
	Zinc
Anodic end (active metals)	Magnesium alloys
	Magnesium

Because titanium is nearly always the cathodic member of any galvanic couple, hydrogen may be evolved at its surface in an amount proportional to the galvanic current flow. This may result in formation of surface hydride films that generally are stable and cause no problems. At temperatures above 75°C (170°F), however, the hydrogen may diffuse into the metal, causing embrittlement. In some environments, titanium hydride is unstable and decomposes, or reacts, causing a loss of metal.

ALLOYING ADDITIONS

General

Alloying titanium with other metals has pronounced effects on its chemical properties. Anodic control of the corrosion reaction predominates when titanium is exposed to reducing acids such as hydrochloric or sulfuric. Alloying with elements that reduce anodic activity therefore should improve corrosion resistance. This can be accomplished by using alloying elements that:

- Shift the corrosion potential of the alloy in the positive direction (cathodic alloying)

- Increase the thermodynamic stability of the alloy and thus reduce the ability of the titanium to dissolve anodically

• Increase the tendency of titanium to passivate

The first group includes noble metals such as platinum, palladium, and rhodium. The second includes nickel, molybdenum, and tungsten. The third group includes zirconium, tantalum, chromium, and, possibly, molybdenum.

Cathodic Alloying

Considerable work has been done on the use of noble metals as alloying additions in titanium. An outgrowth of this work has been the development of Ti-0.2Pd, which has considerably greater resistance to corrosion in reducing environments than that of unalloyed titanium.

Alloying for Thermodynamic Stability

In studying corrosion of titanium in aqueous salt solutions, it was noted that titanium alloys containing nickel, molybdenum, or palladium were more resistant to nonoxidizing acid solutions than commercially pure (CP) titanium. It was concluded that they also should be more resistant to crevice corrosion.

An alloy containing 2% nickel was developed and recommended for service in hot brine environments where crevice corrosion is sometimes a problem. Subsequent studies confirmed that this alloy has much better resistance to crevice corrosion than the unalloyed metal. However, the nickel addition has detrimental effects that diminish its overall value. Ti-2Ni is very susceptible to hydrogen embrittlement and is subject to severe edge cracking during rolling, which makes it difficult to produce.

Passivation Alloying

Various studies have demonstrated that corrosion resistance of titanium is improved by addition of molybdenum. The principal problem with Ti-Mo alloys is the difficulty of obtaining uniform distribution of the molybdenum in large ingots. Because titanium and molybdenum have such widely different melting points, molybdenum is difficult to dissolve and may segregate as high-density inclusions when large amounts are added.

The commercial alloy Ti Code 12, which contains 0.3% Mo and 0.8% Ni, combines some of the favorable properties of nickel and molybdenum additions while avoiding the negative aspects. This alloy has excellent resistance to pitting and crevice corrosion in high-temperature brines, which sometimes attack commercially pure titanium. It also has better resistance to oxidizing environments such as nitric acid. The alloy resists corrosion in reducing environments such as HCl and H_2SO_4 better than commercially pure titanium but not as well as the titanium-palladium alloy.

CREVICE CORROSION

Titanium is subject to crevice corrosion in brine solutions containing oxygen, because the oxygen in the crevice is consumed faster than it can diffuse into the crevice from the bulk solution. As a result, the corrosion potential of metal in the crevice becomes more electronegative than the potential of metal exposed to the bulk solution. Metal in the crevice acts as the anode and dissolves under the influence of the resulting galvanic current. This produces an excess of positive ions at the anode, which is balanced by the migration of chloride ions into the crevice. The titanium chlorides formed in the crevice are unstable and tend to hydrolyze, forming small amounts of HCl. This reaction is very slow at first, but in the very restricted volume of the crevice it can reduce the pH to values as low as 1, which reduces the potential still further until corrosion becomes severe.

Although crevice corrosion of titanium is observed most often in hot chloride solutions, it also occurs in iodide, bromide, and sulfate solutions. Susceptibility increases with increasing temperature, increasing concentration of chloride ions, decreasing concentration of dissolved oxygen and decreasing pH. In solutions with neutral pH, crevice corrosion of titanium has not been observed at temperatures below 120°C (250°F). At lower pH values, crevice corrosion sometimes is encountered at temperatures below 120°C.

PITTING CORROSION

Pitting corrosion is a form of localized corrosion closely related to crevice corrosion. Both are observed mainly on passive metals such as aluminum, stainless steels, and titanium. Pitting initiates at imperfections in the oxide film. (One such imperfection is a grouping of iron particles.) Aggressive ions such as Cl⁻ concentrate at these sites until they are able to displace the oxygen in the passive film. A small crevice is soon formed by an insoluble corrosion product, TiO_2, which fills and covers the pit, thus restricting diffusion into the growing pit and permitting acid conditions to develop.

During fabrication and installation of titanium equipment, the titanium must be handled with enough care to avoid contaminating it with embedded iron particles. Pitting failures of titanium tubing have been traced to scratches in which traces of iron were detected; the failures were attributed to smearing of iron particles into the passive film of TiO_2 until they had penetrated it.

The difference in corrosion potential between low-carbon steel and unalloyed titanium is nearly 0.5 V in saturated brine at temperatures near the boiling point. This difference is sufficient to establish an electrochemical cell in which the iron is consumed at the anode. Before the iron is completely consumed, however, a pit begins to grow in the titanium. Once the pit becomes established, acid conditions develop in it. These conditions prevent the passive film from re-forming, and corrosion continues until the titanium is perforated.

As a final operation in fabrication and installation of titanium equipment, anodizing helps remove extraneous iron particles and thickens the passive film so that corrosion and pitting are much less likely.

EROSION-CORROSION AND CAVITATION

For most materials there are critical velocities beyond which protective films are swept away and accelerated corrosion attack occurs. This accelerated attack is known as erosion-corrosion. The critical velocity differs greatly from one material to another and may be as low as 0.6 to 0.9 m/sec (2 to 3 ft/sec). For titanium, the critical velocity in seawater is high—more than 27 m/sec (90 ft/sec). Numerous erosion-corrosion tests have shown titanium to have outstanding resistance to this form of attack.

Erosion-corrosion can be greatly aggravated by the presence of abrasive particles (such as sand) in a flowing fluid. Titanium exhibited superior resistance to this type of attack in seawater containing fine sand that flowed through conventional titanium condenser tubes at 1.8 m/sec (6 ft/sec).

Cavitation is a phenomenon that occurs in flowing liquids, wherein the relative motion between the liquid and a surface across which it flows is great enough to locally reduce the pressure below the vapor pressure of the liquid. At the reduced pressure, bubbles form in the liquid. When the liquid containing cavitation bubbles flows into a region of higher pressure, the bubbles collapse, inflicting severe, highly localized forces on the surface against which they collapse. This can produce deep, rounded pits in the surface of almost any solid — even glass.

Cavitation resistance tests performed in Boston Harbor proved titanium to be one of the metals most resistant to this type of damage.

STRESS-CORROSION CRACKING

General

Unalloyed titanium generally is immune to stress-corrosion cracking (SSC) unless it has a high oxygen content (0.3% or more). For this reason stress-corrosion cracking is of little concern in the chemical process industries where unalloyed titanium is most commonly used. On the other hand, certain alloys of titanium used principally in the aerospace industry are subject to stress-corrosion cracking, although field stress-corrosion failures are rare.

One of the most important variables affecting susceptibility to stress-corrosion cracking is alloy composition. Aluminum additions increase susceptibility to stress-corrosion cracking; alloys containing more than 6% Al generally are susceptible to stress corrosion. Additions of tin, manganese, and cobalt are detrimental, whereas zirconium appears to be neutral. Beta stabilizers such as molybdenum, vanadium, and niobium are beneficial. Susceptibility to stress-corrosion cracking also can be affected by heat treatment.

A number of environments in which some titanium alloys are susceptible to stress-corrosion cracking along with the temperatures at which cracking has been observed are listed in Table 12.2. Some of these environments are discussed in the following paragraphs.

Hot, Dry Chloride Salts

Hot-salt SCC of titanium alloys is a function of temperature, stress, and time of exposure. In general, hot-salt cracking has not been encountered at temperatures below about 260°C (500°F). The greatest susceptibility occurs at about 290° to 425°C (about 550° to 800°F), based on laboratory tests. Time-to-failure decreases as either temperature or stress level is increased. All commercial alloys, but not unalloyed titanium, have some degree of susceptibility to hot-salt cracking.

Residual salts cause surface pitting and even cracking of certain alloys under high tensile loads. Although rarely encountered in service, cracking of titanium parts due to hot-salt

Table 12.2 Environments and temperatures that may be conducive to SCC of titanium alloys

Environmental	Temperature
Nitric acid, red fuming	Ambient
Hot, dry chloride salts	260-480°C (500-900°F)
Cadmium, solid and liquid	Ambient to 400°C (750°C)
Chlorine	Elevated
Hydrogen chloride	Elevated
Hydrochloric acid, 10%	Ambient to 40°C (100°F)
Nitrogen tetroxide	Ambient to 75°C (165°F)
Methyl and ethyl alcohols	Ambient
Seawater	Ambient
Trichloroethylene	Elevated
Trichlorofluorethane	Elevated
Chlorinated diphenyl	Elevated

corrosion has been encountered by fabricators during stress-relieving operations. Responsibility has been traced to vapors of chlorinated cleaning fluids that were not completely removed prior to thermal processing; to chloride traces from other process fluids (including tap water); and even to salt residues from fingerprints.

Since the original finding that hot halogenated salts can damage titanium, the phenomenon has been studied extensively. Although much has been learned about the reaction, relative susceptibility, together with related variables, it is generally agreed that laboratory tests do not simulate service conditions well and thus do not correctly predict field performance.

The extent of damage by salts is directly related to temperature exposure time and tensile-stress level. Processing history, alloy composition, salt composition, and other environmental conditions also have important effects. Susceptibility to hot-salt corrosion appears to be influenced considerably by processing and alloy additions. Therefore, control of these factors should make it possible to avoid the phenomenon in service.

Hot-salt SCC has been associated with use of silver as an antigallant for titanium. Despite the ordinary resistance of titanium to in-service, low-salt SCC failure, the silver caused hot-salt SCC to occur by trapping chloride ions as $AgCl_2$ and altering the environment seen by the titanium alloy. Silver is no longer used in this application, yet the experience suggests that field failures are possible under certain conditions. Silver may participate in liquid metal embrittlement, as discussed below.

Chlorine, Hydrogen Chloride, Hydrochloric Acid

With respect to environments containing chlorine, hydrogen chloride, or hydrochloric acid, the mechanism of cracking is not completely understood, although it appears that both oxygen and water must also be present for cracking to occur.

Nitrogen Tetroxide

Nitrogen tetroxide (N_2O_4) containing small amounts of dissolved oxygen causes cracking of titanium and some titanium alloys. No cracking occurs if the nitrogen tetroxide contains a small percentage of nitric oxide (NO). The cracking may be transgranular, intergranular or both, depending on alloy composition.

Methyl and Ethyl Alcohol

Methyl and ethyl alcohols containing small amounts of water, chloride, bromide, and iodide promote cracking at ambient temperatures. Greater concentrations of water inhibit cracking. Higher alcohols may induce cracking but to a lesser extent; the longer the chain, the less reactive the alcohol becomes.

ACCELERATED CRACK PROPAGATION IN SEAWATER

Titanium is known to be highly resistant to corrosion by seawater. However, for certain alloys, components containing very sharp notches or cracks exhibit accelerated crack propagation and thus lose resistance to fracture when exposed to seawater. Failure of titanium due to loss of fracture resistance appears to be similar to delayed fracture of high-strength steels containing sharp notches or cracks on exposure to various liquid environments; it is not considered a form of stress-corrosion cracking.

Exposure to seawater does not appear to diminish service life of titanium alloys — such as Ti-8Mn and Ti-5Al-2.5Sn — that exhibit this phenomenon in laboratory testing. These two alloys have been employed successfully in aircraft during the past 15 years without reported failures. Apparently, the conditions leading to accelerated crack propagation (primarily, the existence of a crack) have not been encountered in service.

Accelerated crack propagation in seawater can be avoided by proper alloy selection. Alloys containing more than 6% aluminum are particularly susceptible. Additions of tin, manganese, cobalt, and oxygen are detrimental, whereas beta stabilizers such as molybdenum, niobium, and vanadium tend to reduce or eliminate susceptibility to this phenomenon. Unalloyed titanium is not susceptible unless it contains more than about 0.3% oxygen.

LIQUID-METAL EMBRITTLEMENT

Some titanium alloys crack under tensile stress when in contact with liquid cadmium, mercury, or silver-base brazing alloys. This type of embrittlement differs from stress-corrosion cracking although there are some similarities. Liquid-metal embrittlement appears to result from diffusion along grain boundaries and formation of brittle phases, which in turn produce the loss of ductility.

Titanium also can be embrittled by contact with certain solid metals (cadmium and silver, for example) when the titanium is under tensile stress. The failure mechanism is not completely understood, although many investigators believe it is similar to liquid-metal embrittlement. Service failures have occurred in cadmium-plated titanium alloys at temperatures as low as 65°C (150°F), and in silver-brazed titanium parts at temperatures above 315°C (600°F).

Silver-plated components should not be used in contact with titanium under stress at temperatures above 230°C (450°F). (Refer to the discussion of stress-corrosion cracking, above.) Cadmium-plated parts such as interference-fit fasteners or press-fit bushings should not be used in contact with titanium at any temperature. Other cadmium-plated parts and fasteners should not be used in contact with titanium at temperatures above 230°C (450°F).

HYDROGEN-DAMAGE FAILURES

Most metals and alloys are susceptible to hydrogen damage, and many are susceptible to more than one type of hydrogen damage. Hydrogen damage in titanium alloys results from embrittlement caused by absorbed hydrogen. Hydrogen may be supplied by a number of sources, including water vapor, pickling acids, and hydrocarbons. The amount of absorption depends primarily on the titanium oxide film on the metal surface; an adherent unbroken film can significantly retard hydrogen absorption.

Titanium and its alloys become embrittled by hydrogen at concentrations that produce a hydride phase in the matrix. The exact level of hydrogen at which a separate hydride phase is formed depends on the composition of the alloy and the previous metallurgical history. In commercial, unalloyed material, this hydride phase is normally found at levels of 150 ppm of hydrogen; however, hydride formation has been observed at levels as low as 40 or 50 ppm of hydrogen.

At temperatures near the boiling point of water, the diffusion rate of hydrogen into the metal is relatively slow, and the thickness of the layer of titanium hydride formed on the surface rarely exceeds about 0.4 mm (0.015 in.) because spalling takes place when the hydride layer reaches thicknesses in this range.

Hydride particles form much more rapidly at temperatures above about 250°C (480°F) because of the decrease in hydrogen solubility within the titanium lattice. Under these conditions, surface spalling does not occur; the formation of hydride particles through the entire thickness of the metal results in complete embrittlement and high susceptibility to failure. This type of embrittlement is often seen in material that has absorbed excess hydrogen at elevated temperatures — such as during heat treatment or welding — and subsequently has formed hydride particles during cooling.

There have been instances of localized formation of hydrides in environments where titanium has otherwise given good performance. Investigations of such instances suggest that the formation is the result of impurities in the metal (particularly the iron content) and the amount of surface contamination introduced during fabrication.

There is a strong link between surface iron contamination and formation of hydrides of titanium. Severe hydride formation has been noted in high-pressure, dry, gaseous hydrogen around particles of iron present on the surface. Anodizing in a 10% ammonium sulfate solution removes surface contamination and leads to thickening of the normal oxide film.

In chemical plant service, where temperatures are such that hydrogen can diffuse into the metal if the protective oxide film is destroyed, severe embrittlement may occur. For example, in highly reducing acids where the titanium oxide film is unstable, hydrides can form rapidly. Hydrogen pickup also has been noted under high-velocity conditions where the protective film erodes away as rapidly as it forms.

Hydrogen contents of 100 to 200 ppm may cause severe losses in tensile ductility and notched tensile strength in titanium alloys, and they may even cause brittle, delayed failure under sustained loading conditions. The sensitivity to hydrogen embrittlement from formation of hydrides varies with alloy composition and is reduced substantially by alloying with aluminum.

Care should be taken to minimize hydrogen pickup during fabrication. Welding operations generally require inert-gas shielding to minimize hydrogen pickup.

Chapter 13

Recent and Future Advances

INTRODUCTION

Titanium metallurgy, as it exists today, grew out of the application of titanium to gas turbines. From there the metal was applied to airframes, and then to more sophisticated and critical areas, as noted throughout this volume. Within a relatively short history of development, the record of accomplishment has been remarkable. At the same time comments on recent and future advances in titanium alloys may sound more like a litany of a quarter of a century of "attempts" to get the alloys to perform in an improved manner. For example, cast titanium was itself thought of as a "future" advance for over a decade before it evolved into an acceptable production technique for critical aerospace component applications.

The topics contained in the following discussions actually form a kind of check list to be consulted by the titanium alloy buyer when a sophisticated application need arises. Long-range planning remains a classic style despite the recent fashionable accent on short-term industrial profits. Consequently, the titanium designer or potential user should know what may be possible should the need for special properties or capabilities occur. Such an individual should also understand exactly which projects may require either financial or developmental backing in order to promote the technology into the area of commercial application.

The technologies mentioned below, for the most part, owe their current status to the support of the United States military establishment, particularly the U.S. Air Force Wright Aeronautical Laboratories at Dayton, Ohio. Many initial titanium metallurgy concepts were brought to fruition with government support, including the metal's very early use in military gas turbines and also in supersonic aircraft. (It is of interest that Lockheed's SR-71 spy plane was the very first utilization of a beta titanium alloy for structural aerospace applications.)

Military involvement points to a simple fact: the developmental costs of many technologies today — as in the past — cannot be exclusively supported by the very limited industrial market perceived at that point in time when the technology, or application, is conceived.

Commercial applications of titanium, in most cases, have tracked military developments. The requirements of the steam turbine, chemical, automotive, or other industries

13

have been secondary causes contributing to the enrichment of titanium's technological base. At the same time, a great deal of credit is due to those original titanium-producing companies which expended so many research-and-development dollars in alloy-development projects. Realistically speaking, however, technologies such as powder metallurgy have been, and will continue to be, spurred and funded by the military. It is after all the military establishment where developmental needs are greatest and where perceived needs as far into the future as a decade can find support and funding.

The following discussions cover these broad topics:

- Beta alloys
- Higher-modulus alloys
- Superplastic forming/diffusion bonding
- Advanced powder metallurgy
- Rapid solidification rate processing
- Higher-temperature-capability alloys

BETA ALLOYS

In a sense, beta alloys have been "rediscovered" within recent years. New beta alloys such as Ti-10V-2Fe-3Al are now available. The formability of alloys such as Ti-15V-3Cr-3Al-3Sn interests many users. These alloys are discussed in earlier chapters, yet they are worth mentioning again. They offer an attractive alternative to the alpha-beta alloys because of their increased heat treatability and their formability or forgeability characteristics. These last two characteristics, whichever applies, are intrinsic features of metastable beta structures. Admittedly, the first major use of a metastable beta alloy was in the SR-71 "Blackbird" aircraft over 25 years ago. Subsequent to that first-of-a-kind application, forgeability along with combined toughness and strength have emerged as important design tools within recent years.

In the solution-treated and quenched condition, beta alloys show excellent formability, a characteristic much sought after in early beta alloys. Strength is increased by aging, which disperses the alpha phase in the beta matrix. These alloys have higher density and costs than do the alpha or alpha-beta alloys, because of the greater amounts of alloying elements present. Beta alloys, however, are characterized by a good balance of strength, toughness, and ductility. When the high strength-to-density ratio of beta alloys is coupled with their increased formability or forgeability characteristic, alloy costs per pound may not be a significant factor, relative to the increased performance. (In fact, higher performance can be realized by substituting beta alloys for alpha-betas.)

Beta alloys are commercially available now. The designer must work with the alloy producer and the fabricator in order to incorporate the most appropriate type into the planned application in the most effective and efficient manner.

HIGHER-MODULUS ALLOYS

Higher-modulus alloys may currently appear trivial in value, yet they may become significant in the future. Titanium has a low modulus relative to its density. (It's about 50% of the modulus of steel, but about 60% of the density of steel.) Titanium's modulus can vary with alloy type (beta vs. alpha) from as low as about 13.5×10^6 psi up to about 17.5×10^6 psi. The modulus is, to a degree, also a function of crystal orientation. Thus there exists a continuing hope that a practical process for the texturing of titanium alloys could permit the attainment of higher moduli without a corresponding loss of the strength properties developed in titanium systems by means of historical processing "tricks."

There do not appear to be any low- or intermediate-temperature titanium alloy developments from which such modulus improvement could be derived within the immediate future. In the recent past the working of plate produced texture and improved moduli, yet no transfer of that concept has been made to forgings. It is highly unlikely that any texturing process will be applied to castings in the foreseeable future, although directional solidification — currently precluded by the high reactivity of molten titanium — or directional recrystallization may offer limited promise for modulus control.

SUPERPLASTIC FORMING/DIFFUSION BONDING

Superplastic forming/diffusion bonding (SPFDB) was discussed briefly in Chapter 10. For designs which use reinforced sheet components of the honeycomb type, or for similar applications, the opportunity to combine forming and bonding operations by the SPFDB technique may be of some interest.

Structures which can be accessed internally by gas under pressure are conceptually all that are required to create a complex structure containing ribs, doublers, or other reinforcement. Designers should give serious consideration to the application of superplastic forming/diffusion bonding techniques to new construction.

ADVANCED POWDER METALLURGY

It is difficult to state exactly where the "current" stops and the "advanced" begins with respect to P/M titanium alloys. Assuming that rapid solidification rate processing is a separate category, there are perhaps three areas into which advances may move.

The first area is cost reduction. The second is powder purity improvement and prealloyed (PA) powder. While positive steps have been taken here, there is little likelihood of a major reduction in cost levels very soon. A breakthrough in the purity of blended elemental powders, however, seems more likely. The use of low-chlorine powders and cold isostatic pressing plus sinter plus HIP (CHIP) may result in lower-cost compacts with fatigue properties comparable to current prealloyed compacts.

The third area of possible advance is the continuing development of heat treatment and the corresponding chemical "tricks" used to refine the microstructure of powder titanium alloys. The goal of this activity is to improve fatigue properties beyond the current levels of P/M products — and even beyond those of available wrought products.

RAPID SOLIDIFICATION RATE PROCESSING

The development of powder titanium alloys has been restricted by the difficulty of obtaining a proper dispersion of the alloying element, or precipitate thereof, throughout the matrix. One of the more exciting new technologies for titanium alloys is the application of rapid solidification rate (RSR) techniques to the formation of titanium powders for processing into usable components. RSR involves a number of different schemes for very-rapid-cooling molten titanium alloys. As the result, desired alloy elements which would otherwise be rejected on normal solidification, even by rotating-electrode process (REP) atomization, remain in solution. They may thus be dispersed in a favorable manner in subsequent processing steps.

True RSR powder is essentially limited to one dimension at this time in order to facilitate heat extraction. Cooling rates as high as 10^8 °K/sec have been used to rapid quench titanium alloys. (These contrast with the cooling rate of about 10^2 °K/sec for REP.) Most of the information on RSR titanium alloys is in the exploratory category. Conventional phase diagrams have not been of great help in alloy design, but trial-and-error attempts are proceeding.

Alloy compositions have been studied. At room temperature, strengths of 165 ksi and ductility of 30% have been reported for HIP RSR flake of a Ti-Al-Mo alloy. Alpha-beta and beta alloy compositions have been studied. Tensile strengths of filaments have been measured at over 300 ksi.

The strengths claimed for the Ti-Al-Mo, mentioned above, are thought to represent levels approachable in alpha-beta alloys made by RSR processing. The most significant fact about these strength levels is that ductility, as measured by elongation, is very high. This suggests that the RSR technique may permit the attainment of higher strength along with toughness in conventional alloy systems.

The most comprehensive studies on RSR, however, involve alloy Ti-6Al-4V, as may be expected.

In addition to the more conventional alloys and to alloys based on conventional systems, rapid solidification rate processing offers the potential to incorporate alloy elements which may be otherwise difficult or even impossible to utilize. It is claimed that RSR processing of titanium alloys with rare earth additions produces an ultrafine dispersion of rare earth oxides which may be useful for development of high strengths. In particular, they may support the development of high-temperature strengths in titanium alloys.

RSR processing also appears to be a useful technique for the introduction of high fractions of metalloid (carbon, boron, silicon, gallium) compounds. Even the conventional alpha systems with sufficient aluminum to produce a high-volume fraction of alpha-2 (Ti_3Al) compound may be amenable to RSR processing.

There also is the possibility of applying RSR processing to the titanium aluminide alloys now under development for high-temperature titanium applications. (Refer to a discussion on this topic later in this chapter.)

Most probably, however, the first practical utilization of RSR titanium alloys will result from the use of solid solution alpha, alpha-beta, or beta alloys hardened by constituents introduced and retained by RSR processing itself. An integral part of the development of RSR technology will be the development of powder processing techniques and thermal schedules in order to optimize properties, especially for long-time, high-temperature service. However the surface has been only fractionally penetrated, and it will be some time in the 1990s before commercial RSR titanium alloys will be available for structural design applications.

HIGHER-TEMPERATURE-CAPABILITY Ti ALLOYS

Conventional Alloys

The possible use of titanium alloys at higher temperatures has been studied for many years. In fact, about a quarter a century ago there was a great deal of effort in, and many claims about, the improved creep-rupture strength of the alpha or superalpha alloys such as Ti-5Al-5Sn-5Zr, which demonstrated substantially improved creep resistance over the then current Ti-6Al-4V. For example, typical creep strengths for Ti-6Al-4V, Ti-8Al-1Mo-1V, and Ti-5Al-6Sn-5Zr were quoted as 9.5, 22 and 43 ksi, respectively, at 538°C (1000°F). On a life basis, this is a dramatic variation in high-temperature strength capability. One of the reasons that the superalpha alloys such as Ti-5Al-5Sn-5Zr did not become one of the standard high-temperature materials was that such alloys seemed to be much more susceptible to hot-salt stress-corrosion cracking (SCC), even to the point of cracking during preparation for hot-salt SCC testing.

A second reason for the general lack of application for the alloys was the difficulty of preventing, or at least obviating, the effects of oxygen pickup (scaling and alloy embrittlement). Also involved was a concern for the metallurgical stability of structures employed for long times at the temperatures above 482°C (900°F), where high-temperature titanium alloys were expected to perform.

A third reason for the diminished enthusiasm for superalloys of this type was the introduction of new titanium alpha-beta compositions. They significantly enhanced the creep resistance of such alloys. The appearance of Ti-6Al-2Sn-4Zr-2Mo (a "near-alpha" alloy) and the addition of small amounts of silicon (0.25%) substantially increased the creep capability. In fact the improvement reached the point that Ti-6Al-2Sn-4Zr-2Mo alloy has become the standard higher-temperature alloy. (Alloy Ti-6Al-4V still remains the main component of the lower-temperature compressor sections of gas turbines.)

It has been reported that some Ti-6Al-2Sn-4Zr-2Mo alloy was exposed to temperatures about 538°C (1000°F) for extended periods of time. No failures were observed, yet no postexposure examinations were mentioned, either.

It is, of course, obvious that titanium alloys for this type of application are not really at high temperatures, at least relative to their melting points. To attain really high-temperature levels, like those comparable to superalloys, aluminides are required. (Refer to the following discussion on that topic.)

Some continuing work on conventional titanium alloys has produced higher-temperature capability in other near-alpha alloys. IMI 829, offered by IMI Titanium, Ltd. of England, is an example. (The company specifies the nominal chemical composition as Ti-5.5Al-3.5Sn-3Zr-1Cb-0.3Mo-0.3Si.) Its literature claims high creep resistance up to 550°C (1022°F) for long periods.) Both IMI 829 and Ti-6Al-2Sn-4Zr-2Mo achieve part of their strength increment from beta processing, a standard method of increasing creep strength, although it is done at the expense of certain other properties.

Alloys such as IMI 829 do have the claimed improved creep strength, but the available literature contains no evidence of hot-salt SCC testing. Before actually using these alloys, a designer should obtain some test data from this important aspect of the alloy in order to verify its possible susceptibility, or lack of it, to such problems.

Furthermore, even if this type of alloy can be used to higher temperatures than present conventional alloys, there is no indication that resistance to oxygen pickup has been increased, nor that metallurgical stability has been improved. The industry unfortunately remains in a condition in which the environmental and exposure restrictions limit the upper application temperature point for conventional titanium alloys.

One solution to the problem of surface stability could be the introduction of coatings. They were first considered when it was believed that hot-salt stress-crack corrosion would become a very real service problem. Over the last 20 years measurable advances in coating technology have taken place. Of course, it may be that loss of surface integrity may be prevented by developing new inherently oxidation-resistant alloys. This appears, however, to be an unlikely process, except for aluminides, discussed later in this chapter.

A more likely prospect is the development or application of new coatings for conventional titanium alloys. Noble metal coatings have been found to provide good surface protection without degradation of fatigue resistance, which becomes a problem when coatings are applied to most materials. Ion plating was used to introduce the noble metals and a tungsten-noble metal combination to the surface.

Other more conventional methods of coating application probably will be evaluated, assuming a demand, in order to extend the temperature capability of the alloys. In any event, further development is required. Considerable attention will inevitably be given to costs, not just properties, before any decision is made on their eventual uses. It is unlikely that effective coatings could be available before 1990.

Titanium Aluminides

Extensive research and development has been carried out for the past 15 years in order to answer the following question: Can ductile titanium aluminides be made to compete with the nickel-based superalloys? In fact, there are two candidate aluminide bases: alpha-2 (Ti-3Al) and gamma (Ti-Al). Intermetallics such as these two bases are particularly desirable because they are of even lower density than conventional titanium alloys. They also have high melting temperatures and moduli, and they retain their modulus better with temperature. The titanium aluminides have higher moduli at 816°C (1500°F) than does titanium at room temperature! It is expected that aluminides will cause even better strength retention at higher temperatures than is currently available.

It may be wondered why aluminide-base materials are not already in use. The answer is found in the general tendency of many metal aluminides in all types of systems to have limited room-temperature ductility and poor fracture toughness. Although aluminides have excellent intrinsic oxidation resistance, there is a real limiting tendency for the oxide scales formed on these aluminides at temperatures above 871°C (1600°F) to spall on cooling.

The answer to all of the above problems has been alloy and process development. In fact, the two efforts are inextricably intertwined: alloy developments can enhance processability; the converse is also true.

Alloy developmental efforts have been undertaken on both types of aluminide bases, alpha-2 and gamma. The larger part of the process development has centered on the former. Virtually all of this effort has been sponsored by the U.S. Air Force Wright Aeronautical Laboratories or allied agencies of the Air Force.

Melting, casting and forging have been extensively studied. Other processing technologies such as machining have not been neglected. Precision investment castings have been made from an alpha-2 base alloy, and a ring for a combustor liner has been rolled successfully from it. In one instance a rolled ring of this type of alloy was made into a compressor case component and successfully run for 65 hours.

Other components for high-temperature use have been made and tested. The weight savings in one application alone amounted to 43% when compared to the more conventional superalloy commonly used.

In general, the alpha-2 based titanium alloys have been thought to be comparable in creep strength to the Inconel 713 superalloy. The gamma-base titanium alloys are thought to be capable of being compared with the IN-100 superalloy. Precise comparisons are not possible owing to publication restrictions on both the alloy and the process developmental efforts. It is possible that aluminide alloys will turn out to be somewhat lower in strength levels than the two nickel-base superalloys noted.

The important features which will limit application of these titanium alloys are:

- Fabricability

- Room-temperature handling capability (which is a concern during installation in the machines in which they will be used)

- Resistance to elevated temperature embrittlement during exposure

Fabricability may be enhanced by the future application of RSR powder processing, which also may directly affect alloying flexibility. The susceptibility to embrittlement during exposure can only be evaluated after appropriate alloy compositions are determined.

It would appear that there has been a considerable advance in the capability of aluminides for higher-temperature applications. Still, much work remains to be done. It is unlikely that aluminides will be commercially available much before the mid-1990s. The temperature limits for high-temperature titanium alloys remain, for the near future, those set by the limits of oxygen pickup or embrittlement, oxidation, and the metallurgical stability set for conventional titanium alloys, namely $\pm 538°C$ (1000°F).

Appendix A

Selected References for Additional Reading

INTRODUCTION

There may be a need to pursue individual topics in more depth than is found in the discussions throughout the guide. This appendix is intended to satisfy such a need, although it does so in a brief, nondefinitive manner.

The first source of more detailed data should be handbooks, since these are essentially collections of highly condensed information. A great deal of specific data related to the properties of titanium and titanium alloys is contained in *Metals Handbook*, 9th ed., Vol 9, ASM INTERNATIONAL, 1980, p. 372.

Reference items are divided according to the following groupings:

- Overview and general background

- Physical metallurgy

- Corrosion

- Processing

- Applications and recent developments

(Each reference item is numbered for ease of use.)

Note that Appendix C, "Machining Data," contains a number of references related to tool-life charts.

OVERVIEW AND GENERAL BACKGROUND

1. R. R. Boyer and H. W. Rosenberg, Ed., *Beta Titanium Alloys in the 80's*, The Metallurgical Society of AIME, Warrendale, PA, 1984

2. F. A. Crossley, "Aircraft Applications of Titanium: A Review of the Past and Potential for the Future," *J. Aircft.*, Vol. 18, 1981, p 983-1002

3. M. J. Donachie, Ed., *Titanium Alloys Source Book*, American Society for Metals, 1983

4. M. J. Donachie, Ed.," Titanium," in *Metals Handbook, Desk Edition*, American Society for Metals, 1985, p 9.1-9.12

5. R. M. Duncan and B. Hanson, *The Selection and Use of Titanium*, Oxford University Press, 1980

6. T. W. Farthing, "Introducing a New Material—The Story of Titanium," *Proc. Inst. Mech. Engrs.*, Vol 191, 1977, p 59-73

7. F. H. Froes *et al.*, Ed., *Titanium Technology: Present Status and Future Trends*, Titanium Development Association, Dayton, OH, 1985

8. J. R. B. Gilbert, "The Uses of Titanium," *Mater. Sci. Technol.*, Vol 1 (No. 4), 1985, p 257-262

9. B. H. Hanson, "Present and Future Use of Titanium in Engineering," *Materials and Design*, Vol 7 (No. 6), 1986, p 301-307

10. W. E. Herman and H. Kessler, "Recent Developments in the Titanium Industry," *SAMPE Quarterly.*, Vol 9, Oct 1977, p 40-44

11. R. I. Jaffee, "An Overview on Titanium Development and Application," in *Titanium 80*, The Metallurgical Society of AIME, Warrendale, PA, 1981, p 53-76

12. R. Syre, "Characterization, Selection and Use of Titanium Base Alloys," in *The Characterization and Application of Materials*, Lecture Series No. 51, Advisory Group for Aerospace Research and Development and North Atlantic Treaty Organization (AGARD-NATO), Neuilly-Sur-Seine, France, May 1971, p 7.24-7.46

13. R. A. Wood, "Titanium Utilization and Availability," *Current Awareness Bulletin*, Metals and Ceramics Information Center, Battelle Mem. Inst., Columbus, OH, Bulletin No. 102, Aug 31, 1981 and Bulletin No. 103, Sept 25, 1981

14. R. A. Wood and R. J. Favor, *Titanium Alloys Handbook MCIC-HB-02*, Metals and Ceramics Information Center, Battelle Mem. Inst., Columbus, OH, 1972

15. "Introduction to Titanium and Its Alloys," in *Properties and Selection: Stainless Steels, Tool Materials and Special Purpose Metals, Metals Handbook*, Vol 3, 9th ed., American Society for Metals, 1980, p 353-360

16. "Spotlight on Titanium," *Mets. and Mats.*, July-Aug 1979, p 27-32; Mets. and Mats., Oct 1979, p 31-38; *Mets. and Mats.*, Dec. 1979, p 43-50

17. *Titanium: Past, Present and Future*, NMAB-392, National Materials Advisory Board, Washington, DC, 1983

PHYSICAL METALLURGY

18. M. Blackburn, "Some Aspects of Phase Transformations in Titanium Alloys" in *The Science, Technology and Application of Titanium*, Pergamon 1970, p 633-643

19. C. Hammond and J. Nutting, "The Physical Metallurgy of Superalloys and Titanium Alloys," in *Forging and Properties of Aerospace Materials*, The Metals Society, London, 1977, p 75-102. (See also *Met. Technol.*, Vol 4, 1977, p 474-490.)

20. B. S. Hickman, "The Formation of Omega Phase in Titanium and Zirconium Alloys: A Review," *J. Mater. Sci.*, Vol. 4, 1969, p 544-562

21. H. Margolin and P. Farrar, "The Physical Metallurgy of Titanium Alloys," *Ocean Engng.*, Vol 1, 1969, p 329-343

22. Y. Murakami, "Critical Review: Phase Transformations and Heat Treatment," in *Titanium 80*, The Metallurgical Society of AIME, Warrendale, PA, 1981, p 153-167

23. J. C. Williams, "Precipitation in Titanium-Base Alloys," in *Precipitation Processes in Solids*, The Metallurgical Society of AIME, New York, 1978, p 191-221

PROPERTIES AND MICROSTRUCTURE

24. L. J. Bartlo, "Effect of Microstructure on the Fatigue Properties of Ti-6Al-4V Bar," in *Fatigue at High Temperature*, STP No. 459, ASTM, Philadelphia, 1969, p 144-154

25. H. B. Bomberger et al., "Environmental Effects on Titanium and Its Alloys," in *Titanium Science and Technology*, Deutsch Gesell. f. Metallk., 1984, p 2435-2534

26. A. W. Bowen, "On the Importance of Crystallographic Texture in the Characterization of Alpha Based Titanium Alloys," *Sc. Met.*, Vol II, 1976, p 17-21

27. R. R. Boyer et al., "Comparative Evaluation on Ti-10V-2Fe-3Al Cast, PM and Wrought Product Forms," *Powd. Met Intl*, Vol 17 (No. 5), 1985, p 239-240

28. H. Conrad, "Effect of Interstitial Solutes on the Strength and Ductility of Titanium," in *Prog. Mats. Sc.*, Vol 26, Pergamon Press, 1981, p 123-403

29. F. A. Crossley and R. H. Jeal, "Fatigue and Fracture Behavior of the High Hardenability Martensitic Transage Titanium Alloys," *J. Aircft.*, Vol 18, 1981, p 683-686

30. W. J. Cruchlow and T. Lunde, "High Cycle Fatigue Properties of Titanium in Aircraft Application," in *Titanium Science and Technology*, Plenum Press, 1973, p 1257-1270

31. D. Eylon, "Fatigue Crack Initiation in Hot Isostatically Pressed Ti-6Al-4V Castings," *J. Mater. Sci.*, Vol 14, 1979, p 1914-1923

32. J. M. Fitzpatrick et al., "Texture Strengthening of Ti-6Al-4V," *Met. Eng. Quart.*, Vol 12, 1972, p 27-31

33. P. J. Fopiano and C. F. Hickey, "Comparison of the Heat Treatment Responses of Three Commercial Titanium Alloys," *J. Test Eval.*, Vol 1, 1973, p 514-519

34. F. H. Froes et al., "Relationship of Fracture Toughness and Ductility to Microstructure and Fractographic Features in Advanced Deep Hardenable Titanium Alloys," in

Toughness and Fracture Behavior of Titanium, STP No. 651, ASTM, Philadelphia, 1978, p 115-153

35. D. I. Golland and C. J. Beevers, "The Effect of Temperature on the Fatigue Response of Alpha-Titanium," *Met. Sci. J.*, Vol 5, 1971, p 174-177

36. J. A. Hall *et al.*, "Property-Microstructure Relationships in the Ti-6Al-2Sn-4Zr-6Mo Alloy," *Mater. Sci. Eng.*, Vol 9, 1972, p 197-210

37. J. A. Hall and C. M. Pierce, "Improved Properties of Ti-6Al-6V-2Sn Alloy through Microstructure Modifications," *Mater. Sci. Eng.*, Vol 8, 1971, p 121-122

38. C. F. Hickey and P. J. Fopiano, "Heat Treatment Effects on the Mechanical Properties in Ti-6Al-6V-2Sn," *J. Test Eval.*, Vol 1, 1973, p 166-169

39. J. P. Hirth and F. H. Froes, "Interrelation Between Fracture Toughness and Other Mechanical Properties in Titanium Alloys," *Metall. Trans.*, Vol 8A, 1977, p 1749-1761

40. R. W. Judy *et al.*, "Fracture Toughness of Titanium Alloys," in *Titanium Science and Technology*, Deutsche Gesell. f. Metallk., 1985, p 1925-1943

41. A. Lasalmonie and M. Loubradon, "Age Hardening Effect in Ti-6Al-4V due to ω Precipitation and α Precipitation in the β Grains," *J. Mater. Sci.*, Vol 14, 1979, p 2589-2596

42. H. A. Lipsitt et al., "The Deformation and Fracture of TiAl at Elevated Temperatures," *Metall. Trans.*, Vol 6A, 1975, p 1991-1996

43. H. Margolin *et al.*, "A Review of the Fracture and Fatigue Behavior of Ti Alloys," in *Titanium 80*, The Metallurgical Society of AIME, Warrendale, PA, 1981, p 169-216

44. N. Paton, "Low-Temperature Hydrogen Embrittlement of Titanium Alloys," in *Titanium Science and Technology*, Deutsche Geselle. f. Metallk., 1985, p 2519-2526

45. N. E. Paton *et al.*, "The Effects of Microstructure on the Fatigue and Fracture of Commercial Titanium Alloys," in *Alloy Design for Fatigue and Fracture Resistance*, AGARD-NATO Conf. Proc. No. 185, Advisory Group for Aerospace Research and Development and North Atlantic Treaty Organization, Neuilly-Sur-Seine, France, 1975, p 4.1-4.14

46. C. G. Rhodes and N. E. Paton, "The Influence of Microstructure on Mechanical Properties in Ti-3Al-4V-6Cr-4Mo-4Zr (Beta-C)," *Metall. Trans.*, Vol 8A, 1977, p 1749-1761

47. S. R. Seagle *et al.*, "High Temperature Properties of Ti-6Al-2Zr-2Mo-0.09Si," *Met. Eng. Quart.*, Vol 15, Feb 1975, p 48-54

48. G. Sridhar *et al.*, "The Influence of Heat Treatment on the Structure and Properties of a Near-Alpha Titanium Alloy," *Metall. Trans.*, Vol 18A, No. 5, 1987, p 877-891

49. R. M. Tchorzewski and W. B. Huln, "Effect of Texture on Fatigue Crack Path in Titanium 6Al-4V," *Met. Sci.*, Vol 12, 1978, p 109-112

50. M. A. Umam *et al.*, "Corrosion Fatigue Properties of the Ti-6Al-4V Alloy," in *Environmental Degradation of Engineering Materials*, Virginia Polytech Institute, Blacksburg, VA, 1977, p 783-790

51. R. P. Wei and D. L. Ritter, "The Influence of Temperature on Fatigue Crack Growth in a Mill Annealed Ti-6Al-4V Alloy," *J. Mats.*, Vol 7, 1972, p 240-250

52. D. N. Williams and R. A. Wood, "Effects of Surface Condition on the Mechanical Properties of Titanium and Its Alloys," MCIC-71-01, Metals and Ceramics Information Center, Battelle Mem. Inst., Columbus, OH, 1971

53. J. C. Williams *et al.*, "The Effect of Omega Phase on the Mechanical Properties of Titanium Alloys," *Met. Trans.*, Vol 2, 1971, p 1913-1919

54. G. R. Yoder *et al.*, "Improvement of Environmental Crack Propagation Resistance in Ti-8Al-1Mo-1V through Microstructural Modification," in *Advances in Materials Technology in the Americas — 1980*, Vol 2 — Materials Processing and Performance, ASME, 1980, p 135-140

CORROSION

55. M. J. Blackburn *et al.*, "Stress Corrosion Cracking of Titanium Alloys," in *Advances in Corrosion Science and Technology*, Plenum Press, 1973, p 67-292

56. M. J. Blackburn and W. H. Smyrl, "Critical Review — Stress Corrosion and Hydrogen Embrittlement," in *Titanium Science and Technology*, Plenum Press, 1973, p 2577-2609

57. H. B. Bomberger, "The Corrosion Resistance of Titanium," in *Titanium — 1966*, DMIC Memo. No. 215, Defense Metals Information Center, Battelle Mem. Inst., Columbus, OH, 1966, p 35-43

58. B. Champin *et al.*, "Corrosion Resistance of Titanium and Its Alloys in Marine Environments," *J. Less-Comm. Mets.*, Vol 69, 1980, p 149-162

59. B. Champin *et al.*, "Oxidation of Titanium Alloys at Temperatures Used in Turbine Engines," *J. Less-Comm. Mets.*, Vol 69, 1980, p 163-183

60. J. B. Cotton, "Critical Review — General Corrosion and Oxidation: Coatings," in *Titanium Science and Technology*, Plenum Press, 1973, p 2363-2372

61. L. C. Covington, "Titanium Solves Corrosion Problems in Petroleum Processing," *Met. Prog.*, Vol III, Feb 1977, p 38-45

62. W. P. Danesi, *et al.*, "The Effects of Microstructural Variations on Stress Corrosion Cracking Susceptibility of Ti-8Al-1Mo-1V Alloy on Martinized Gas Turbines," Paper 67-GT-5, ASME, New York, 1967

63. J. M. Ferguson, "The Oxidation and Contamination of Titanium and Its Alloys," DMIC Memo. No. 238, Defense Metals Information Center, Battelle Mem. Inst., Columbus, OH, 1968

64. H. R. Gray, "Effects of Hot Salt Stress Corrosion on Titanium Alloys," *Met. Eng. Quart.*, Vol 12, Nov 1972, p 10-17

65. H. R. Gray and J. R. Johnston, "Effect of Exposure Cycle on Hot-Salt Stress-Corrosion of a Titanium Alloy," Memo TMX-3145, National Aeronautics and Space Administration, Cleveland, OH, 1974

66. J. D. Jackson and W. K. Boyd, "Corrosion of Titanium," DMIC Memo No. 218, Defense Metals Information Center, Battelle Mem. Inst., Columbus, OH, 1968

67. M. W. Mahoney and A. S. Tetelman, "The Effect of Microstructure on the Hot Salt Stress Corrosion Susceptibility of Titanium Alloys," *Metall. Trans.*, Vol 7A, 1976, p 1549-1558

68. R. J. Solar, "Corrosion Resistance of Titanium Surgical Implant Alloys: A Review," in *Corrosion and Degradation of Implant Materials*, STP No. 684, ASTM, Philadelphia, 1979, p 259-273

PROCESSING

69. H. L. Black and R. O. Kaufman, "Isothermal Rolling of Superalloy and Ti Precision Shapes," *Ind. Heating*, Vol 43, Oct 1976, p 30-36

70. J. A. Burger and D. K. Hanink, "Heat Treating Titanium and Its Alloys," *Met. Prog.*, Vol 91, June 1967, p 70-75

71. C. C. Chen, "Role of Processing Variables and Microstructures in Producing Dual Property Titanium Forgings," in *Process Modeling—Fundamentals and Applications to Metals*, American Society for Metals, 1980, p 365-386

72. R. H. Colley, "Engineer's Guide to Titanium Castings," in *Titanium Science and Technology*, Deutsche Gesell. f. Metallk., 1985, p 151-156

73. C. W. Corti and G. H. Gessinger, "Isothermal Forging of Complex Components in Ti-6Al-4V Alloys," in Forging Properties of Aerospace Materials, The Metals Society, London, 1978, p 266-278

74. J. E. Coyne, "Microstructural Control in Ti- and Ni-Base Forgings — An Overview," in *Forging Properties of Aerospace Materials*, The Metals Society, London, 1978, p 234-247. (See also *Mets. Technol.*, Vol 4, 1977, p 337-345.)

75. J. E. Coyne *et al.*, "The Effect of Beta Forging on Several Titanium Alloys," *Met. Eng. Quart.*, Vol 8, Aug 1968, p 10-15

76. D. S. Duvall and W. A. Owczarski, "Fabrication and Repair of Titanium Engine Components by Welding," in *Materials and Processes for the 70's — Cost Effectiveness and Reliability*, SAMPE, Azusa, CA, 1973, p 472-485

77. D. Eylon et al., "Status of Titanium Powder Metallurgy," in *Industrial Applications of Titanium and Zirconium: Third Conference*, STP 830, ASTM, Philadelphia, 1984, p 48-65

78. F. H. Froes, "Developments in Titanium Powder Metallurgy," *J. Mets.*, Vol 32, Feb 1980, p 47-54

79. F. H. Froes and D. Eylon, "Powder Metallurgy of Titanium Alloys—A Review," *Powd. Met. Int.*, Vol 17 (No. 4), 1985, p 163-168; *Powd. Met. Int.*, Vol 17 (No. 5), 1985, p 235-239

80. F. H. Froes and D. Eylon, Ed., *Titanium Net Shape Technologies*, The Metallurgical Society of AIME, Warrendale, PA, 1984

81. D. F. Hasson and C. H. Hamilton, Ed., *Advanced Processing Methods for Titanium*, The Metallurgical Society of AIME, Warrendale, PA, 1982

82. H. Heitman *et al.*, "The Effect of Alpha + Beta and Beta Forging on the Fracture Toughness of Several High-Strength Titanium Alloys," *Met. Eng. Quart.*, Vol 8, Aug 1968, p 15-18

83. R. E. Keith, "Adhesive Bonding of Ti and Its Alloys," in *Handbook of Adhesive Bonding.*, McGraw-Hill, 1973

84. J. F. Kenney and C. Deminet, "Low Pressure Diffusion Bonding of Ti," in *Structures*, American Institute of Aeronautics and Astronautics, New York, 1978, p 126-131

85. H. D. Kessler, "Critical Review—Consolidation, Primary and Secondary Fabrication," in *Titanium Science and Technology*, Plenum Press, 1973, p 303-318

86. R. E. Key *et al.*, "Titanium Structural Brazing," *Weld. J.*, Vol 53, Oct 1974, p 426s-431s

87. E. J. Kubel, "Titanium NNS Technology Shaping Up," *Adv. Mater. Proc.*, Vol 131 (No. 2), 1987, p 46-50

88. E. J. Kubel, "Titanium Alloy Technology Update," *Met. Prog.*, Vol 129 (No. 7), 1986, p 41-42; 47-51

89. G. Kuhlman and T. B. Gurganus, "Optimizing Thermomechanical Processing of Ti-10V-2Fe-3Al Forgings," *Met. Prog.*, Vol 118, July 1980, p 30-35

90. P. Lowenstein, "Extruding Titanium and Titanium Alloys," Tech. Paper No. MF71-139, Society of Tool and Manufacturing Engineers, Dearborn, MI, 1971

91. J. Magnuson, "Repair of Titanium Airframe Casting by Hot Isostatic Pressing," *Metallography,* Vol 10, 1977, p 223-233

92. L. J. Maidment and H. Paweletz, "An Evaluation of Vacuum Centrifuged Titanium Castings for Helicopter Components," *Metallwissenscaft und Tech.*, Vol 35, 1981, p 137-140

93. L. Parsons *et al.*, "Titanium P/M Comes of Age," *Met. Prog.*, Vol 126 (No. 4), 1984, p 83-84, 86, 89, 91, 93, 94

94. V. C. Petersen *et al.*, "Hot Isostatic Pressing of Large Titanium Shapes," in *Powder Metallurgy of Titanium Alloys,* The Metallurgical Society of AIME, Warrendale, PA, 1980, p 243-254

95. J. V. Scanlan and G. J. G. Chambers, "Forgings in Titanium Alloys," in *The Science Technology and Application of Titanium,* Pergamon Press, 1970, p 79-95

96. R. H. Witt and A. L. Ferreri, "Titanium Near Net Shape Components for Demanding Airframe Applications," *SAMPE Quart.*, Vol 17 (No. 3), 1986, p 55

97. R. H. Witt and W. T. Highberger, "Hot Isostatic Pressing of Near Net Titanium Structural Parts," in *Powder Metallurgy of Titanium Alloys*, The Metallurgical Society of AIME, Warrendale, PA, 1980, p 255-266

APPLICATIONS AND RECENT DEVELOPMENTS

98. J. E. Allison *et al.*, "Titanium in Engine Valve Systems," *J. Mets.*, Vol 39 (No. 3), 1987, p 15-17

99. B. P. Bannon *et al.*, "Emerging Applications for Titanium-Based Alloys," in *Industrial Applications of Titanium and Zirconium: Third Conference*, STP 830, ASTM, Philadelphia, 1984, p 66-85

100. R. G. Berryman *et al.*, "A New Cost Effective Titanium Alloy With High Fracture Toughness," *Met. Prog.*, Vol 112, Dec 1977, p 40-45

101. P. A. Blenkinsop, "Developments in High Temperature Titanium Alloys," in *Titanium Science and Technology*, Deutsche Gesell. f. Metallk., 1985, p 2323-2338

102. R. R. Boyer, "Design Properties of a High Strength Titanium Alloy, Ti-10V-2Fe-3Al," *J. Mets.*, Vol 32, Mar 1980, p 61-65

103. R. W. Broomfield *et al.*, "Application of Advanced Powder Process Technology to Titanium Aerospace Components," *Powd. Met. Int.*, Vol 28 (No. 1), 1985, p 27-34

104. S. P. Cooper and G. H. Whitley, "Titanium in the Powder Generation," *Mater. Sci. Technol.*, Vol 3 (No. 2), 1987, p 91-96

105. F. A. Crossley, "The Martensitic Transage Alloys for Improved Structural Efficiency and Reduced Cost," in *Materials and Processes—Continuing Innovations*, SAMPE, Anaheim, CA, 1983, p 1352-1367

106. D. Eylon *et al.*, "Titanium Alloys for High Temperature Applications—A Review," *High Temp. Mater. Proc.*, Vol 6 (No. 1 and 2), 1984, p 81-91

107. D. Eylon *et al.*, "Manufacture of Cost Affordable High Performance Titanium Components for Advanced Air Force Systems," in *Materials 1980*, SAMPE, Azusa, CA, 1980, p 356-367

108. T. W. Farthing, "Applications of Titanium and Titanium Alloys," in *Titanium Science and Technology*, Deutsche Gesell. f. Metallk., Vol 9, 1985, p 39-54

109. F. H. Froes and D. Eylon, *Titanium Rapid Solidification Technology*, The Metallurgical Society of AIME, Warrendale, PA, 1986

110. N. F. Harpur, "Applications for Titanium in Airframes," *Metall. Mater. Technol.*, Vol 13, 1981, p 259-263

111. R. D. Kane and W. K. Boyd, "Use of Titanium and Zirconium in Chemical Environments," in *Industrial Applications of Titanium and Zirconium*, STP No. 728, ASTM, Philadelphia, PA, 1981, p 3-8

112. R. L. Kane, "Titanium in Seawater Piping," *J. Mets.*, Vol 39 (No. 3), 1987, p 10-11

113. H. A. Lipsitt, "Titanium Aluminides—Future Turbine Materials," in *Advanced High-Temperature Alloys: Processing and Properties*, American Society for Metals, 1986, p 157-164

114. H. W. Rosenberg, "Beta Titanium Alloys for Structural Efficiency," in *Materials and Processes—Continuing Innovations*, Vol 28, SAMPE, Anaheim, CA, 1983, p 1223-1230

115. T. M. Rust and W. G. Steltz, "Titanium for Steam Turbine Blades," *J. Mets.*, Vol 34 (No. 9), 1982, p 42-53

116. T. M. Rust *et al.*, "Operating Experience with Titanium Steam Turbine Blades," *Met. Prog.*, Vol 116, July 1979, p 62-66

117. R. W. Schutz, "Performance and Application of Titanium Alloys in Geothermal Service," *Mater. Perform.*, Vol 24 (No. 1), 1985, p 39-48

118. R. W. Schutz and L. C. Covington, "Titanium Applications in the Energy Industry," Paper No. A 81-53, The Metallurgical Society of AIME, Warrendale, PA, 1981

119. S. R. Seagle *et al.*, "Titanium Alloy Springs," *Mater. Sci. Technol.*, Vol 3s (No. 2), 1987, p 97-100

120. S. G. Steinemann *et al.*, "Titanium Alloys as Metallic Biomaterials," in *Titanium Science and Technology*, Gesell. f. Metallk., 1985, p 1327-1334

Appendix B

Glossary

INTRODUCTION

This glossary is composed of a selection of technical words relevant to a discussion of titanium and its alloys. It is included to aid the reader who may not be familiar with all aspects of titanium, its production, or its uses. This glossary is not all inclusive. (Refer also to the various volumes ASM INTERNATIONAL's *Metals Handbook*, 9th edition, which frequently have extensive glossaries covering the entire field of metal production and use.)

A

AC. Air cooled.

Acicular alpha. A product of nucleation and growth from beta to the lower-temperature allotrope alpha phase. It may have a needlelike appearance in a photomicrograph and may have needle, lenticular or flattened bar morphology in three dimensions. Its typical aspect ratio is about 10:1.

Activation. The changing of a passive surface of a metal to a chemically active state. (Contrast with *passivation*.)

Age hardening. Hardening by aging, usually after rapid cooling or cold working. (See also *aging*.)

Aging. A change in the properties of certain metals and alloys that occurs with time at ambient or moderately elevated temperatures after working or a heat treatment (natural or artificial aging) or after a cold working operation (strain aging). The change in properties is often, but not always, due to a phase change (precipitation), but it never involves a change in chemical composition of the metal or alloy. (See *overaging*.)

Allotropy. The property by which certain elements may exist in more than one crystal structure. An allotrope is a specific crystal structure of the metal.

Alloy. A substance having metallic properties and being composed of two or more chemical elements of which at least one is a metal.

Alloying element. An element, added to and remaining in a metal, that changes structure and properties.

Alloy powder, alloyed powder. A metal powder consisting of at least two constituents that are partially or completely alloyed with each other.

Alloy system. A complete series of compositions produced by mixing in all proportions any group of two or more components, at least one of which is a metal.

Alpha. The low-temperature allotrope of titanium with a hexagonal, close-packed crystal structure. (It occurs below the beta transus.)

Alpha-beta structure. A microstructure containing α and β as the principal phases at a specific temperature. (See also *beta*.)

Alpha case. The oxygen-, nitrogen-, or carbon-enriched, α-stabilized surface resulting from elevated temperature exposure. (See also *alpha stabilizer*.)

Alpha double prime (orthorhombic martensite). A supersaturated, nonequilibrium orthorhombic phase formed by a diffusionless transformation of the β phase in certain alloys.

Alpha prime (hexagonal martensite). A supersaturated, nonequilibrium hexagonal α phase formed by a diffusionless transformation of the β phase. It is often difficult to distinguish from acicular α, although the latter is usually less well defined and frequently has curved, instead of straight, sides.

Alpha stabilizer. An alloying element which dissolves preferentially in the alpha phase and raises the alpha-beta transformation temperature.

Alpha transus. The temperature that designates the phase boundary between the α and $\alpha + \beta$ fields.

Alpha 2 structure. A structure consisting of the ordered alpha phase, Ti_3Al_1.

Annealed powder. A powder that is heat treated to render it soft and compactible.

Annealing. A generic term denoting a treatment, consisting of heating to, and holding at, a suitable temperature followed by cooling at a suitable rate. It is used primarily to soften metallic materials, but also to simultaneously produce desired changes in other properties or in microstructure. The purpose of such changes may be, but is not confined to: improvement of machinability, facilitation of cold work; improvement of mechanical or electrical properties, and/or increase in stability of dimensions. When the term is used without qualification, full annealing is implied. When applied only for the relief of stress, the process is properly called stress relieving or stress-relief annealing. In nonferrous alloys, annealing cycles are designed to: (1) remove part or all of the effects of cold working (recrystallization may or may not be involved); (2) cause substantially complete coalescence of precipitates from solid solution in relatively coarse form; or (3) both, depending on composition and material condition. Specific process names in commercial use are final annealing, full annealing, intermediate annealing, partial annealing, recrystallization annealing, stress-relief annealing, anneal to temper. (See *multiple annealing*.)

atm. Atmosphere (pressure).

Atomic number. The number of protons in an atomic nucleus; determines the individuality of the atom as a chemical element.

Atomic percent. The number of atoms of an element in a total of 100 representative atoms of a substance.

Atomization. The disintegration of a molten metal into particles by a rapidly moving gas or liquid stream or by other means.

AWG. American wire gage.

AWS. American Welding Society.

B

B. Bar.

Bake (verb). To remove gases from a powder at low temperatures.

Banded structure. A segregated structure consisting of alternating, nearly parallel bands of different composition, typically aligned in the direction of primary hot working.

Barstock. Same as bar.

Basal plane. A plane perpendicular to the principal axis (c axis) in a tetragonal or hexagonal structure.

Basketweave. Alpha platelets with or without interleaved β platelets that occur in colonies in a Widmanstätten structure.

Batch. The total output of one mixing of powder metal; sometimes called a lot.

Batch sintering. Presintering or sintering in such a manner that compacts are sintered and removed from the furnace before additional unsintered compacts are placed in the furnace.

Bauschinger effect. For both single-crystal and polycrystalline metals, any change in stress-strain characteristics that can be ascribed to changes in the microscopic stress distribution within the metal, as distinguished from changes caused by strain hardening. In the narrow sense, the process whereby plastic deformation in one direction causes a reduction in yield strength when stress is applied in the opposite direction.

Beta. The high-temperature allotrope of titanium with a body-centered cubic crystal structure that occurs above the β transus.

Beta eutectoid stabilizer. An alloying element that dissolves preferentially in the β phase, lower the α-β to β transformation temperature, and results in β decomposition to α plus a compound.

Beta fleck. Alpha-lean region in the α-β microstructure significantly larger than the primary α width. This β-rich area has a β transus measurably below that of the matrix. Beta flecks have reduced amounts of primary α which may exhibit a morphology different from the primary α in the surrounding α-β matrix.

Beta isomorphous stabilizer. An alloying element that dissolves preferentially in the β phase, lowers the α-β to β transformation temperature without a eutectoid reaction, and forms a continuous series of solid solutions with β-titanium.

Beta-STOA. Beta solution overaging. (See *solution heat treatment* and *overaging*.)

Beta transus. The minimum temperature above which equilibrium α does not exist. For β eutectoid additions, the β transus ordinarily is applied to hypoeutectoid compositions or those that lie to the left of the eutectoid composition.

Billet. (1) A solid, semifinished round or square product that has been hot worked by forging, rolling or extrusion; usually smaller than a bloom. (2) A general term for wrought starting stock used in making forgings or extrusions.

Binder. A substance added to the powder to: (1) increase the strength of the compact; and (2) cement together powder particles that alone would not sinter into a strong object.

Blank. A pressed, presintered, or fully sintered compact, usually in the unfinished condition, to be machined or otherwise processed to final shape or condition.

Body-centered cubic lattice structure. A unit cell which consists of atoms arranged at cube corners with one atom at the center of the cube.

Braze. A joint produced by heating an assembly to suitable temperatures and by using a filler metal having a liquidus above 450°C (840°F) and below the solidus of the base metal. The filler metal is distributed between the closely fitted faying surfaces of the joint by capillary action.

Brazeability. The capacity of a metal to be brazed under the fabrication conditions imposed into a specific suitably designed structure and to perform satisfactorily in the intended service.

Brazing. A group of processes that join solid materials together by heating them to a suitable temperature and by using a filler metal having a liquidus above about 450°C (840°F) and below the solidus of the base materials. The filler metal is distributed between the closely fitted surfaces of the joint by capillary attraction.

Brazing filler metal. A nonferrous filler metal used in brazing and braze welding.

Braze welding. A method of welding by using a filler metal having a liquidus above 450°C (840°F) and below the solidus of the base metals. Unlike brazing, in braze welding, the filler metal is not distributed in the joint by capillary attraction.

Brinell hardness number (HB). A number related to the applied load and to the surface area of the permanent impression made by a ball indenter.

Brinell hardness test. A test for determining the hardness of a material by forcing a hard steel or carbide ball of specified diameter into it under a specific load. The result is expressed as the Brinell hardness number, which is the value obtained by dividing the applied load in kilograms by the surface area of the resulting impression in square millimeters.

Brittle. Permitting little or no plastic (permanent) deformation prior to fracture.

Brittle fracture. Separation of a solid accompanied by little or no macroscopic plastic deformation. Typically, brittle fracture occurs by rapid crack propagation with less expenditure of energy than for ductile fracture.

Brittleness. The tendency of a material to fracture without first undergoing significant plastic deformation. (Contrast with *ductility*.)

B&S. Brown and Sharpe (gage).

C

Cake. A coalesced mass of unpressed metal powder.

CAP. See *consolidation by atmospheric pressure*.

Cast or casting. To fabricate an item by pouring molten metal into a shaped cavity and permitting the metal to solidify. A cast can relate to the item or may be a synonym for heat, that is an identifiable chemistry lot.

Charpy test. An impact test in which a V-notched, keyhole-notched, or U-notched specimen, supported at both ends, is struck behind the notch by a striker mounted at the lower end of a bar that can swing as a pendulum. The energy that is absorbed in fracture is calculated from the height to which the striker would have risen had there been no specimen and the height to which it actually rises after fracture of the specimen.

CHM. Chemical milling, a machining technique.

Chemical vapor deposition (CVD). The precipitation of a metal from a gaseous compound onto a solid or particulate substrate.

CHIP. CIP plus sinter plus HIP. A 3-stage P/M process. (See *CIP*.)

CIP. The acronym representing the words cold isostatic pressing.

Close-packed. A geometric arrangement in which a collection of equally sized spheres (atoms) may be packed together in a minimum total volume.

Coarse grains. Grains larger than normal for the particular wrought metal or alloy or of a size that produces a surface roughening known as orange peel or alligator skin in wrought alloys.

Cold pressing. The forming of a compact from powder at, or below, room temperature.

Cold working. Deforming metal plastically under conditions of temperature and strain rate that induce strain hardening. Usually, but not necessarily, conducted at room temperature. (Contrast with *hot working*.)

Cold-worked structure. A microstructure resulting from plastic deformation of a metal or alloy below its recrystallization temperature.

Colonies. Regions within prior beta grains with alpha platelets having nearly identical orientations. In commercially pure titanium, colonies often have serrated boundaries. Colonies arise as transformation products during cooling from the beta fields at cooling rates that induce platelet nucleation and growth.

Compact. An object produced by the compression of metal powder, generally while confined in a die, with or without the inclusion of nonmetallic constituents.

Compact, compacting, compaction. The operation or process of producing a compact; sometimes called pressing.

Consolidation by atmospheric pressure (CAP). A P/M consolidation process.

Consumable electrode. A general term for any arc welding electrode made chiefly of filler metal.

Consumable electrode remelting. A process for refining metals in which an electric current passes between an electrode made of the metal to be refined.

Corrosion. The deterioration of a metal by a chemical or electrochemical reaction with its environment.

Corrosion embrittlement. The severe loss of ductility of a metal resulting from corrosive attack, usually intergranular and often not visually apparent.

Corrosion fatigue. Cracking produced by the combined action of repeated or fluctuating stress and a corrosive environment at lower stress levels or fewer cycles than would be required in the absence of a corrosive environment.

Corrosive wear. Wear in which chemical or electrochemical reaction with the environment is significant.

cph. Close-packed hexagonal.

Creep. Time-dependent strain occurring under stress. The creep strain occurring at a diminishing rate is called primary, or transient, creep; that occurring at a minimum and almost constant rate, secondary, or steady-rate creep; that occurring at an accelerating rate, tertiary creep. These rates are frequently represented graphically as 1, 2, and 3, or as I, II, and III.

Creep limit. (1) The maximum stress that will cause less than a specified quantity of creep in a given time. (2) The maximum nominal stress under which the creep strain rate decreases continuously with time under constant load and at constant temperature. Sometimes used synonymously with creep strength.

Creep rate. The slope of the creep-time curve at a given time determined from a Cartesian plot.

Creep recovery. Time-dependent strain after release of load in a creep test.

Creep-rupture strength. The stress that will cause fracture in a creep test at a given time in a specified constant environment. Also known as stress-rupture strength.

Creep-rupture test. Same as stress-rupture test.

Creep strain. The time-dependent total strain (extension plus initial gage length) produced by applied stress during a creep test.

Creep strength. The stress that will cause a given creep strain in a creep test at a given time in a specified constant environment.

Creep stress. The constant load divided by the original cross-sectional area of the specimen.

Crevice corrosion. A type of concentration cell corrosion; corrosion caused by the concentration or depletion of dissolved salts, metal ions, oxygen, or other gases, and such, in crevices or pockets remote from the principal fluid stream, with a resultant building up of differential cells that ultimately cause deep pitting. Localized corrosion of a metal surface at, or immediately adjacent to, an area that is shielded from full exposure to the environment because of close proximity between the metal and the surface of another material.

Critical stress intensity factor (K_{IC}). A measure of fracture toughness. Increased K_{Ic} indicates greater resistance to fracture. K_{Ic} is a common means, but not the only one, of describing quantitatively the fracture resistance of an alloy. Stress intensity factors vary with loading conditions. The I refers to the loading condition, i.e. plane strain. The c refers to critical condition, above which the stress (load) need no longer be increased to cause fracture. K_{Ic} varies with material chemistry and processing history. It is a function of temperature, and generally decreases as temperature decreases.

Crystal. A solid composed of atoms, ions, or molecules arranged in a pattern that is periodic in three dimensions.

Crystallization. (1) The separation, usually from a liquid phase on cooling, of a solid crystalline phase. (2) Sometimes erroneously used to explain fracturing that actually has occurred by fatigue.

Curing. The processing of a mold to obtain desired characteristics.

CVD. See *chemical vapor disposition*.

D

DA. See *duplex annealing*.

da/dN. See *fatigue crack growth rate*.

DBTT. Ductile-to-brittle transition temperature.

Deformation. A change in the form of a body due to stress, thermal change, change in moisture, or other causes. Measured in units of length.

Descaling. Removing the thick layer of oxides formed on some metals at elevated temperatures.

DFB. See *diffusion brazing*.

DFW. See *diffusion welding*.

diam. Diameter.

Diffusion brazing (DFB). A brazing process that joins two or more components by heating them to suitable temperatures and by using a filler metal or an *in situ* liquid phase. The filler metal may be distributed by capillary attraction or may be placed or formed at the faying surfaces. The filler metal is diffused with the base metal to the extent that the joint properties have been changed to approach those of the base metal. Pressure may or may not be applied.

Diffusion welding (DFW). A high-temperature, solid-state welding process that permanently joins faying surfaces by the simultaneous application of pressure and heat. The process does not involve macroscopic deformation, melting, nor relative motion of parts. A solid filler metal (diffusion aid) may or may not be inserted between the faying surfaces.

Double aging. Employment of two different aging treatments to control the type of precipitate formed from a supersaturated matrix in order to obtain the desired properties. The first aging treatment, sometimes referred to as intermediate or stabilizing, is usually carried out at higher temperature than the second.

Ductile fracture. Fracture characterized by tearing of metal accompanied by appreciable gross plastic deformation and expenditure of considerable energy.

Ductility. The ability of a material to deform plastically before fracturing. Measured by elongation or reduction of area in a tension test, by height of cupping in a cupping test or by the radius or angle of bend in a bend test. (Contrast with *brittleness*; see also *plastic deformation*.)

Duplex annealing (DA). See *multiple annealing*.

Duplexing. Any two-furnace melting or refining process. Also called duplex melting or duplex processing.

E

EBC. See *electron beam cutting*.

EBW. See *electron beam welding*.

ECM. Electrochemical machining.

Electrode. (1) The isolated sponge, master alloy and/or revert used in consumable vacuum arc melting. (2) The solidified ingot in cases when it is to be remelted again in double and triple melting operations.

Electron beam cutting (EBC). A cutting process which uses the heat obtained from a concentrated beam composed primarily of high-velocity electrons which impinge upon the workpieces to be cut; it may or may not use an externally supplied gas.

Electron beam welding (EBW). A welding process which produces coalescence of metals with the heat obtained from a concentrated beam composed primarily of high-velocity electrons impinging upon the surfaces to be joined.

ELI. Extra-low interstitial.

Elongated alpha. A fibrous type of structure brought about by unidirectional metal-working. It may be enhanced by the prior presence of blocky and/or grain boundary alpha.

Elongated grain. A grain with one principal axis significantly longer than either of the other two.

Elongation. A term used in mechanical testing to describe the amount of extension of a test piece when stressed. In tensile testing, the increase in the gage length, measured after fracture of the specimen within the gage length, usually expressed as a percentage of the original gage length.

Embrittlement. The severe loss of ductility and/or toughness of a material, usually a metal or alloy.

Endurance limit. The maximum stress below which a material can presumably endure an infinite number of stress cycles. If the stress is not completely reversed, the value of the mean stress, the minimum stress or the stress ratio also should be stated. (Compare with *fatigue limit*.)

Equiaxed structure. A polygonal or spheroidal microstructural feature having approximately equal dimensions in all directions. In alpha-beta titanium alloys, such a term commonly refers to a microstructure in which most of the minority phase appears spheroidal.

Equilibrium. A dynamic condition of physical, chemical, mechanical, or atomic balance, where the condition appears to be one of rest rather than change.

Erosion. Destruction of metals or other materials by the abrasive action of moving fluids, usually accelerated by the presence of solid particles or matter in suspension. When corrosion occurs simultaneously, the term erosion-corrosion is often used.

Erosion-corrosion. A conjoint action involving corrosion and erosion in the presence of a moving corrosive fluid, leading to the accelerated loss of material.

Eutectic. (1) An isothermal reversible reaction in which a liquid solution is converted into two or more intimately mixed solids upon cooling; the number of solids formed equals the number of components in the system. (2) An alloy having the composition indicated by the eutectic point on an equilibrium diagram. (3) An alloy structure of intermixed solid constituents formed by a eutectic reaction.

Eutectic melting. Melting of localized microscopic areas whose composition corresponds to that of the eutectic in the system.

Eutectoid. (1) An isothermal, reversible transformation in which a solid solution is converted into two or more intimately mixed solids. The number of solids formed equals the number of components in the system. (2) An alloy having the composition indicated by the eutectoid point on an equilibrium diagram. (3) An alloy structure of intermixed solid constituents formed by a eutectoid transformation.

Eutectoid point. The composition of a solid phase that undergoes univariant transformation into two or more other solid phases upon cooling.

F

Face-centered cubic lattice structure. A unit cell which consists of atoms arranged at cube corners with one atom at the center of each cube face.

Fatigue. The phenomenon leading to fracture under repeated or fluctuating stresses having a maximum value less than the tensile strength of the material. Fatigue fractures are progressive, beginning as minute cracks that grow under the action of the fluctuating stress.

Fatigue crack growth rate (da/dN). The rate of crack extension caused by constant-amplitude fatigue loading.

Fatigue failure. Failure that occurs when a specimen undergoing fatigue completely fractures into two parts, or has softened, or been otherwise significantly reduced in stiffness by thermal heating or cracking. Fatigue failure generally occurs at loads which, if applied statically, would produce little perceptible effect. Fatigue failures are progressive, beginning as minute cracks that grow under the action of the fluctuating stress.

Fatigue life. The number of cycles of stress that can be sustained prior to failure for a stated test condition.

Fatigue limit. The maximum stress that presumably leads to fatigue fracture in a specified number of stress cycles. If the stress is not completely reversed, the value of the mean stress, the minimum stress, or the stress ratio also should be stated. (Compare with _endurance limit_.)

Fatigue ratio. The fatigue limit under completely reversed flexural stress divided by the tensile strength for the same alloy and condition.

Fatigue strength. The maximum stress that can be sustained for a specified number of cycles without failure, the stress being completely reversed within each cycle unless otherwise stated.

FC. Furnace cooled.

fcc. Face-centered cubic.

FCP. Fatigue crack propagation.

Filler metal. The metal to be added in making a welding, brazed, or soldered joint.

Fissure. A small crack-like discontinuity with only slight separation (opening displacement) of the fracture surfaces. The prefixes macro or micro indicate relative size.

Flake. Powder of an essentially two-dimensional nature.

Flash welding (FW). A resistance welding process which produces coalescence simultaneously over the entire area of abutting surfaces, by the heat obtained from resistance to electric current between the two surfaces, and by the application of pressure after heating is substantially completed. Flashing and upsetting are accompanied by expulsion of molten metal from the joint.

Fluid die process. A P/M consolidation process. Also known as rapid omnidirectional compaction (ROC) process.

Foil. Among many definitions is: a flat-rolled product 0.127 mm (0.005 in.), or less, in thickness, regardless of width. (Any flat-rolled product thicker than this dimension is not considered foil.) Only thickness, not width, is a factor in determining foil.

Forgeability. Term used to describe the relative ability of material to flow under a compressive load without rupture.

Forged structure. The macrostructure through a suitable section of a forging that reveals direction of working.

Forging. (1) Plastically deforming metal, usually hot, into desired shapes with compressive force, with or without dies. (2) Reshaping a billet or ingot by hammering. (3) The process of placing a powder in a container, removing the air from the container, and sealing it. This is followed by conventional forging of the powder and container to the desired shape.

Forging stock. A rod, bar, billet, or other section used to make forgings.

Formability. The relative ease with which a metal can be shaped through plastic deformation.

Forming. The shaping of a component or part by bending, stretching, etc.

Fracture. (1) The irregular surface produced when a piece of metal is broken. (2) To separate a metal or alloy into two or more pieces by an applied load.

Fracture stress. (1) The maximum principal true stress at fracture. Usually refers to unnotched tensile specimens. (2) The (hypothetical) true stress that will cause fracture without further deformation at any given strain.

Fracture toughness. See *stress-intensity factor*.

FRW. Friction welding.

Friction welding. A solid-state process in which materials are welded by the heat obtained from rubbing together surfaces that are held against each other under pressure.

FW. See *flash welding*.

G

Gall. (1) To damage the surface of a component by momentary adhesion between facing surfaces. (2) Damage to a compact or die part, caused by adhesion of powder to the die cavity wall or a punch surface.

Galling. A condition whereby excessive friction between high spots results in localized welding with subsequent spalling and a further roughening of the rubbing surfaces of one or both of two mating parts.

Gamma structure. An ordered structure of titanium-aluminum compound with a stoichiometry of Ti-Al.

Gas carbon arc welding (CAW-G). A carbon arc welding process variation which produces coalescence of metals by heating them with an electric arc between a single carbon electrode and the work. Shielding is obtained from a gas or gas mixture.

Gas metal arc welding (GMAW). An arc welding process which produces joining of metals by heating them with an arc between a continuous filler metal (consumable) electrode and the work. Shielding is obtained entirely from an externally supplied gas or gas mixture. Some variations of this process are called MIG or CO_2 welding (nonpreferred terms).

Gas shielding arc welding. A general term used to describe gas metal arc welding, gas tungsten arc welding, and flux cored arc welding when gas shielding is employed.

Gas tungsten arc welding (GTAW). An arc welding process which produces coalescence of metals by heating them with an arc between a tungsten (nonconsumable) electrode and the work. Shielding is obtained from a gas or gas mixture. At times called tungsten inert gas (TIG) welding.

GAW-G. See *gas carbon arc welding*.

General corrosion. A form deterioration that is distributed more or less uniformly over a surface. (See also *corrosion*.)

GMAW. See *gas metal arc welding*.

Grain boundary alpha. Primary alpha outlining prior beta grain boundaries. It may be continuous unless broken up by subsequent work. Also may accompany blocky alpha. Occurs by slow cooling from the beta field into the alpha-beta field and is associated with insufficient deformation in working.

Grindability. Relative ease of grinding. (Analogous to machinability).

Grinding. Removing material from a workpiece with a grinding wheel or abrasive belt.

Grinding cracks. Shallow cracks formed in the surface of relatively hard materials because of excessive grinding heat or the high sensitivity of the material. (See *grinding sensitivity*.)

Grinding sensitivity. Susceptibility of a material to surface damage such as grinding cracks; it can be affected by such factors as hardness, microstructure, hydrogen content, and residual stress.

Grinding stress. Residual stress, generated by grinding, in the surface layer of work. It may be tensile, compressive or both.

GTAW. Gas tungsten arc welding. Often called TIG (tungsten inert gas).

H

HAD. See *high-aluminum defect*.

Hardener. (1) An alloy element introduced to produce strength. (2) An alloy, rich in one or more alloying elements, added to a melt to permit closer composition control than possible by addition of pure metals or to introduce refractory elements not readily alloyed with the base metal. Sometimes called master alloy or rich alloy.

Hardening. Increasing hardness by suitable chemical, thermal, or mechanical treatment, usually involving heating and cooling. (See also *age hardening, case hardening, induction hardening, precipitation hardening* and *quench hardening*.)

Hardness. A measure of the resistance of a material to surface indentation or abrasion; may be thought of as a function of the stress required to produce some specified type of surface deformation. There is no absolute scale for hardness; therefore, to express hardness quantitatively, each type of test has its own scale of arbitrarily defined hardness. Indentation hardness may be measured by Brinell, Knoop, Rockwell, Scleroscope and Vickers hardness tests.

HAZ. See *heat-affected zone*.

HB. See *Brinell hardness number*.

Heat-affected zone. That portion of the base metal which has not been melted, but whose mechanical properties or microstructure have been altered by the heat of welding, brazing, soldering, or cutting.

Heat treatment. Heating and cooling a solid metal or alloy in such a way as to obtain desired conditions or properties. Heating for the sole purpose of hot working is excluded from the meaning of this definition.

Hexagonal close-packed lattice structure. A unit cell which consists of a hexagonal arrangement of atoms in a plane and surrounding an atom followed by three atoms in the next horizontal plane. This last plane is offset from the initial plane atoms, followed by an identical planar location of atoms above this. If the first plane is A and the second B, then the repetitive arrangement of atom planes is A-B-A-B-A-B etc.

HID. See *high-interstitial defect*.

High-aluminum defect (HAD). An alpha-stabilized region containing an abnormally large amount of aluminum which may extend across a large number of beta grains. It contains an inordinate fraction of primary alpha but has a microhardness only slightly higher than the adjacent matrix. These are also known as Type II defects.

High-interstitial defect. Interstitially stabilized α-phase region in titanium of substantially higher hardness than surrounding material. It arises from very high local nitrogen or oxygen concentrations that increase the β transus and produce the high-hardness, often brittle α phase. Such a defect is often accompanied by a void resulting from thermomechanical working. Also termed Type I or low-density interstitial defects, although they are not necessarily low density.

Higher modulus alloys. Alloys in which the modulus values exceed the customary average. These are achieved by texture control and, possibly, by chemistry adjustments.

HIP. See *hot isostatic pressing*.

Hogging. Machining a part from barstock, plate, or a simple forging in which much of the original stock is removed.

Hot isostatic pressing. (1) A process for simultaneously heating and forming a powder metallurgy compact in which metal powder, contained in a sealed flexible mold, is subjected to equal pressure from all directions at a temperature high enough for sintering to take place. (2) A process similar to the one explained in (1), but applied to castings in order to close internal porosity.

Hot pressing. Forming a powder metallurgy compact at a temperature high enough to have concurrent sintering.

Hot quenching. An imprecise term used to cover a variety of quenching procedures in which a quenching medium is maintained at a prescribed temperature above 70°C (160°F).

Hot-worked structure. The structure of a material worked at a temperature higher than the recrystallization temperature.

Hot working. Deforming metal plastically at such a temperature and strain rate that recrystallization takes place simultaneously with the deformation, thus avoiding any strain hardening.

HR. See *Rockwell hardness number*.

HRA. Rockwell A hardness.

HRB. Rockwell B hardness.

HRC. Rockwell C hardness.

HV. See *Vickers hardness number*.

Hydride descaling. Descaling by action of a hydride in a fused alkali.

Hydride phase. The phase TiH_x formed in titanium when the hydrogen content exceeds the solubility limit, generally locally due to some special circumstance.

Hydrogen embrittlement. A condition of low ductility in metals resulting from the absorption of hydrogen.

I

Immersion cleaning. Cleaning where the work is immersed in a liquid solution.

Impingement attack. Corrosion associated with turbulent flow of liquid. May be accelerated by entrained gas bubbles. (See also *erosion-corrosion*.)

Impurities. Undesirable elements or compounds in a material.

Inclusion. A particle of foreign material in a metallic matrix. The particle is usually a compound (such as an oxide, sulfide, or silicate), but may be of any substance that is foreign to (and essentially insoluble in) the matrix. Inclusions are usually considered undesirable, although in some cases — such as in free-machining metals — manganese sulfides, phosphorus, selenium, or tellurium may be deliberately introduced to improve machinability.

Ingot. A casting of simple shape, suitable for hot working or remelting.

Intergranular beta. Beta phase situated between alpha grains. It may be at grain corners as in the case of equiaxed alpha-type microstructures in alloys having low beta stabilizer contents.

Intermetallic compound. A phase in an alloy system having a restricted solid solubility range. Nearly all are brittle and of stoichiometric composition.

Interstitial element. An element with a relatively small atom which can assume a position in the interstices of the titanium lattice. Common examples are oxygen, nitrogen, hydrogen, and carbon.

Interstitial solid solution. A type of solid solution that sometimes forms in alloy systems having two elements of widely different atomic sizes. Elements of small atomic size, such as carbon, hydrogen, oxygen, and nitrogen, often dissolve in solid metals to form this solid solution. The space lattice is similar to that of the pure metal, and the atoms of carbon, hydrogen, oxygen, and nitrogen occupy the spaces or interstices between the metal atoms.

Investment casting. (1) Casting metal into a mold produced by surrounding (investing) an expendable pattern with a refractory slurry that sets at room temperature after which the wax, plastic, or frozen mercury pattern is removed through the use of heat. The pattern is then fixed at high temperature. Also called precision casting or lost-wax process. (2) A part made by the investment casting process.

Investment compound. A mixture of a graded refractory filler, a binder, and a liquid vehicle, used to make molds for investment casting.

Isothermal forging. Forging of a material while maintaining it at an essentially constant forging temperature.

Isostatic pressing. A process for forming a powder metallurgy compact by applying pressure equally from all directions to metal powder contained in a sealed flexible mold. (See also *hot isostatic pressing* and *cold isostatic pressing*.)

J

J. Joules.

K

K$_{Ic}$. See *critical stress intensity factor*.

Kroll process. A process for the production of metallic titanium sponge by the reduction of titanium tetrachloride with a more active metal, such as magnesium. The sponge is further processed to granules or powder.

K$_t$. Stress-concentration factor.

L

Laser beam cutting (LBC). A cutting process which severs materials with the heat obtained from the application of a concentrated coherent light beam impinging upon the workpiece to be cut. The process can be used with or without an externally supplied gas.

Laser beam welding (LBW). A welding process which produces joining of materials with the heat obtained from the application of a concentrated coherent light beam impinging upon the members to be joined.

LBC. See *laser beam cutting*.

LBW. See *laser beam welding*.

Liquidus. In a constitution or equilibrium diagram, the locus of points representing the temperatures at which the various compositions in the system begin to freeze on cooling or finish melting on heating. (See also *solidus*.)

Longitudinal direction. Usually, the direction parallel to the direction of working in wrought alloys or the direction of crystal growth in directionally solidified or single-crystal cast alloys. Commonly, it corresponds to the direction parallel to the direction of maximum elongation in a worked material. (See also *normal direction* and *transverse direction*.)

M

Machinability. The relative ease of machining a metal.

Machining. Removing material from a metal part, usually using a cutting tool, and usually using a power-driven machine.

Macrostructure. The structure of metals as revealed by macroscopic examination of a specimen. The examination may be carried out using an as-polished or a polished and etched specimen.

Martensite. (1) The alpha product resulting from cooling from the beta phase region at rates too high to permit transformation by nucleation and growth. Martensite is supersaturated with beta stabilizer. Also called martensitic alpha. (2) A generic term for microstructures formed by diffusionless phase transformation in which the parent and product phases have a specific crystallographic relationship. Martensite is characterized by an acicular pattern in the microstructure in ferrous and nonferrous alloys. The amount of high-temperature phase that transforms to martensite upon cooling depends to a large extent on the lowest temperature attained, there being a distinct starting temperature (M_s) and a temperature at which the transformation is essentially complete (M_f), which is the martensite finish temperature. (See also *transformation temperature*.)

Martensitic. A platelike constituent having an appearance and a mechanism of formation similar to that of martensite.

Martensitic transformation. A reaction that takes place in some metals on cooling, with the formation of an acicular structure called martensite.

Master alloy. An alloy, rich in one or more desired addition elements, that is added to a melt to raise the percentage of a desired constituent.

Matrix. The constituent which forms the continuous or dominant phase of a two-phase microstructure.

Mechanical properties. The properties of a material that reveal its elastic and inelastic (plastic) behavior when force is applied, thereby indicating its suitability for mechanical (load-bearing) applications. Examples are elongation, fatigue limit, hardness, modulus of elasticity, tensile strength, and yield strength. (Compare with *physical properties*.)

Melting point. The temperature at which a pure metal, compound, or eutectic changes from solid to liquid; the temperature at which the liquid and the solid are in equilibrium.

Metastable. Refers to a state of pseudoequilibrium that has a higher free energy than the true equilibrium state.

Metastable beta. A β phase composition that can be partially or completely transformed to martensite, α, or eutectoid decomposition products with thermal or strain-energy during subsequent processing or service exposure.

M_f. The temperature at which the martensite reaction is complete.

MIG welding. See preferred terms *gas metal arc welding* and *flux cored arc welding*.

Mill forms, products. Shaped metal not produced in the design of a part or component; e.g. strip, sheet, plate, tubing, etc. (A forged disk, however, is not considered a mill product.)

M_o. The maximum temperature at which alpha double prime (orthorhombic martensite) begins to form from the beta on cooling.

Modulus. See *modulus of elasticity*.

Modulus of elasticity. A measure of rigidity or stiffness of a metal; the ratio of stress, below the proportional limit, to the corresponding strain. Specifically, the modulus obtained in tension or compression is Young's modulus (E), stretch modulus, or modulus of extensibility; the modulus obtained in torsion or shear is modulus of rigidity, shear modulus (G), or modulus of torsion; the modulus covering the ratio of the mean normal stress to the change in volume per unit volume is the bulk modulus.

Modulus of rigidity. See *modulus of elasticity*.

Modulus of rupture. Nominal stress at fracture in a bend test or torsion test. In bending, modulus of rupture is the bending moment at fracture divided by the section modulus. In torsion, modulus of rupture is the torque at fracture divided by the polar section modulus.

M_s. The maximum temperature at which the alpha prime martensite reaction begins upon cooling from the beta phase.

Multiple annealing. The process of giving two or more heat treatments to a titanium alloy to enhance ductility and toughness at the expense of modest decreases in strength. STOA results from multiple annealing of an alloy. Historically, multiple annealing has been called duplex or triplex annealing. It occurs whenever a high-temperature solution heat treatment is followed by a second or third thermal treatment. An example might be a heat treatment of Ti-6Al-4V alloy: at 954°C (1750°F) for 2 hours and water quenched; plus 593°C (1100°F) for 2 hours and air cooled; plus 704°C (1300°F) for 2 hours and air cooled.

N

NNS. See *near-net shape*.

Near-net shape (NNS). A quality of P/M and investment casting techniques.

O

OD. Outside diameter.

ODS. See *oxide dispersion strengthening*.

Omega phase. A nonequilibrium, submicroscopic phase that forms as a nucleation and growth product; often thought to be a transition phase during the formation of α from β. It occurs in metastable β alloys and can lead to severe embrittlement. It typically occurs during aging at low temperature, but can also be induced by high hydrostatic pressures.

OQ. Oil quenched.

Ordered structures. That crystal structure of a solid solution in which the atoms of different elements seek preferred lattice positions.

Overaging. Aging under conditions of time and temperature greater than those required to obtain maximum change in a certain property. (See *aging*.)

Overheating. Heating a metal or alloy to such a high temperature that its properties are impaired. When the original properties cannot be restored by further heat treating, by mechanical working, or by a combination of working and heat treating, the overheating is known as burning.

Oxidation. (1) A reaction in which there is an increase in valence resulting from a loss of electrons. (Contrast with *reduction*.) (2) A corrosion reaction in which the corroded metal forms an oxide; usually applied to reaction with a gas containing elemental oxygen, such as air.

Oxide dispersion strengthening (ODS). (1) Said of P/M alloys. (2) Alloys strengthened by the uniform dispersion of refractory oxide particles throughout the matrix.

P

P. Plate.

PA. See *prealloyed powder*.

PAW. See *plasma-arc welding*.

Passivation. The changing of a chemically active surface of a metal to a much less reactive state. (Contrast with *activation*.)

Passivity. A condition in which a piece of metal, because of an impervious covering of oxide or other compound, has a potential much more positive than when the metal is in the active state.

Pickling. Removal of the oxide film on a casting by a chemical process; pickling is sometimes used solely to show up defects.

Physical properties. Properties of a metal or alloy that are relatively insensitive to structure and can be measured without the application of force; for example, density, electrical conductivity, coefficient of thermal expansion, magnetic permeability, heat capacity, and lattice parameter. Does not include chemical reactivity. (Compare with *mechanical properties*.)

Plasma-arc welding (PAW). An arc-welding process that produces coalescence of metals by heating them with a constricted arc between an electrode and the workpiece (transferred arc) or the electrode and the constricting nozzle (nontransferred arc). Shielding is obtained from hot, ionized gas issuing from an orifice surrounding the electrode and may be supplemented by an auxiliary source of shielding gas, which may be an inert gas or a mixture of gases. Pressure may or may not be used, and filler metal may or may not be supplied.

Plastic deformation. The permanent (inelastic) distortion of metals under applied stresses.

Plate. A flat-rolled metal product of some minimum thickness and width—at times less than 610 mm (24 in.). (It is relatively thick when compared with sheet.)

Platelet alpha. Alpha phase arranged in plates, often in colonies or domains in Widmanstätten structures. Beta may or may not be present.

Platelet alpha structure. Acicular alpha of a coarser variety, usually with low aspect ratios. This microstructure arises from cooling alpha or alpha-beta alloys from temperatures at which a significant fraction of beta phase exists.

Platelets. Grains of a phase existing with essentially a two-phase shape; similar to plate.

P/M. Powder metallurgy.

Poisson's ratio. The absolute value of the ratio of the transverse strain to the corresponding axial strain in a body subjected to uniaxial stress; usually applied to elastic conditions.

Powder. Particles of a solid characterized by small size, nominally within the range of 1 to 1000 μm.

Powder lubricant. An agent mixed with or incorporated in a powder to facilitate the pressing and ejecting of a powder metallurgy compact.

Powder metallurgy (P/M). The technology and art of producing metal powders and of the utilization of metal powders for the production of massive materials and shaped objects.

Powder metallurgy forging. Plastically deforming a powder metallurgy compact or preform into a fully dense finished shape using compressive force; usually done hot, and usually within closed dies.

Powder metallurgy part. A shaped object that has been formed from metal powders and sintered by heating below the melting point of the major constituent. A structural or mechanical component made by the powder metallurgy process.

Powder production. The process by which a powder is produced, such as machining, milling, atomization, condensation, reduction, oxide decomposition, carbonyl decomposition, electrolytic deposition, or precipitation from a solution.

Powder technology. A broad term encompassing the production and utilization of both metal and nonmetal powders.

ppm. Parts per million.

Prealloyed powder. A metallic powder composed of two or more elements that are alloyed in the powder manufacturing process and in which the particles are of the same nominal composition throughout.

Precipitation. (1) Separation of a new phase from solid or liquid solution, usually with changing conditions of time, temperature, and stress. (2) The removing of a metal from a solution caused by the addition of a reagent by displacement. (3) The removal of a metal from a gas by displacement.

Precipitation hardening. Hardening caused by the precipitation of a constituent from a supersaturated solid solution. (See also *age hardening* and *aging*.)

Precipitation heat treatment. Artificial aging in which a constituent precipitates from a supersaturated solid solution.

Precision casting. A metal casting of reproducible accurate dimensions regardless of how it is made.

Precision part, precision sintered part. A powder metallurgy part that is compacted and sintered, closely conforming to specified dimensions without a need for substantial finishing.

Precision sheet. A section of flat-rolled metal in a short length. The width is considered to be 610 mm (24 in.), or greater. The thickness is less than 0.381 mm (0.015 in.), but greater than 0.127 mm (0.005 in.).

Precision strip. A section of flat-rolled metal in a short length. The width is considered to be less than 610 mm (24 in.). The thickness is less than 0.381 mm (0.015 in.), but greater than 0.127 mm (0.005 in.).

Preform. An initially pressed compact to be subjected to repressing or forging.

Preforming. (1) The initial pressing of a metal powder to form a compact that is to be subjected to a subsequent pressing operation other than coining or sizing. Also, the preliminary shaping of a refractory metal compact after presintering and before the final sintering. (2) Preliminary forming operations, especially for impression die forging.

Preheat. An early stage in the sintering procedure when, in a continuous furnace, lubricant or binder burnoff occurs without atmosphere protection prior to actual sintering in the protective atmosphere of the high heat chamber.

Premium grade. A term used to describe titanium alloys used for jet engines. Equivalent of aircraft quality, flight quality.

Premium quality. See *premium grade*.

Premix (noun). A uniform mixture of components prepared by a powder producer for direct use in compacting.

Premix (verb). A term sometimes applied to the preparation of a premix. (See preferred term mixture.)

Presintered blank. A compact sintered at a low temperature but at a long enough time to make it sufficiently strong for metal working. (See *presintering*.)

Presintering. Heating a compact to a temperature below the final sintering temperature, usually to increase the ease of handling or shaping of a compact or to remove a lubricant or binder (burnoff) prior to sintering.

Primary alpha. Alpha phase in a crystallographic structure that is retained from the last high-temperature α-β working or heat treatment. The morphology of α is influenced by the prior thermomechanical history.

Principal stress (normal). The maximum or minimum value of the normal stress at a point in a plane considered with respect to all possible orientations of the considered plane. On such principal planes the shear stress is zero. There are three principal stresses on three mutually perpendicular planes. The state of stress at a point may be: (1) uniaxial, a state of stress in which two of the three principal stresses are zero; (2) biaxial, a state of stress in which only one of the three principal stresses is zero; or (3) triaxial, a state of stress in which none of the principal stresses is zero. Multiaxial stress refers to either biaxial or triaxial stress.

Process annealing. (1) An imprecise term denoting various treatments used to improve workability. For the term to be meaningful, the condition of the material and the time-temperature cycle used must be stated. (2) A heat treatment used to soften metal for further cold working. In ferrous sheet and wire industries, heating to a temperature close to but below the lower limit of the transformation range and subsequently cooling for working. In the nonferrous industries, heating above the recrystallization temperatures at a time and temperature sufficient to permit the desired subsequent cold working.

Progressive aging. Aging by increasing the temperature in steps or continuously during the aging cycle.

Prior-beta grain size. Size of β grains established during the most recent β field excursion. Grains may be distorted by subsequent subtransus deformation. Beta grain boundaries may be obscured by a superimposed α-β microstructure and detectable only by special techniques.

Q

Quench aging. Aging induced from rapid cooling after solution heat treatment.

Quench hardening. Hardening suitable alpha-beta alloys by solution treating and quenching to develop a martensite-like structure.

Quenching. Rapid cooling. When applicable, the following more specific terms should be used: direct quenching, fog quenching, hot quenching, interrupted quenching, selective quenching, spray quenching, and time quenching.

Quench time. (1) In resistance welding, the time from the end of weld time to the beginning of temper time. (2) In heat treatment, the time from the removal of an object from a heat source until the object reaches the desired temperature.

R

RD. Rolling direction.

Reactive. Capable of interacting with other elements, most usually with gases such as oxygen or liquids.

Recrystallization. (1) Formation of new, strain-free grain structure from the structure existing in cold-worked metal. (2) A change from one crystal structure to another, such as that occurring upon heating or cooling through critical temperature.

Reducing atmosphere. A chemically active protective atmosphere which at elevated temperature will reduce metal oxides to their metallic state. (Reducing atmosphere is a relative term and such an atmosphere may be reducing to one oxide but not to another oxide.)

Reduction. (1) In cupping and deep drawing, a measure of the percentage decrease from blank diameter to cup diameter, or of diameter reduction in redraws. (2) In forging, rolling and drawing, either the ratio of the original to final cross-sectional area or the percentage decrease in cross-sectional area. (3) A reaction in which there is a decrease in valence resulting from a gain in electrons. (Contrast with *oxidation*.)

Reduction in area. (1) Commonly, the difference, expressed as a percentage of original area, between the original cross-sectional area of a tensile test specimen and the minimum cross-sectional area measured after complete separation. (2) The difference, expressed as a percentage of original area, between original cross-sectional area and that after straining the specimen.

Regrowth alpha. Alpha that grows on pre-existing alpha during cooling.

rem. Remainder.

REP atomization. Formation of powder, usually of a reactive metal, by an arc struck between a rotating consumable ingot source and a tungsten electrode in a vacuum or a low-reactivity atmosphere.

Residual stress. Stress remaining in a structure or member as a result of thermal or mechanical treatment or both. Stress arises in fusion welding primarily because the weld metal contracts on cooling from the solidus to room temperature.

Resistance brazing. Brazing by resistance heating, the joint being part of the electrical circuit.

Resistance seam welding (RSEW). A resistance welding process which produces coalescence at the faying surfaces by the heat obtained from resistance to electric current through the work parts held together under pressure by electrodes. The resulting weld is a series of overlapping resistance spot welds made progressively along a joint by rotating the electrodes.

Resistance welding. Welding with resistance heating and pressure, the working being part of the electrical circuit. Examples: resistance spot welding, resistance seam welding, projection welding, and flash butt welding.

Revert. Reclaimed Ti and/or Ti alloy scrap.

ROC. Rapid omnidirectional compaction process. (See *fluid die process*.)

Rockwell hardness number (HR). A number derived from the net increase in the depth of impression as the load on an indenter is increased from a fixed minor load to a major load and then returned to the minor load. Rockwell hardness numbers are always quoted with a scale symbol representing the penetrator, load, and dial used.

Rotor grade. Titanium alloy material approved for the most stringent rotating-part applications in gas turbine engines.

Rotor quality turnings. Titanium alloy machining scraps which are carefully segregated and controlled to prevent contamination or, subsequently, inclusions.

Rupture stress. The stress at failure. Also known as breaking stress or fracture stress.

RSR. Rapid solidification rate.

RSEW. See *resistance seam welding*.

RT. Room temperature.

S

SAW. Submerged arc welding.

Scaling. (1) Forming a thick layer of oxidation products on metals at high temperature. (2) Depositing water-insoluble constituents on a metal surface, as in cooling tubes and water boilers.

SCC. Stress-corrosion cracking. Often associated with hot-salt-induced cracking (HSSCC).

Seam. On the surface of metal, an unwelded fold or lap that appears as a crack, usually resulting from a discontinuity obtained in casting or in a wrought workpiece.

Sheet. A flat-rolled metal product of some maximum thickness and minimum width arbitrarily dependent on the type of metal. It is thinner than plate and has a width-to-thickness ratio greater than about 50.

Shot-blasting. A stream of iron shot or grit being directed onto the surface of a casting for the purpose of dislodging foreign materials not removed during removal of mold.

Shrink cavity. A defect caused by the solidification of metal. In titanium and zirconium this cavity is caused by the outside surface solidifying before the inside; as the inside hardens and shrinks, it causes a void in the metal.

Shrinkage. As molten metal solidifies during cooling it contracts, which in turn makes the finished part a little smaller than the mold.

Slab. A piece of metal, intermediate between ingot and plate, with the width at least twice the thickness.

SMAW. Shielded metal arc welding.

Solidification. The change in state from liquid to solid on cooling through the melting temperature or melting range.

Solidification shrinkage. The reduction in volume of metal from beginning to end of solidification.

Solidification shrinkage crack. A crack that forms, usually at elevated temperature, because of the internal (shrinkage) stresses that develop during solidification of a metal casting. Also termed hot crack.

Solid shrinkage. The reduction in volume of metal from the solidus to room temperature.

Solid solution. A solid crystalline phase containing two or more chemical species in concentrations that may vary between limits imposed by phase equilibrium.

Solid solution strengthening. A mechanism for strengthening the alloy by dissolved elements in solid solution.

Solidus. In a constitution or equilibrium diagram, the locus of points representing the temperatures at which various compositions finish freezing on cooling or begin to melt on heating. (See also *liquidus*.)

Solidification range. The temperature range between the liquidus and the solidus.

Solute. The component of a liquid or solid solution that is present to the lesser or minor extent; the component that is dissolved in the solvent.

Solution. A phase existing over a range of composition.

Solution heat treatment. A heat treatment in which an alloy is heated to a suitable temperature, held at that temperature long enough to cause one or more constituents to enter into solid solution, then cooled rapidly enough to hold these constituents in solution.

Solution heat treatment and overaging (STOA). A process whereby an alloy is "solutioned" high in the alpha-beta range and, after quenching from that temperature, is aged above the standard aging temperature. Used to produce incremented ductility and toughness, but little change in strength. (See *multiple annealing*.)

Spatter. The metal particles expelled during arc or gas welding. They do not form part of the weld.

SPF/DB. Superplastic forming/diffusion bonding.

Splat. Used in P/M to describe a particle coating occurring during the manufacturing process when a solid particle collides with a liquid particle. Associated with RSR in some instances.

Sponge. (1) Also called "virgin sponge," so designated because of its porous, sponge-like texture. (2) A form of metal characterized by a porous condition that is the result of the decomposition or reduction of a compound without fusion. The term is applied to one particular form of titanium.

Sponge titanium powder. Ground and sized titanium sponge. (See *Kroll process*.)

Spot welding. Welding of lapped parts in which fusion is confined to a relatively small circular area. It is generally resistance welding, but may also be gas tungsten-arc, gas metal-arc, or submerged-arc welding.

Springback. (1) The elastic recovery of metal after cold forming. (2) The degree to which metal tends to return to its original shape or contour after undergoing a forming operation. (3) In flash, upset, or pressure welding, the deflection in the welding machine caused by the upset pressure.

ST. See *solution heat treatment*.

STA. Solution heat treated and aged.

std. Standard.

STDA. Solution treated and double aged.

STOA. See solution heat treated and overaged.

Stopoff. A material used on the surfaces adjacent to the joint to limit the spread of soldering or brazing filler metal.

STQ. Solution treated and quenched.

Strain. A measure of the relative change in the size or shape of a body. Linear strain is the change per unit length of a linear dimension. True strain (or natural strain) is the natural logarithm of the ratio of the length at the moment of observation to the original gage length. Conventional strain is the linear strain over the original gage length. Shearing strain (or shear strain) is the change in angle (expressed in radians) between two lines originally at right angles. When the term "strain" is used alone, it usually refers to the linear strain in the direction of applied stress.

Strain-age embrittlement. A loss in ductility accompanied by an increase in hardness and strength that occurs with some alloys aged, following plastic deformation. The degree of embrittlement is a function of aging time and temperature, occurring in a matter of minutes at higher temperatures but requiring a few hours to year at room temperature. (Not normally a great concern in nonferrous alloys.)

Strain hardening. An increase in hardness and strength caused by plastic deformation at temperatures below the recrystallization range.

Strain rate. The time rate of straining for the usual tensile test. Strain as measured directly on the specimen gage length is used for determining strain rate. Because strain is dimensionless, the units of strain rate are reciprocal time.

Stress. The intensity of the internally distributed forces or components of forces that resist a change in the volume or shape of a material that is or has been subjected to external forces. Stress is expressed in force per unit area and is calculated on the basis of the original dimensions of the cross section of the specimen. Stress can be either direct (tension or compression) or shear. Usually expressed in pounds per square inch (psi) or megapascals (MPa).

Stress-corrosion cracking. Failure of metals by cracking under combined action of corrosion and stress, residual or applied. In brazing, the term applies to the cracking of stressed base metal due to the presence of a liquid filler metal.

Stress-relief heat treatment. Uniform heating of a structure or a portion thereof to a sufficient temperature to relieve the major portion of the residual stresses, followed by uniform cooling.

Stress-relief cracking. Intergranular cracking in the heat-affected zone or weld metal that occurs during the exposure of weldments to elevated temperatures during postweld heat treatment or high temperature service.

Stress relieving. Heating to a suitable temperature, holding long enough to reduce residual stresses, and then cooling slowly enough to minimize the development of new residual stresses.

Stress-rupture test. A method of evaluating elevated-temperature durability in which a tension-test specimen is stressed under constant load until it breaks. Data recorded commonly include: initial stress, time to rupture, initial extension, creep extension, reduction of area at fracture. Also known as creep-rupture test.

Striation. A fatigue fracture feature often observed in electron micrographs that indicates the position of the crack front after each succeeding cycle of stress. The distance between striations indicates the advance of the crack front across that crystal during one stress cycle, and a line normal to the striation indicates the direction of local crack propagation.

Substrate. The layer of metal underlying a coating, regardless of whether the layer is base metal.

Subsurface corrosion. Formation of isolated particles of corrosion products beneath a metal surface. This results from the preferential reaction of certain alloy constituents by inward diffusion of oxygen, nitrogen or sulfur.

Superalpha. Refers to titanium alloys in the α-β region but which have chemistries very close to the α region and which contain a very large amount of α.

Surface hardening. A generic term covering several processes applicable to a suitable ferrous alloy that produces, by quench hardening only, a surface layer that is harder or more wear resistant than the core. There is no significant alteration of the chemical composition of the surface layer. The processes commonly used are induction hardening, flame hardening, and shell hardening. Use of the applicable specific process name is preferred.

Surface stability. Ability of a surface to resist interaction with an atmosphere either at low or high temperatures.

T

TA. See *multiple annealing*.

TCP. Topologically close-packed.

TCR. Temperature coefficient of resistance.

TD. Transverse direction.

Temper. In nonferrous alloys the hardness and strength produced by mechanical or thermal treatment, or both, and characterized by a certain structure, mechanical properties, or reduction in area during cold working.

Tensile strength. In tensile testing, the ratio of maximum load to original cross-sectional area. Also called ultimate strength. (Compare with *yield strength*.)

Tension. The force or load that produces elongation.

Tension testing. A method of determining the behavior of materials subjected to uniaxial loading, which tends to stretch the metal. A longitudinal specimen of known length and diameter is gripped at both ends and stretched at a slow, controlled rate until rupture occurs. Also known as tensile testing.

Tickle. Slang for Ti-Cl$_4$.

Toll melting. Contracting with a company to melt a metal.

Transformed beta. A local or continuous structure comprised of decomposition products arising by nucleation and growth from beta; typically consists of alpha platelets which may or may not be separated by beta phase.

Transverse direction. Literally "across." Usually signifying a direction or plane perpendicular to the direction of working. In rolled plate or sheet, the direction across the width is often called long transverse, and the direction through the thickness, short transverse. (See also *longitudinal direction* and *normal direction*.)

Triplex annealing. See *multiple annealing*.

TS. Tensile strength.

U

Ultimate strength. The maximum stress (tensile, compressive or shear) a material can sustain without fracture, determined by dividing maximum load by the original cross-sectional area of the specimen. Also known as nominal strength or maximum strength.

Ultrasonic welding (USW). A solid state process in which materials are welded by locally applying high-frequency vibratory energy to a joint held together under pressure.

Unit cell. A unit of atoms arranged such that its repetitive occurrence constitutes a grain of metal.

UNS. Unified Numbering System.

Upset. (1) The localized increase in cross-sectional area of a workpiece or weldment resulting from the application of pressure during mechanical fabrication or welding. (2) Bulk deformation resulting from the application of pressure in welding. The upset may be measured as a percent increase in interfacial area, a reduction in length or a percent reduction in thickness (for lap joints).

Upset forging. A forging obtained by upset of a suitable length of bar, billet, or bloom.

Upsetting. Working metal so that the cross-sectional area of a portion or all of the stock is increased.

Upset weld. A weld made by upset welding.

Upset welding (UW). A resistance welding process which produces coalescence simultaneously over the entire area of abutting surfaces or progressively along a joint by the heat obtained from resistance to electric current through the area where those surfaces are in contact. Pressure is applied before heating is started and is maintained throughout the heating period.

USW. See *ultrasonic welding*.

UW. See *upset welding*.

V

Vacuum arc remelting (VAR). A consumable electrode remelting process in which heat is generated by an electric arc between the electrode and the ingot. The process is performed inside a vacuum chamber. Exposure of the droplets of molten metal to the reduced pressure reduces the amount of dissolved gas in the metal.

Vacuum induction melting (VIM). A process for remelting and refining metals in which the metal is melted inside a vacuum chamber by induction heating. The metal may be melted in a crucible, then poured into a mold. The process may also be operated in a configuration similar to that used in consumable electrode remelting except that the heat is supplied by an induction heating coil rather than from the passage of electric current through the electrode.

Vacuum melting. Melting in a vacuum to prevent contamination from air, as well as to remove gases already dissolved in the metal; the solidification may also be carried out in a vacuum or at low pressure.

VAR. See *vacuum arc remelting*.

VHP. Vacuum hot processing.

Vickers hardness number (HV). A number related to the applied load and the surface area of the permanent impression made by a square-based pyramidal diamond indenter having included face angles of 136°.

Vickers hardness test. An indentation hardness test employing a 136° diamond pyramid indenter (Vickers) and variable loads, enabling the use of one hardness scale for all ranges of hardness—from very soft lead to tungsten carbide. Also known as diamond pyramid hardness test.

VIM. See *vacuum induction melting*.

W

Weldability. A specific or relative measure of the ability of a material to be welded under a given set of conditions. Implicit in this definition is the ability of the completed weldment to fulfill all service designed into the part.

Widmanstätten structure. A structure characterized by a geometrical pattern resulting from the formation of a new phase along certain crystallographic planes of the parent solid solution. The orientation of the lattice in the new phase is related crystallographically to the orientation of the lattice in the parent phase. The structure was originally observed in meteorites, but is readily produced in many alloys, such as titanium, by appropriate heat treatment.

Wire. A thin, flexible, continuous length of metal, usually of circular cross section, and usually produced by drawing through a die.

Work hardening. Same as strain hardening.

WQ. Water quenched.

Wrought. A metal or alloy which has been deformed plastically one or more times and which exhibits little or no evidence of cast structure. (See *cast*.)

wt. Weight.

Y

Yield. (1) As a noun: the weight of a finished casting divided by the total weight of metal needed to produce it, including running systems, etc. (2) As a verb: to deform plastically on the first instance in a tensile or compressive test. (See *yield point*.)

Yield point. The first stress in a material, usually less than the maximum attainable stress, at which an increase in strain occurs without an increase in stress. Only certain metals exhibit a yield point. If there is a decrease in stress after yielding, a distinction may be made between upper and lower yield points.

Yield strength. The stress at which a material exhibits a specified deviation from proportionality of stress and strain. An offset of 0.2% is used for many metals. (Compare with *tensile strength*.)

YS. Yield strength.

Appendix C

Machining Data

APPENDIX SUMMARY

The paragraphs which follow contain highly specific data related to a number of titanium machining operations. The topics include:

- Sawing
- Turning
- Drilling
- Reaming
- Tapping
- Broaching

- Face milling
- End milling: slotting
- End milling: peripheral
- Surface grinding
- Thermal cutting

(Refer to Section 6, "Machining," earlier in this volume for a general overview of the topic.)

SPECIFIC DATA LISTINGS

Data contained in these tables must be considered as typical and representative. Although all specifics were compiled from widespread experience by two major producers, the data may be seen only as a starting point both for individual operating conditions and for a given manufacturer's particular grade of titanium. Fabricators must begin building a documented database for each type of machining operation encountered. Work closely with suppliers, and request more exact data from them.

(Some of the following material is reprinted with the generous permission of **RMI Company**. Other material was contributed by **Industrial Titanium Corporation**. The data represent a synthesis of experiences and recommendations on the topic of machining titanium and its alloys.)

SAWING

Titanium may be sawed using abrasive cutting wheels, hacksaws, or bandsaws. Chip formation is the most important gage in determining the success of any sawing operation. The formation of nearly invisible, flake-like chips can cause early blade failure. Fortunately, preventive measures are available to control the formation of these chips. Liberal use of proper coolants, ones providing good lubricity and anti-chip-weld characteristics, is recommended. In addition, the use of special brush attachments on cutting apparatus will help prevent accumulation of chips as the work passes through the saw.

ABRASIVE CUTTING

Titanium is easily sawed with abrasive wheels when proper wheel compositions are employed and the work is flooded with coolant.

Abrasive grains should fracture to expose fresh, sharp cutting edges. The wheel should be of the proper hardness to prevent rapid wear (too soft) or loading up (too hard).

Cutting rates will probably vary from 6.5 to 19.5 cm^2/min (1 to 3 in.^{2}/min).

Wheels

Only silicon carbide wheels are recommended for titanium cutout operations. Aluminum oxide wheels are not satisfactory. Table C.1 lists recommended wheel types.

Manufacturers provide extensive technical data, which should be requested before initiating the kind of operations described here. Table C.2 is a typical example. Note that it varies in small details with Table C.1, thereby indicating that a close understanding with a supplier is necessary.

Table C.1 Recommended cutoff wheel material characteristics; Ti abrasive cutting

WHEEL CHARACTERISTICS

Abrasive material	Silicon carbide
Wheel width, in.	1/8 to 3/16
Abrasive type	37C(a)
Abrasive size	60
Wheel hardness	L
Structure	6
Bond	Rubber

(a)Norton Code

Machine Setting	Bar Diameter	
	Up to 3.00″	Over 3.00″
Feed, sq. in/min.	2-4	5-6
Speed, sfpm	7000-12,000	6000-7000
Cutting motion	Oscillating wheel	Oscillating wheel & work rotation
Coolant:	1. 10% water solution of rust inhibitor (Nitrite-amine types)	
	2. 10% water solution of soluble oil	

Table courtesy of Industrial Titanium Corp.

Table C.2 Hack sawing, band sawing data (a)

MATERIAL	Brinell Hardness Number	Condition	POWER HACK SAW — Pitch—Teeth Per Inch Minimum Thickness of Material—Inches				Speed Strokes/ Minute	Feed Inches/ Stroke	POWER BAND SAW — Pitch—Teeth Per Inch Minimum Thickness of Material—Inches			Speed (fpm)
			¼ under	¼-¾	¾-2	2 Over			¼ under	¼-1½	1½ over	
Grade 1	110-170	Annealed	10	6	6	4	180	.009	8-10	6-8	3-6	200
Grade 2 Grade 3 Ti-Pd	140-200	Annealed	10	6	6	4	150	.009	8-10	6-8	3-6	175
Grade 4 3Al-2.5V	200-275	Annealed	10	6	6	4	120	.009	8-10	6-8	3-6	130
5Al-2.5Sn 5Al-2.5Sn ELI 6Al-2Cb-1Ta-1Mo 4Al-3Mo-1V	300-340	Annealed	10	6	6	4	70	.006	8-10	6-8	3-6	100
6Al-4V 6Al-4V ELI 8Mn	310-350	Annealed	10	6	6	4	60	.006	8-10	6-8	3-6	90
7Al-4Mo 8Al-1Mo-1V 6Al-6V-2Sn	320-370	Annealed	10	6	6	4	50	.005	8-10	6-8	3-6	80
1Al-8V-5Fe	320-375	Annealed	10	6	6	4	30	.003	8-10	6-8	3-6	70
6Al-4V	350-375	Solution Treated & Aged	10	6	6	4	30	.003	8-10	6-8	3-6	70
13V-11Cr-3Al	310-350	Solution Treated	10	6	6	4	25	.003	8-10	6-8	3-6	60

(a) Use of a high-speed steel blade is assumed.

Table courtesy of RMI Company

Data presented in Table C.2 are based on average conditions, and are intended only as typical starting points. Higher titanium speeds and feeds are being realized in plants using high-power tools and advanced technology methods.

Setup

Machines with cutting heads capable of oscillating and plunging motions are best. For work with a cross-sectional area greater than 7.62 cm (3 in.), work rotation is necessary to minimize wheel wear, wheel breakage, and workpiece burning.

Work rotation and use of oscillating heads minimize the amount of titanium surface being cut, thereby greatly reducing the tendency for the cutoff wheel to load up.

Generous amounts of a water solution of nitrite rust inhibitor must be applied to the work during cutting. This is necessary to keep the cutting temperatures to a minimum, and to avoid burning or heat checking the titanium.

Although a 10% solution of nitrite rust inhibitor in water is recommended, soluble oils may be used if necessary. However, some soluble oils have a tendency to foam and lose a good part of their cooling power.

Hacksawing

Hacksawing of titanium metal is basically a roughing operation. Problems encountered from galling, smearing, and high temperatures generated at the cutting edge of the teeth have, however, largely been overcome.

Heavier, more rugged machine designs (for example, the Marvel Heavy-duty Number 6 or 9 machine), blades and full recognition of the value of the proper coolant — sulfo-chlorinated oil — have doubled blade life in the last five years.

As in most titanium machining operations, surface speed must be kept low, and continual positive feed rates are to be employed. Coarse-pitched (3, 4, or 6 teeth/in.), high-speed steel blades have proved most successful. Use 3 or 4 teeth/in. on solid stock, and 6 teeth/in. for tubing and shapes. Blade tension should be maintained at about 20 kN (4500 lbf).

Successful blade characteristics are shown in Table C.3.

Table C.3 Successful blade characteristics

Teeth per in., number	3, 4 and 6
Blade width, in.	1½ (up to 10″ capacity machine)
	2 (over 10″ capacity machine)
Blade thickness, in.	0.072-0.075 (Up to 10″ capacity machine)
	0.100 (over 10″ capacity machine)

Table courtesy of Industrial Titanium Corp.

The positive feed mechanism should govern the feed rate. The friction or variable feed should be kept medium.

On alloy and heat-treatable grades, cutting rates of 6.5 to 13 cm^2/min (1 to 2 in.^{2}/min) can be obtained. Total area cut before a blade change is required will be about 1300 to 1936 cm^2 (200 to 300 in.^{2}). On commercially pure titanium grades, blade life and cutting rate is increased 50 to 100%. (See Table C.4.)

Table C.4 Operational information

Machine Setting	Work Size	Commercially-pure	All Allow Grades	
			Annealed	Heat-Treated
Speed, strokes/min	—	90-100	60.90	30-60
Feed, inches/stroke	4-6	0.012	0.009	0.009
" " "	6-8	0.009	0.006	0.006
" " "	8-10	0.006	0.003	0.003
" " "	10 & Over	0.003	0.003	0.003

Table courtesy of Industrial Titanium Corp.

Forging Skin Caution

Initial cutting should never be attempted on scale or forging skin since a forging skin thickness of less than 0.025 mm (0.001 in.) may completely ruin a blade in a few strokes of the sawframe. Forging skin is best removed by turning or pickling. Satisfactory results have also been obtained by use of portable grinders to remove the skin.

Some shops have used the expedient of trying an old blade to get under the skin, and then substituting a new blade to complete the work. In general, this approach is never recommended since the danger of work-hardening is ever-present.

Bandsawing

Bandsawing of titanium — on horizontal or vertical machines — is generally a freer operation than bandsawing heat-resistant materials such as stainless steel.

Further, if equipment is available, the bandsaw offers certain advantages over the hacksaw in cut-off operations on commercially pure grades of titanium.

Four basic recommendations apply to bandsawing operations, regardless of whether vertical or horizontal equipment is used. These are:

- Work thickness determines the number of blade teeth/mm (teeth/in.): 3 to 10 for inches. In general, the thicker the workpiece, the less the number of teeth recommended.

- The blade should be eased into the work before the feed is fully increased. This will provide the greatest accuracy of cuts and prolong blade life.

- Positive feeds must be maintained once the work has been started.

- Use of coolant is required.

Saws employed for cutting stock up to 12.7 cm (5 in.) of cross section (for example, the Do All Company's C-57 or C-58 automatic indexing model) have a blade life of 3900 to 6500 cm^2 (600 to 1000 in.^{2}). On equipment employed for larger stock (for example, the Do All C-24), the blade life is estimated at 6500 to 11 700 cm^2 (1000 to 1800 in.^{2}).

Horizontal Bandsawing Machines (Cut-off Sawing)

Pitch selection is determined by the work thickness involved. As a general rule, the greater the work thickness, the coarser the pitch.

For example, for work thicknesses ranging from 10 to 15 cm (4 to 6 in.), coarse-pitched (6 teeth/in.), high-speed steel blades 2.5 cm (1 in.) wide, employed at speeds of 24 to 27 m^2/min (80 to 90 ft^2/min) have yielded the best results.

Optimum cutting rates, providing a balance of good tool life and quality of cut have been established at 6.5 sq cm/min (1 sq in./min). With rigid setups, thin slabs are easily cut from barstock with an accuracy within 0.05 mm (0.002 in.) thickness variation per mm (in.) of work area.

Effective feed force ranges from medium to medium-heavy.

A summary of blade characteristics found effective for bandsawing either commercially pure or alloy grades of titanium is shown in Table C.5.

Table C.5 Recommended blade characteristics

Blades	High Speed Steel
Saw Gage	0.042″
Saw Set	0.065″ raker
Pitch, teeth per inch	6
Width	1″
Velocity, surface feet/min	80
Lubricant	Sulfochlorinated Oil
Feed Force	Medium to Medium-Heavy
Cutting Rate, sq. in/min	1.00

Table courtesy of Industrial Titanium Corp.

Reference also Table C.2 for typical manufacturer's data.

Vertical Bandsawing Machines

Vertical bandsawing machines are employed for either cut-off or contour operations.

As in horizontal bandsawing, coarse-pitched (6 teeth/in.) high-speed steel blades are employed at speed of about 24 to 27 m^2/min (80 to 90 ft^2/min). Blade width is 25.4 mm (1 in.) and the thickness is 0.889 mm (0.035 in.). Medium to medium-heavy feeding force is used. (Metric here is the approximate equivalent.)

Blade life is estimated at about 3870 to 6452 cm^2 (600 to 1000 in^2.).

Equipment is constantly being improved, and a wide variety of refinements is available.

Contour Sawing

Conventional contour sawing of titanium is an effective method for cutting shapes of all kinds—including bevelling, recessing, etc.

Fine-pitched (10 teeth/in.), high-speed steel bands have met with success in shaping material up to 19.1 mm (0.750 in.) thick. Blades of 6 to 8 teeth/in. are generally advised for material from 19.1 to 101.6 mm (3/4 to 4 in.) thick. Heavier work requires 2, 3, and 4 pitch blades.

Medium-to-heavy feeding force is required with saw velocity geared to about 24 to 27 m^2/min (80 to 90 ft^2/min). Life expectancy of vertical band saw blades is about 3870 to 6452 cm^2 (600 to 1000 in.^{2}).

TURNING

Commercially pure and alloyed titanium can be turned with little difficulty. Carbide tools of the throwaway type should be used, wherever possible, for turning and boring titanium and titanium alloys, because of the higher production rates attainable with them.

The "straight" tungsten, carbide grades of standard designations C1-C4, such as Metal Carbides C-91 and similar types, give the best results. Cobalt-type, high-speed steels appear to be the best of the many types of high-speed steel available. Cast-alloy tools may be used when carbide is not available and when the cheaper high-speed steels are not satisfactory.

Overhang should be kept to a minimum in all cases to avoid deflection, thereby reducing the tendency of titanium to smear on the tool flank.

A heavy stream of cutting fluid should be applied constantly to the tool. Live centers must be used in turning, since seizing occurs on dead centers.

Manufacturers provide extensive technical data, which should be requested before initiating the kind of operations described here. Data presented in Table C.6 are based on average conditions, and are intended only as typical starting points. (Higher titanium speeds and feeds are being realized in plants using high-power tools and advanced technology methods.)

DRILLING

Successful drilling of titanium and its alloys requires the use of low surface speeds, heavy cuts (with controlled down feed if available), sharp tools of the correct geometry, rigid setups, and liberal amounts of coolant.

These procedures reduce the amount of heat generated during cutting operations, thereby providing protection against dulling of the drill and galling and smearing on the cutting faces and margins.

The major factors in successful titanium drilling are: chip flow, clogging, and point smearing. These factors are determined by the depth of the hole drilled which, in turn, determines drill life.

Table C.6 Single point and box tools (a)

MATERIAL	Condition	Brinell Hardness Number	Depth of Cut (Inches)	HIGH SPEED STEEL TOOL				CARBIDE TOOL				
				Speed (fpm)	Feed (ipr)	Tool Matl.	Tool* Geometry	SPEED (fpm)		Feed (ipr)	Tool Matl.	Tool* Geometry
								Brazed Tool	Throw-Away Tool			
Grade 1	Annealed	110-170	.250	175	.015	T-15, or M-3	SR:5° BR:0° SCEA:15° ECEA:15° Relief: 5° NR: .030"	400	450	.015	C-2	SR:-5° BR:-5°
			.100	200	.010			450	500	.010	C-2	SR:-5° BR:-5°
			.025	250	.005			500	550	.005	C-3	SR:-5° BR:-0° SCEA:15°, ECEA:15° Relief:5°, NR:.030"
Grade 2 Grade 3 Ti-Pd	Annealed	140-200	.250	140	.015	T-15, or M-3	SR:5° BR:0° SCEA:15° ECEA:15° Relief: 5° NR: .030"	325	375	.015	C-2	SR:-5° BR:-5°
			.100	160	.010			375	425	.010	C-2	SR:-5° BR:-5°
			.025	180	.005			425	475	.005	C-3	SR:5° BR:0° SCEA:15°, ECEA:15° Relief:5°, NR:.030"
Grade 4	Annealed	200-275	.250	90	.015	T-15, or M-3	SR:5° BR:0° SCEA: 15° ECEA: 15° Relief: 5° NR: .030"	225	275	.015	C-2	SR:-5° BR:-5°
			.100	100	.010			250	310	.010	C-2	SR:-5° BR:-5°
			.025	110	.005			275	350	.005	C-3	SR:5° BR:0° SCEA:15°, ECEA:15° Relief:5°, NR:.030"

(Continued)

Table C.6 (Continued)

MATERIAL	Condition	Brinell Hardness Number	Depth of Cut (Inches)	HIGH SPEED STEEL TOOL				CARBIDE TOOL				
				Speed (fpm)	Feed (ipr)	Tool Matl.	Tool* Geometry	SPEED (fpm) Brazed Tool	SPEED (fpm) Throw-Away Tool	Feed (ipr)	Tool Matl.	Tool* Geometry
5Al-2.5Sn 5Al-2.5Sn ELI** 6Al-2Cb-1Ta-1Mo 4Al-3Mo-1V	Annealed	300-340	.250 .100 .025	60 70 80	.015 .010 .005	T-15, or M-3	SR:5° BR:0° SCEA:15° ECEA:15° Relief:5° NR:.030"	150 180 215	185 220 250	.015 .010 .005	C-2 C-2 C-3	SR:-5° BR:-5° SR:-5° BR:-5° SR:5° BR:0° SCEA:15°, ECEA:15° Relief:5°, NR:.030"
6Al-4V 6Al-4V ELI**	Annealed	310-350	.250 .100 .025	50 60 70	.015 .010 .005	T-15, or M-3	SR:5° BR:0° SCEA:15° ECEA:15° Relief:5° NR:.030"	125 150 170	160 180 210	.015 .010 .005	C-2 C-2 C-3	SR:-5° BR:-5° SR:-5° BR:-5° SR:5° BR:0° SCEA:15°, ECEA:15° Relief:5°, NR:.030"
7Al-4Mo 8Al-1Mo-1V 6Al-6V-2Sn	Annealed	320-370	.250 .100 .025	40 50 60	.015 .010 .005	T-15, or M-3	SR:5° BR:0° SCEA:15° ECEA:15° Relief:5° NR:.030"	110 130 155	150 165 185	.015 .010 .005	C-2 C-2 C-3	SR:-5° BR:-5° SR:-5° BR:-5° SR:5° BR:0° SCEA:15°, ECEA:15° Relief:5°, NR:.030"
1Al-8V-5Fe	Annealed	320-380	.250 .100 .025	15 20 30	.015 .010 .005	T-15, or M-3	SR:5° BR:0° SCEA:15° ECEA:15° Relief:5° NR:.030"	70 90 115	90 110 135	.015 .010 .005	C-2 C-2 C-3	SR:-5° BR:-5° SR:-5° BR:-5° SR:5° BR:0° SCEA:15°, ECEA:15° Relief:5°, NR:.030"
6Al-4V	Solution Treated and Aged	350-400	.250 .100 .025	45 55 65	.015 .010 .005	T-15, or M-3	SR:5° BR:0° SCEA:15° ECEA:15° Relief:5° NR:.030"	100 120 145	140 160 185	.015 .010 .005	C-2 C-2 C-3	SR:-5° BR:-5° SR:-5° BR:-5° SR:5° BR:0° SCEA:15°, ECEA:15° Relief:5°, NR:.030"

(Continued)

Table C.6 (Continued)

MATERIAL	Condition	Brinell Hardness Number	Depth of Cut (Inches)	HIGH SPEED STEEL TOOL				CARBIDE TOOL				
				Speed (fpm)	Feed (ipr)	Tool Matl.	Tool* Geometry	SPEED (fpm) Brazed Tool	SPEED (fpm) Throw-Away Tool	Feed (ipr)	Tool Matl.	Tool* Geometry
6Aʟ-6V-2Sɴ 7Aʟ-4Mo 4Aʟ-3Mo-1V	Solution Treated and Aged	375-420	.250 .100 .025	30 40 50	.010 .010 .005	T-15	SR:15° BR:0° SCEA:15° ECEA:15° Relief:5° NR: .030"	80 100 120	100 120 150	.010 .010 .005	C-2 C-2 C-3	SR:-5° BR:-5° SR:-5° BR:-5° SR:5° BR:0° SCEA:15°, ECEA:15° Relief:5°, NR:.030"
1Aʟ-8V-5Fᴇ	Solution Treated and Aged	375-440	.250 .100 .025	20 25 35	.010 .010 .005	T-15	SR:15° BR:0° SCEA:45° ECEA:10° Relief:5° NR: .030"	60 80 100	80 100 120	.010 .010 .005	C-2 C-2 C-3	SR:-5° BR:-5° SR:-5° BR:-5° SR:5° BR:0° SCEA:15°, ECEA:15° Relief:5°, NR:.030"
13V-11Cʀ-3Aʟ	Solution Annealed	310-350	.250 .100 .025	20 25 35	.015 .010 .005	T-15, or M-3	SR:5° BR:0° SCEA:15° ECEA:15° Relief:5° NR: .030"	80 100 125	100 120 150	.015 .010 .005	C-2 C-2 C-3	SR:-5° BR:-5° SR:-5° BR:-5° SR:5° BR:0° SCEA:15°, ECEA:15° Relief:5°, NR:.030"
13V-11Cʀ-3Aʟ	Solution Treated and Aged	375-440	.250 .100 .025	20 25 35	.010 .010 .005	T-15	SR:15° BR:0° SCEA:45° ECEA:10° Relief:5° NR: .030"	60 80 100	80 100 120	.010 .010 .005	C-2 C-2 C-3	SR:-5° BR:-5° SR:-5° BR:-5° SR:5° BR:0° SCEA:15°, ECEA:15° Relief:5°, NR:.030"

(a) Nominal Tool Life to be Expected for the Recommended Turning Conditions.
High Speed Steel or Brazed Carbide Tool Bits—60 min.
Throwaway Carbide Inserts —30 min.

(b) When 15° SCEA and ECEA are specified, it is assumed that throwaway tooling providing this geometry can be applied.

Table courtesy of RMI Company

Successful drilling can be accomplished with ordinary high-speed steel drills. One of the most important factors in drilling titanium is the length of the unsupported section of the drill. This portion of the drill should be no longer than necessary to drill the required depth, yet still allow the chips to flow unhampered through the flutes and out of the hole. This permits application of maximum cutting pressure, as well as rapid removal and re-engagement to clear chips, without drill breakage.

Chip Flow

It is very important that drills be kept sharp and clean. (Titanium alloys may be difficult to drill unless correct cutting conditions are employed.) A dull drill, or one with smeared lips and margins, impedes the flow of chips along the flutes.

A freshly ground drill will produce good chips; no drilling difficulty will be apparent. As the drill starts to become dull, chips flow with increasing difficulty, and some titanium remains stuck to the drill lips and margins. Soon the chips become packed and wedged in the flutes. The drill will fail, either by overheating or by seizing in the hole.

Careful attention to chip formation in drilling of titanium should provide an accurate gage of the condition of the drill employed.

Uniform, smooth chips are produced by a sharp drill. When chips become shirred or feathered, the drill has dulled. When chips become discolored and irregular, the tool has failed. If the hole being drilled is to be tapped subsequently, generation of discolored chips may mean the workpiece has been ruined.

Recommended Drilling Tools

Cobalt (T-4 or T-5) or molybdenum (M-10) high-speed steel drills have been found suitable for most drilling operations, while carbide drills seem especially useful for deep hole drilling on long production runs.

Stub-type screw machine drills or the shortest drills with shortest possible length of flute should be used. If the required hole depth makes their use impractical, drill jigs should be used to align the drill and prevent deflection. Without this protection, drills have a tendency to drift or break off in the hole. Increasing the web thickness will also increase the rigidity of the drill and help prevent deflection that causes chatter and uneven cutting action.

Flutes should be large enough to prevent chips from clogging the drill. Drills should be long enough to provide unrestricted chip flow through the hole.

All drill angles should be machine ground to assure correct tool geometry. (Hand grinding is not recommended.)

Point angle is dependent on the hole diameter and operation involved (deep holes, sheet drilling, etc.). Blunt points (140°) are better suited to smaller diameter holes (6.35 mm [1/4 in.] and less), while sharp points (90°) are more suited to large-diameter holes where higher pressures are employed to feed the drill. Double-point angle drills are recommended over single-point angle drills for larger size holes. The increased cutting surface on the double-point drill, which thins the chip and distributes the cutting load over a greater area of the cutting lip, provides approximately 25% greater drill life over single-point angle drills.

Relief angles are important to drill life. Too small a relief angle causes smearing and galling on the cutting lips, while too high an angle causes tool failure by chipping on the cutting edge.

Sheet drilling is best accomplished by a drill with no body clearance. The same general rules apply to the drilling of sheet as deep holes; i.e., use drills only long enough to accomplish the required job, and use positive feeds. In order to reduce the required pressure to keep the drill cutting, it is sometimes wise to clear the web to the center leaving no chisel point.

Manufacturers provide extensive technical data, which should be requested before initiating the kind of operations described here. Data presented in Table C.7 are based on average conditions, and are intended only as typical starting points. (Higher titanium speeds and feeds are being realized in plants using high-power tools and advanced technology methods.)

Table C.7 arranges drilling data according to specific titanium grades and/or alloys. A more general approach, one organized according to operations, is given in Table C.8.

Spiral-Point Drills

Grinding of spiral points on drills produces a tool with a marked superiority for most titanium drilling operations over the conventional chisel edge. Equipment is available for grinding the spiral point on any two-fluted, high-speed steel drill up to 50.8 mm (2 in.) diameter.

Companies employing spiral-point drills report these type drills reduce the large negative rake angle of the chisel-edge drill; provide a proper clearance angle along the entire surface of the cutting edge; reduce thrust loading about 30%; and — because the spiral point terminates at its center in a sharp point — automatically center themselves on the axis of the drill when first engaging the workpiece. Use of spiral point geometry has also made it largely unnecessary for employment of guide bushings to maintain proper location of a hole.

When spiral point drilling sheet, 180° point angles are recommended; the standard 118° angle is used for round or bar stock.

Drills should be either chrome plated or oxide coated to resist galling and smearing on the drill margin.

Setup

As in all machining operations, rigid machines and holding fixtures are a necessity when drilling titanium. In production operations, the use of drill jigs is recommended to prolong drill life and to prevent out-of-tolerance holes.

When sheets are stack-drilled they should be clamped firmly together for optimum results.

Table C.7 Drilling Data (a)

MATERIAL	Condition	Brinell Hardness Number	Speed (fpm)	FEED—INCHES/REVOLUTION NOMINAL HOLE DIAMETER—INCHES								Tool Material	Tool Geometry
				⅛	¼	½	¾	1	1½	2	3		
Grade 1	Annealed	110-170	100	.0005	.002	.006	.007	.008	.010	.013	.015	M-1, M-10 M-2	Stub Length Drill PA: 118° Lip Cl: 7° PG: Crankshaft
Grade 2 Grade 3 Ti-Pd	Annealed	140-200	80	.0008	.003	.006	.007	.008	.010	.013	.015	M-1, M-10 M-2	Stub Length Drill PA: 118° Lip Cl: 7° PG: Crankshaft
Grade 4	Annealed	200-275	50	.002	.005	.006	.007	.008	.010	.013	.015	M-1, M-10 M-2	Stub Length Drill PA: 118° Lip Cl: 7° PG: Crankshaft
3Al-2.5V	Annealed	200-260	50	.002	.005	.006	.008	.009	.010	.012	.013	M-1, M-10 M-2	Stub Length Drill PA: 118° Lip Cl: 7° PG: Crankshaft
5Al-2.5Sn 5Al-2.5Sn ELI 6Al-2Cb-1Ta-1Mo 4Al-3Mo-1V	Annealed	300-340	40	.002	.005	.006	.007	.008	.010	.011	.012	M-1, M-10 M-2	Stub Length Drill PA: 118° Lip Cl: 7° PG: Crankshaft

(Continued)

Table C.7 (Continued)

MATERIAL	Condition	Brinell Hardness Number	Speed (fpm)	FEED—INCHES/REVOLUTION NOMINAL HOLE DIAMETER—INCHES								Tool Material	Tool Geometry
				⅛	¼	½	¾	1	1½	2	3		
6Aʟ-4V 8Mɴ	Annealed	310-350	30	.002	.005	.006	.007	.008	.009	.010	.011	M-1, M-10 M-2	Stub Length Drill PA: 118° Lip Cl: 7° PG: Crankshaft
7Aʟ-4Mo 8Aʟ-1Mo-1V 6Aʟ-6V-2Sɴ	Annealed	320-370	20	.002	.005	.006	.007	.008	.009	.010	.011	M-1, M-10 M-2	Stub Length Drill PA: 118° Lip Cl: 7° PG: Crankshaft
1Aʟ-8V-5Fᴇ	Annealed	320-380	15	.002	.004	.005	.006	.007	.008	.009	.010	T-15, M-33	Stub Length Drill PA: 118° Lip Cl: 7° PG: Crankshaft
6Aʟ-4V	Solution Treated & Aged	350-400	25	.001	.002	.004	.005	.006	.007	.008	.008	T-15, M-33	Stub Length Drill PA: 118° Lip Cl: 7° PG: Crankshaft

(Continued)

Table C.7 (Continued)

MATERIAL	Condition	Brinell Hardness Number	Speed (fpm)	FEED—INCHES/REVOLUTION NOMINAL HOLE DIAMETER—INCHES								Tool Material	Tool Geometry
				⅛	¼	½	¾	1	1½	2	3		
6Aʟ-6V-2Sɴ 7Aʟ-4Mo 4Aʟ-3Mo-1V	Solution Treated & Aged	375-420	20	.001	.002	.003	.004	.004	.005	.005	.005	T-15, M-33	Stub Length Drill PA: 118° Lip Cl: 7° PG: Crankshaft
1Aʟ-8V-5Fᴇ	Solution Treated & Aged	375-440	15	.0005	.001	.0015	.0015	.002	.002	.003	.004	T-15, M-33	Stub Length Drill PA: 118° Lip Cl: 7° PG: Crankshaft
13V-11Cʀ-3Aʟ	Solution Annealed	310-350	20	.001	.003	.004	.005	.006	.007	.008	.009	M-1, M-10 M-2	Stub Length Drill PA: 118° Lip Cl: 7° PG: Crankshaft
13V-11Cʀ-3Aʟ	Solution Treated & Aged	375-440	15	.0005	.001	.0015	.0015	.002	.002	.003	.004	T-15, M-33	Stub Length Drill PA: 118° Lip Cl: 7° PG: Crankshaft

(a) Nominal Tool Life to be Expected for the Recommended Drilling Conditions using High Speed Steel Drills—75 holes, for 2 to 1 depth to diameter ratio

Table courtesy of RMI Company.

Table C.8 Recommended tool materials, angles, machine settings for drilling

TOOL MATERIALS

Type of Operation	Tool Material
General Drilling	T-4 or T-5 H.S.S.
Deep Holes, Low Production	T-5 H.S.S.
Deep Holes, High Production	C-1 or C-2 carbide
Sheet, Power Drilling	T-4, T-5, or M-10, H.S.S.
Sheet, Hand Drilling	M-10, T-4 or T-5 H.S.S.

TOOL GEOMETRY

Operation	Tool Angle	H.S.S.	Carbide
General and Deep Hole Drilling	Point Angle, °		
	Less than ¼ dia.	140	Single lip
	¼ to ½ dia.	90 or double angle	Gun drill
	Helix Angle, °	28-35	— —
	Relief Angle, °	9-10	6-8
	Cutting Angle, °	0	— —
	Body Clearance	Yes	— —
Sheet, Power Drilling	Point Angle, °		
	Less than ¼ dia.	135	
	¼ to ½ dia.	118	
	Helix Angle, °	15	Not
	Relief Angle, °	12-15	Recommended
	Cutting Angle, °	0	
	Body Clearance	Yes	
Sheet, Hand Drilling	Point Angle, °		
	Less than ¼ dia.	150	
	¼ to ½ dia.[a]	135	
	Helix Angle, °	15	Not
	Relief Angle, °	12-15	Recommended
	Cutting Angle, °	0	
	Body Clearance	No	

[a] Freehand drilling not recommended over 5/16 dia.

(Continued)

Operation

A positive feed must be maintained to prevent smearing and galling of the cutting lips, which produce rapid drill failure. Although hand- drilling operations are necessary in some cases, they should be kept to a minimum. Generally, hand drilling yields only 20 to 30% of the tool life that machine drilling yields.

When drilling holes over one diameter deep, the drill should be retracted frequently to clear the drill flutes and holes of chips. The deeper the hole drilled, the shorter the drill life. For deep, small-diameter holds, sulfo-chlorinated oils are the best coolant, and they should be supplied freely at the point of contact between work and tool. Oil-feeding drills may be used for larger diameter holes. Some shops report that no lubricant is required for drilling sheet when the hole depth is no greater than twice the drill diameter.

Table C.8 (Continued)

OPERATIONAL INFORMATION

	Commercially Pure Titanium	Titanium— All Alloy Grades	
		Annealed	Heat-Treated
GENERAL AND DEEP HOLE DRILLING WITH HIGH-SPEED STEEL			
Speed, sfpm	40-60	20-50	5-40
Feed, ipr			
less than 1/8 dia.	0.0015	0.0015	0.0015
1/8 to 1/4 dia.	0.002-0.005	0.002-0.005	0.002-0.005
1/4 to 1/2 dia.	0.005-0.009	0.005-0.009	0.005-0.009
DRILLING DEEP HOLES WITH CARBIDE DRILLS			
Speed, sfpm	200	100-170	75-145
Feed, ipr	0.0005	0.0005	0.0005
SHEET DRILLING WITH HIGH-SPEED STEEL			
Speed, sfpm	15-40	20-30	10.25
Feed, ipr	0.002-0.005[a]	0.002-0.005[a]	0.002-0.005[a]

[a] Hand drilling titanium requires approximately twice the axial force for drilling aluminum.

We gratefully acknowledge TIMET for their cooperation in preparing this TechReport.

Table courtesy of Industrial Titanium Corp.

At a depth-to-diameter ratio of 5:1, or greater, drill life may be reduced to 30% of that obtained at depth-to-diameter ratios of 3:1, or less. For this reason, designing of holes as shallow as possible is recommended.

REAMING

Holes drilled or bored for the reaming of titanium and titanium alloys should be 0.254 to 0.058 mm (0.010 to 0.020 in.) undersize. Standard high-speed steel and carbide reamers perform satisfactorily, except that clearance on the chamfer should be 10°. To provide maximum tooth space for chip clearance, reamers with the minimum number of flutes for a given size should be selected.

Manufacturers provide extensive technical data, which should be requested before initiating the kind of operations described here. Data presented in Table C.9 are based on average conditions, and are intended only as typical starting points. (Higher titanium speeds and feeds are being realized in plants using high-power tools and advanced technology methods.)

TAPPING

It is essential that straight, clean holes be drilled to assure good tapping results, since variations in diameter and tapered holes are detrimental to this work. Sound threads can be assured by reducing the tendency of the titanium to smear on the lands of the tap and by providing for a free flow of chips in the flutes. Failure to do this will result in poor threads, undersize holes, seizures and, consequently, broken taps.

Best results in tapping titanium have been with a 65% thread.

Table C.9 Reaming Data (a)

MATERIAL	Brinell Hardness Number	HIGH SPEED STEEL TOOL									CARBIDE TOOL								
		Speed (fpm)	Feed—(ipr) Nominal Hole Dia.—In.						Tool Matl.	Tool Geometry	Speed (fpm)	Feed—(ipr) Nominal Hole Dia.—In.						Tool Matl.	Tool Geometry
			⅛	¼	½	1	1½	2				⅛	¼	½	1	1½	2		
Grade 1 Annealed	110-170	100	.003	.006	.009	.012	.015	.020	M-1 or M-2	Right Hand Helix: 10° CA: 45° Per. Relief: 10°	375	.003	.006	.009	.012	.015	.020	C-2	RR: 6° Helix: 7° Per. Relief: 10° CA: 45° Lead Angle 2° x 3/16''
Grade 2 Grade 3 Ti-Pd Annealed	140-200	80	.003	.006	.009	.012	.015	.020	M-1 or M-2	Right Hand Helix: 10° CA: 45° Per. Relief: 10°	300	.003	.006	.009	.012	.015	.020	C-2	RR: 6° Helix: 7° Per. Relief: 10° CA: 45° Lead Angle 2° x 3/16''
Grade 4 Annealed	200-275	70	.003	.005	.008	.011	.014	.016	M-2	Right Hand Helix: 10° CA: 45° Per. Relief: 10°	250	.003	.005	.008	.011	.014	.016	C-2	RR: 6° Helix: 7° Per. Relief: 10° CA: 45° Lead Angle 2° x 3/16''
3Al-2.5V Annealed	200-260	60	.003	.005	.007	.009	.012	.015	M-1 or M-2	Right Hand Helix: 10° CA: 45° Per. Relief: 10°	250	.003	.005	.007	.009	.012	.015	C-2	RR: 6° Helix: 7° Per. Relief: 10° CA: 45° Lead Angle 2° x 3/16''
5Al-2.5Sn 5Al-2.5Sn ELI 6Al-2Cb-1Ta-1Mo 4Al-3Mo-1V Annealed	300-340	45	.002	.005	.007	.009	.012	.015	M-2	Right Hand Helix: 10° CA: 45° Per. Relief: 10°	175	.002	.005	.007	.009	.012	.015	C-2	RR: 6° Helix: 7° Per. Relief: 10° CA: 45° Lead Angle 2° x 3/16''

(Continued)

Table C.9 (Continued)

MATERIAL	Brinell Hardness Number	HIGH SPEED STEEL TOOL										CARBIDE TOOL									
		Speed (fpm)	Feed—(ipr) Nominal Hole Dia.—In.						Tool Matl.	Tool Geometry		Speed (fpm)	Feed—(ipr) Nominal Hole Dia.—In.						Tool Matl.	Tool Geometry
			⅛	¼	½	1	1½	2					⅛	¼	½	1	1½	2		
6Al-4V 6Al-4V ELI 8Mn Annealed	310-350	35	.002	.005	.007	.009	.012	.015	M-2	Right Hand Helix: 10° CA: 45° Per. Relief: 10°		150	.002	.005	.007	.009	.012	.015	C-2	RR: 6° Helix: 7° Per. Relief: 10° CA: 45° Lead Angle: 2° x 3/16''
7Al-4Mo 8Al-1Mo-1V 6Al-2Sn Annealed	320-370	30	.002	.005	.007	.009	.012	.015	M-2	Right Hand Helix: 10° CA: 45° Per. Relief: 10°		120	.002	.005	.007	.009	.012	.015	C-2	RR: 6° Helix: 7° Per. Relief: 10° CA: 45° Lead Angle: 2° x 3/16''
1Al-8V-5Fe Annealed	320-380	30	.002	.005	.007	.009	.012	.015	M-2	Right Hand Helix: 10° CA: 45° Per. Relief: 10°		120	.002	.005	.007	.009	.012	.015	C-2	RR: 6° Helix: 7° Per. Relief: 10° CA: 45° Lead Angle: 2° x 3/16''
6Al-4V Solution Treated & Aged	350-400	30	.002	.005	.007	.009	.012	.015	M-2	Right Hand Helix: 10° CA: 45° Per. Relief: 10°		120	.002	.005	.007	.009	.012	.015	C-2	RR: 6° Helix: 7° Per. Relief: 10° CA: 45° Lead Angle: 2° x 3/16''

(Continued)

Table C.9 (Continued)

MATERIAL	Brinell Hardness Number	HIGH SPEED STEEL TOOL									CARBIDE TOOL								
		Speed (fpm)	Feed—(ipr) Nominal Hole Dia.—In.						Tool Matl.	Tool Geometry	Speed (fpm)	Feed—(ipr) Nominal Hole Dia.—In.						Tool Matl.	Tool Geometry
			⅛	¼	½	1	1½	2				⅛	¼	½	1	1½	2		
6Aʟ-6V-2Sɴ 7Aʟ-4Mo 4Aʟ-3Mo-1V Solution Treated & Aged	375-420	25	.002	.005	.007	.009	.012	.015	M-2	Right Hand Helix: 10° CA: 45° Per. Relief: 10°	100	.002	.005	.007	.009	.012	.015	C-2	RR: 6° Helix: 7° Per. Relief: 10° CA: 45° Lead Angle: 2° x ³⁄₁₆″
1Aʟ-8V-5Fᴇ Solution Treated & Aged	375-440	25	.002	.004	.006	.008	.010	.012	M-2	Right Hand Helix: 10° CA: 45° Per. Relief: 10°	100	.002	.004	.006	.008	.010	.012	C-2	RR: 6° Helix: 7° Per. Relief: 10° CA: 45° Lead Angle: 2° x ³⁄₁₆″
13V-11Cʀ-3Aʟ Solution Annealed	310-350	30	.002	.005	.007	.009	.012	.015	M-2	Right Hand Helix: 10° CA: 45° Per. Relief: 10°	150	.002	.005	.007	.009	.012	.015	C-2	RR: 6° Helix: 7° Per. Relief: 10° CA: 45° Lead Angle: 2° x ³⁄₁₆″
13V-11Cʀ-3Aʟ Solution Treated & Aged	375-440	25	.002	.004	.006	.008	.010	.012	M-2	Right Hand Helix: 10° CA: 45° Per. Relief: 10°	100	.002	.004	.006	.008	.010	.012	C-2	RR: 6° Helix: 7° Per. Relief: 10° CA: 45° Lead Angle: 2° x ³⁄₁₆″

(a) Nominal Tool Life to be Expected for the Recommended Reaming Conditions, High Speed Steel and Carbide Reamers—75 holes, for 2 to 1 depth to diameter ratio.

Table courtesy of RMI Company

One common problem is the smear of titanium on the land of the tap, which can result in the tap freezing or binding in the hole. An activated cutting oil such as a sulfurized and chlorinated oil is helpful in avoiding this. The use of nitrided taps also helps to reduce adherence of titanium to the lands of the tap. Relieving the land or the use of an interrupted tap also helps minimize the smearing tendency.

Chip removal is a problem which makes tapping one of the more difficult machining operations. Chip clogging is reduced by the use of spiral pointed taps, which push the chips ahead of the tool. More chip clearance can be provided by sharply grinding away the trailing edges of the flutes. To give proper clearance, two-fluted, spiral point taps are recommended for diameters up to 5/16 in.; use three-fluted taps for larger sizes.

Manufacturers provide extensive technical data, which should be requested before initiating the kind of operations described here. Data presented in Table C.10 are based on average conditions, and are intended only as typical starting points. (Higher titanium speeds are being realized in plants using high-power tools and advanced technology methods.)

BROACHING

To assure a top quality broaching job, it is essential that the entire machine tool setup and the titanium component be rigid. It is also recommended that broaches be wet ground to improve the finish of the tool, thereby giving better tool performance. During the broaching operation, vapor blasting with the coolant helps to lengthen broach life and to reduce smearing.

There is a tendency for titanium chips to weld to the tool on an interrupted cut such as broaching. This tendency increases as the wearland develops. Both the broach and broach slots should be examined regularly for signs of smearing in order to avoid poor finish, more rapid tool wear and loss of tolerances.

Manufacturers provide extensive technical data, which should be requested before initiating the kind of operations described here. Data presented in Table C.11 are based on average conditions, and are intended only as typical starting points. (Higher titanium speeds are being realized in plants using high-power tools and advanced technology methods.)

MILLING

Face Milling

The life of face milling cutters can be lengthened through the use of a "climb milling" setup, with an anti-backlash device on the table feed screw. Titanium chips tend to weld to the edge of the milling cutter and, when knocked off on re-entering the metal, carry a portion of the cutting edge with them. This is especially true of carbide cutters. Climb milling produces a thin chip as the cutter teeth leave the work, thus reducing the tendency of the chip to weld to the cutting edge.

As in all titanium machining work, sharp tools must be used to reduce galling and welding tendencies. Relief or clearance angles for face milling cutters should be greater than those used for steel. The use of a water base coolant is recommended.

Table C.10 Tapping Data (a)

MATERIAL	Condition	Brinell Hardness Number	Speed (fpm)	HSS Tool Material	Tool Geometry
Grade 1	Annealed	110-170	50	Nitrided M-1, M-10	2 Flute Spiral Point—⁵⁄₁₆″ Tap & Smaller 3 Flute Spiral Point—Larger than ⁵⁄₁₆″ Tap
Grade 2 Grade 3 Ti-Pd	Annealed	140-200	40	Nitrided M-1, M-10	2 Flute Spiral Point—⁵⁄₁₆″ Tap & Smaller 3 Flute Spiral Point—Larger than ⁵⁄₁₆″ Tap
Grade 4	Annealed	200-275	30	Nitrided M-1, M-10	2 Flute Spiral Point—⁵⁄₁₆″ Tap & Smaller 3 Flute Spiral Point—Larger than ⁵⁄₁₆″ Tap
5Al-2.5Sn 5Al-2.5Sn ELI 6Al-2Cʙ-1Tᴀ-1Mo 4Al-3Mo-1V	Annealed	300-340	25	Nitrided M-1, M-10	2 Flute Spiral Point—⁵⁄₁₆″ Tap & Smaller 3 Flute Spiral Point—Larger than ⁵⁄₁₆″ Tap

(Continued)

Table C.10 (Continued)

MATERIAL	Condition	Brinell Hardness Number	Speed (fpm)	HSS Tool Material	Tool Geometry
6Aʟ-4V 6Aʟ-4V ELI	Annealed	310-350	20	Nitrided M-1, M-10	2 Flute Spiral Point— ⁵⁄₁₆'' Tap & Smaller 3 Flute Spiral Point— Larger than ⁵⁄₁₆'' Tap
7Aʟ-4Mo 8Aʟ-1Mo-1V 6V-2Sɴ	Annealed	320-370	15	Nitrided M-1, M-10	2 Flute Spiral Point— ⁵⁄₁₆'' Tap & Smaller 3 Flute Spiral Point— Larger than ⁵⁄₁₆'' Tap
1Aʟ-8V-5Fᴇ	Annealed	320-380	10	Nitrided M-1, M-10	2 Flute Spiral Point— ⁵⁄₁₆'' Tap & Smaller 3 Flute Spiral Point— Larger than ⁵⁄₁₆'' Tap
6Aʟ-4V	Solution Treated & Aged	350-400	10	Nitrided M-1, M-10	2 Flute Spiral Point— ⁵⁄₁₆'' Tap & Smaller 3 Flute Spiral Point— Larger than ⁵⁄₁₆'' Tap

(Continued)

Table C.10 (Continued)

MATERIAL	Condition	Brinell Hardness Number	Speed (fpm)	HSS Tool Material	Tool Geometry
6Aʟ-6V-2Sɴ 7Aʟ-4Mo 4Aʟ-3Mo-1V	Solution Treated & Aged	375-420	10	Nitrided M-1, M-10	2 Flute Spiral Point—5/16″ Tap & Smaller 3 Flute Spiral Point—Larger than 5/16″ Tap
1Aʟ-8V-5Fє	Solution Treated & Aged	375-440	7	Nitrided M-1, M-10	2 Flute Spiral Point—5/16″ Tap & Smaller 3 Flute Spiral Point—Larger than 5/16″ Tap
13V-11Cʀ-3Aʟ	Solution Annealed	310-350	15	Nitrided M-1, M-10	2 Flute Spiral Point—5/16″ Tap & Smaller 3 Flute Spiral Point—Larger than 5/16″ Tap
13V-11Cʀ-3Aʟ	Solution Treated & Aged	375-440	7	Nitrided M-1, M-10	2 Flute Spiral Point—5/16″ Tap & Smaller 3 Flute Spiral Point—Larger than 5/16″ Tap

(a) Nominal Tool Life to be Expected for the Recommended Tapping Conditions, High Speed Steel Taps—75 Holes, for 2 to 1 depth to diameter ratio

Table courtesy of RMI Company

Table C.11 Broaching Data (a)

MATERIAL	Condition	Brinell Hardness Number	Type of Cut	HIGH SPEED STEEL TOOL			
				Speed (fpm)	Chip Load (In./Tooth)	Tool Matl.	Tool Geometry
Grade 1	Annealed	110-170	Roughing	35	.005—.008	T-5	Hook Angle: 8°-10° Cl: 3°-4°
			Finishing	55	.002—.005	T-5	Hook Angle: 8°-10° Cl: 2°-3°
Grade 2 Grade 3 Ti-Pd	Annealed	140-200	Roughing	30	.005—.008	T-5	Hook Angle: 8°-10° Cl: 3°-4°
			Finishing	45	.002—.005	T-5	Hook Angle: 8°-10° Cl: 2°-3°
Grade 4	Annealed	200-275	Roughing	20	.004—.007	T-5	Hook Angle: 8°-10° Cl: 3°-4°
			Finishing	30	.002—.004	T-5	Hook Angle: 8°-10° Cl: 2°-3°
5Al-2.5Sn 5Al-2.5Sn ELI 6Al-2Cb-1Ta-1Mo 4Al-3Mo-1V	Annealed	300-340	Roughing	15	.003—.006	T-5	Hook Angle: 8°-10° Cl: 3°-4°
			Finishing	22	.0015—.003	T-5	Hook Angle: 8°-10° Cl: 2°-3°
6Al-4V 6Al-4V ELI 8Mn	Annealed	310-350	Roughing	12	.003—.006	T-5	Hook Angle: 8°-10° Cl: 3°-4°
			Finishing	18	.0015—.003	T-5	Hook Angle: 8° 10° Cl: 2°-3°
7Al-4Mo 8Al-1Mo-1V 6Al-6V-2Sn	Annealed	320-370	Roughing	10	.003—.006	T-5	Hook Angle: 8°-10° Cl: 3°-4°
			Finishing	16	.0015—.003	T-5	Hook Angle: 8°-10° Cl: 2°-3°

(Continued)

Table C.11 (Continued)

MATERIAL	Condition	Brinell Hardness Number	Type of Cut	HIGH SPEED STEEL TOOL			
				Speed (fpm)	Chip Load (In./Tooth)	Tool Matl.	Tool Geometry
1AL-8V-5FE	Annealed	320-380	Roughing	8	.002—.005	T-5	Hook Angle: 8°-10° Cl: 3°-4°
			Finishing	14	.0015—.0025	T-5	Hook Angle: 8°-10° Cl: 2°-3°
6AL-4V	Solution Treated & Aged	350-400	Roughing	8	.002—.005	T-15	Hook Angle: 8°-10° Cl: 3°-4°
			Finishing	12	.001—.002	T-15	Hook Angle: 8°-10° Cl: 2°-3°
6AL-6V-2SN 7AL-4MO 4AL-3MO-1V	Solution Treated & Aged	375-420	Roughing	7	.002—.004	T-15	Hook Angle: 8°-10° Cl: 3°-4°
			Finishing	10	.001—.002	T-15	Hook Angle: 8°-10° Cl: 2°-3°
1AL-8V-5FE	Solution Treated & Aged	375-440	Roughing	6	.002—.004	T-15	Hook Angle: 8°-10° Cl: 3°-4°
			Finishing	9	.001—.002	T-15	Hook Angle: 8°-10° Cl: 2°-3°
13V-11CR-3AL	Solution Annealed	310-350	Roughing	11	.003—.006	T-5	Hook Angle: 8°-10° Cl: 3°-4°
			Finishing	17	.0015—.003	T-5	Hook Angle: 8°-10° Cl: 2°-3°
13V-11CR-3AL	Solution Treated & Aged	375-440	Roughing	6	.002—.004	T-15	Hook Angle: 8°-10° Cl: 3°-4°
			Finishing	9	.001—.002	T-15	Hook Angle: 8°-10° Cl: 2°-3°

(a) Due to the complexity of most broaching tools and the configurations machined, general predictions of broach life are not practical.

Table courtesy of RMI Company

In milling titanium, when the cutting edge fails, it is usually because of chipping. Thus the results with carbide tools are often less satisfactory than with cast-alloy tools. The increase of 20 to 30% in cutting speeds (which is possible with carbide tools as contrasted with cast-alloy tools) does not always compensate for the additional tool grinding costs. Consequently, it is advisable to try both cast-alloy and carbide tools to determine the better of the two for each milling job.

For slab milling, the work should move in the same direction as the cutting teeth. For face milling, the teeth should emerge from the cut in the same direction as the work is fed.

Manufacturers provide extensive technical data, which should be requested before initiating the kind of operations described here. Data presented in Table C.12 are based on average conditions, and are intended only as typical starting points. (Higher titanium speeds and feeds are being realized in plants using high-power tools and advanced technology methods.)

End Milling: Slotting

Titanium end milling is, for the most part throughout industry, done with high-speed steel cutters. Cutters as short as possible are used. In this machining method, the tooth length-to-diameter ratio is high; thus considerable tool deflection takes place. This condition is extremely critical with carbide cutters. Cutters should have sufficient flute space to prevent chip clogging and subsequent tool failure. Cutters up to 25.4-mm (1-in.) diameter should have 3 to 4 flutes.

Manufacturers provide extensive technical data, which should be requested before initiating the kind of operations described here. Data presented in Table C.13 are based on average conditions, and are intended only as typical starting points. (Higher titanium speeds and feeds are being realized in plants using high-power tools and advanced technology methods.)

End Milling: Peripheral

High-speed steel cutters are preferred throughout industry for peripheral work, since the lack of rigidity inherent in this method is critical for carbide cutters. Here, too, cutters should be as short as possible to reduce tool deflection. Cutters should have sufficient flute space to prevent chip clogging and early cutter failure. Cutters up to 25.4-mm (1-in.) diameter should have three to four flutes.

Climb milling should also be employed on peripheral milling. This produces thinner chips as the cutter teeth leave the work, thereby reducing the tendency of chips to weld the cutter and then to break off portions of the cutting edge as they re-enter the work. Climb milling thus lengthens cutter life.

Manufacturers provide extensive technical data, which should be requested before initiating the kind of operations described here. Data presented in Table C.14 are based on average conditions, and are intended only as typical starting points. (Higher titanium speeds and feeds are being realized in plants using high-power tools and advanced technology methods.)

Table C.12 Face Milling Data (a)

MATERIAL	Condition	Brinell Hardness Number	Depth of Cut (Inches)	HIGH SPEED STEEL TOOL				CARBIDE TOOL			
				Speed (fpm)	Feed (In./Tooth)	Tool Matl.	Tool Geometry	Speed (fpm)	Feed (In./Tooth)	Tool Matl.	Tool Geometry
Grade 1	Annealed	110-170	.250 .050	125 175	.008 .004	M-2	AR: 0° RR: 0° ECEA: 6° CI: 12° CA: 30°	400 500	.008 .004	C-2 C-2	AR: 0° RR: 0° ECEA: 10° CI: 10° CA: 45°
Grade 2 Grade 3 Ti-Pd	Annealed	140-200	.250 .050	100 140	.006 .004	M-2	AR: 0° RR: 0° ECEA: 6° CI: 12° CA: 30°	300 400	.006 .004	C-2 C-2	AR: 0° RR: 0° ECEA: 10° CI: 10° CA: 45°
Grade 4	Annealed	200-275	.250 .050	75 110	.006 .004	M-2	AR: 0° RR: 0° ECEA: 6° CI: 12° CA: 30°	200 300	.006 .004	C-2 C-2	AR: 0° RR: 0° ECEA: 10° CI: 10° CA: 45°
5Al-2.5Sn 5Al-2.5Sn ELI 6Al-2Cb-1Ta-1Mo 4Al-3Mo-1V	Annealed	300-340	.250 .050	50 60	.006 .004	M-3 or T-15	AR: 0° RR: 0° ECEA: 6° CI: 12° CA: 30°	170 210	.006 .004	C-2 C-2	AR: 0° RR: 10° ECEA: 12° CI: 12° CA: 30°

(Continued)

Table C.12 (Continued)

MATERIAL	Condition	Brinell Hardness Number	Depth of Cut (Inches)	HIGH SPEED STEEL TOOL				CARBIDE TOOL			
				Speed (fpm)	Feed (In./Tooth)	Tool Matl.	Tool Geometry	Speed (fpm)	Feed (In./Tooth)	Tool Matl.	Tool Geometry
6Aʟ-4V 6Aʟ-4V ELI 8Mɴ	Annealed	310-350	.250 .050	40 50	.006 .004	M-3 or T-15	AR: 0° RR: 0° ECEA: 6° CI: 12° CA: 30°	130 170	.006 .004	C-2 C-2	AR: 0° RR: 10° ECEA: 12° CI: 12° CA: 30°
7Aʟ-4Mo 8Aʟ-1Mo-1V 6Aʟ-6V-2Sɴ	Annealed	320-370	.250 .050	30 40	.006 .004	T-15 or M-3	AR:0° RR: 0° ECEA: 6° CI: 12° CA: 30°	110 150	.006 .004	C-2 C-2	AR: 0° RR: −10° ECEA: 12° CI: 12° CA: 30°
1Aʟ-8V-5Fᴇ	Annealed	320-380	.250 .050	20 30	.006 .004	T-15	AR: 10° RR: 0° ECEA: 10° CI: 10° CA: 45°	90 120	.006 .004	C-2 C-2	AR: 0° RR: −10° ECEA: 12° CI: 12° CA: 30°
6Aʟ-4V	Solution Treated & Aged	350-400	.250 .050	35 45	.007 .004	T-15	AR: 10° RR: 0° ECEA: 10° CI: 10° CA: 45°	110 130	.006 .004	C-2 C-2	AR: 10° RR: 0° ECEA: 10° CI: 10° CA: 45°

(Continued)

Table C.12 (Continued)

MATERIAL	Condition	Brinell Hardness Number	Depth of Cut (Inches)	HIGH SPEED STEEL TOOL				CARBIDE TOOL			
				Speed (fpm)	Feed (In./Tooth)	Tool Matl.	Tool Geometry	Speed (fpm)	Feed (In./Tooth)	Tool Matl.	Tool Geometry
6Aʟ-6V-2Sɴ 7Aʟ-4Mo 4Aʟ-3Mo-1V	Solution Treated & Aged	375-420	.250 .050	25 35	.007 .004	T-15	AR: 10° RR: 0° ECEA: 10° CI: 10° CA: 45°	80 100	.006 .004	C-2 C-2	AR: 10° RR: 0° ECEA: 10° CI: 10° CA: 45°
1Aʟ-8V-5Fᴇ	Solution Treated & Aged	375-440	.250 .050	20 25	.007 .004	T-15	AR:10° RR: 0° ECEA: 10° CI: 10° CA: 45°	60 70	.006 .004	C-2 C-2	AR: 10° RR: 0° ECEA: 10° CI: 10° CA: 45°
13V-11Cʀ-3Aʟ	Solution Annealed	310-350	.250 .050	25 35	.006 .004	T-15	AR: 10° RR: 0° ECEA: 10° CI: 10° CA: 45°	100 130	.006 .004	C-2 C-2	AR: 0° RR: -10° ECEA: 12° CI: 12° CA: 30°
13V-11Cʀ-3Aʟ	Solution Treated & Aged	375-440	.250 .050	20 25	.007 .004	T-15	AR: 10° RR: 0° ECEA: 10° CI: 10° CA: 45°	60 70	.006 .004	C-2 C-2	AR: 10° RR: 0° ECEA: 10° CI: 10° CA: 45°

(a) Nominal Tool Life to be Expected for the Recommended Face Milling Conditions — High Speed Steel and Carbide Cutters—50 inches/tooth

Table courtesy of RMI Company

Table C.13 End milling; slotting data (a)

MATERIAL	Brinell Hardness Number	Depth of Cut (Inches)	HIGH SPEED STEEL TOOL Speed (fpm)	Feed—Inches/Tooth Cutter Diameter—Inch ½	¾	1	1 To 2	Tool Matl.	Tool Geometry	CARBIDE TOOL Speed (fpm)	Feed—Inches/Tooth Cutter Diameter—Inch ½	¾	1	1 To 2	Tool Matl.	Tool Geometry
Grade 1 Annealed	110-170	.250 .125 .050 .015	80 100 125 150	— — .0005 .0005	.0015 .002 .003 .003	.004 .005 .006 .006	.006 .006 .007 .007	M-2 M-2 M-2 M-2	Helix: 30° RR: 10° End Cl: 3° Per. Cl: 7° ECEA: 3° CA: 45° x .060″ or Radius As Required	200 250 325 375	— — .001 .001	.0015 .0015 .002 .002	.003 .003 .004 .004	.005 .005 .006 .006	C-2 C-2 C-2 C-2	Helix: 15° RR: 0° End Cl: 12° Per. Cl: 12° ECEA: 3° CA: 45° x .040″
Grade 2 Grade 3 Ti-Pd Annealed	140-200	.250 .125 .050 .015	70 90 120 150	— — .0005 .0005	.0015 .002 .003 .003	.004 .005 .006 .006	.006 .006 .007 .007	M-2 M-2 M-2 M-2	Helix: 30° RR: 10° End Cl: 3° Per. Cl: 7° ECEA: 3° CA: 45° x .060″ or Radius As Required	175 225 300 375	— — .001 .001	.0015 .0015 .002 .002	.003 .003 .004 .004	.005 .005 .006 .006	C-2 C-2 C-2 C-2	Helix: 15° RR: 0° End Cl: 12° Per. Cl: 12° ECEA: 3° CA: 45° x .040″
Grade 4 Annealed	200-275	.250 .125 .050 .015	40 50 60 75	— — .0005 .0007	.0007 .001 .002 .003	.003 .003 .004 .005	.004 .004 .005 .006	M-2 M-2 M-2 M-2	Helix: 30° RR: 10° End Cl: 3° Per. Cl: 7° ECEA: 3° CA: 45° x .060″ or Radius As Required	100 125 160 190	— — .0005 .0005	.0007 .0015 .002 .003	.003 .003 .005 .006	.005 .005 .007 .008	C-2 C-2 C-2 C-2	Helix: 15° RR: 0° End Cl: 12° Per. Cl: 12° ECEA: 3° CA: 45° x .040″
5Al-2.5Sn 5Al-2.5Sn ELI 6Al-2Cb-1Ta-1Mo 4Al-3Mo-1V Annealed	300-340	.250 .125 .050 .015	35 45 55 70	— — .0005 .0007	.0007 .001 .002 .003	.003 .003 .004 .005	.004 .004 .005 .006	M-2 M-2 M-2 M-2	Helix: 30° RR: 10° End Cl: 3° Per. Cl: 7° ECEA: 3° CA: 45° x .060″ or Radius As Required	90 110 140 175	— — .0005 .0005	.0007 .0015 .002 .003	.003 .003 .005 .006	.005 .005 .007 .008	C-2 C-2 C-2 C-2	Helix: 15° RR: 0° End Cl: 12° Per. Cl: 12° ECEA: 3° CA: 45° x .040″

(Continued)

Table C.13 (Continued)

MATERIAL	Brinell Hardness Number	Depth of Cut (Inches)	HIGH SPEED STEEL TOOL Speed (fpm)	Feed—Inches/Tooth Cutter Diameter—Inch ⅛	½	¾	1 To 2	Tool Matl.	Tool Geometry	CARBIDE TOOL Speed (fpm)	Feed—Inches/Tooth Cutter Diameter—Inch ⅛	½	¾	1 To 2	Tool Matl.	Tool Geometry
6AL-4V 6AL-4V ELI 8MN Annealed	310-350	.250 .125 .050 .015	30 40 50 65	— — .0005 .0007	.0007 .001 .002 .003	.003 .003 .004 .005	.004 .004 .005 .006	M-2 M-2 M-2 M-2	Helix: 30° RR: 10° End Cl: 3° Per. Cl: 7° ECEA: 3° CA: 45° x .060" or Radius As Required	75 100 125 165	— — .0005 .005	.0007 .0015 .002 .003	.003 .003 .005 .006	.005 .005 .007 .008	C-2 C-2 C-2 C-2	Helix: 15° RR: 0° End Cl: 12° Per. Cl: 12° ECEA: 3° CA: 45° x .040"
7AL-4MO 8AL-1MO-1V 6AL-6V-2SN Annealed	320-370	.250 .125 .050 .015	30 35 45 55	— — .0005 .0007	.0007 .001 .002 .003	.003 .003 .004 .005	.004 .004 .005 .006	M-2 M-2 M-2 M-2	Helix: 30° RR: 10° End Cl: 3° Per. Cl: 7° ECEA: 3° CA: 45° x .060" or Radius As Required	75 90 115 140	— — .0005 .0005	.0007 .0015 .002 .003	.003 .003 .005 .006	.005 .005 .007 .008	C-2 C-2 C-2 C-2	Helix: 15° RR: 0° End Cl: 12° Per. Cl: 12° ECEA: 3° CA: 45° x .040"
1AL-8V-5FE Annealed	320-380	.250 .125 .050 .015	25 30 35 45	— — .0005 .0007	.0007 .001 .002 .003	.003 .003 .004 .005	.004 .004 .005 .006	M-2 M-2 M-2 M-2	Helix: 30° RR: 10° End Cl: 3° Per. Cl: 7° ECEA: 3° CA: 45° x .060" or Radius As Required	60 75 90 115	— — .0005 .001	.0007 .0015 .002 .003	.003 .003 .005 .006	.005 .005 .007 .008	C-2 C-2 C-2 C-2	Helix: 15° RR: 0° End Cl: 12° Per. Cl: 12° ECEA: 3° CA: 45° x .040"
6AL-4V Solution Treated & Aged	350-400	.250 .125 .050 .015	25 30 35 45	— — .0006 .001	.0006 .001 .002 .003	.002 .003 .003 .004	.003 .004 .004 .006	M-2 M-2 M-2 M-2	Helix: 30° RR: 10° End Cl: 3° Per. Cl: 7° ECEA: 3° CA: 45° x .060" or Radius As Required	60 75 90 115	— — .0006 .001	.0006 .001 .002 .003	.003 .003 .004 .005	.004 .005 .006 .007	C-2 C-2 C-2 C-2	Helix: 15° RR: 0° End Cl: 12° Per. Cl: 12° ECEA: 3° CA: 45° x .040"

(Continued)

Table C.13 (Continued)

MATERIAL	Brinell Hardness Number	Depth of Cut (Inches)	HIGH SPEED STEEL TOOL — Speed (fpm)	Feed ⅛	Feed ⅜	Feed ¾	Feed 1 To 2	Tool Matl.	Tool Geometry	CARBIDE TOOL — Speed (fpm)	Feed ⅛	Feed ⅜	Feed ¾	Feed 1 To 2	Tool Matl.	Tool Geometry
6Al-6V-2Sn 7Al-4Mo 4Al-3Mo-1V Solution Treated Aged &	375-420	.250	25	—	.0005	.001	.002	M-2	Helix: 30° RR: 10° End Cl: 3° Per. Cl: 7° ECEA: 3° CA: 45° x .060" or Radius As Required	60	—	.0006	.003	.004	C-2	Helix: 15° RR: 0° End Cl: 12° Per. Cl: 12° ECEA: 3° CA: 45° x .040"
		.125	30	—	.001	.002	.003	M-2		75	—	.001	.003	.004	C-2	
		.050	35	.0006	.0015	.003	.004	M-2		90	.0005	.0015	.005	.006	C-2	
		.015	45	.0008	.003	.004	.006	M-2		115	.001	.003	.005	.007	C-2	
1Al-8V-5Fe Solution Treated & Aged	375-440	.250	20	—	.0004	.001	.002	M-2	Helix: 30° RR: 10° End Cl: 3° Per. Cl: 7° ECEA: 3° CA: 45° x .060" or Radius As Required	50	—	.0004	.001	.002	C-2	Helix: 15° RR: 0° End Cl: 12° Per. Cl: 12° ECEA: 3° CA: 45° x .040"
		.125	25	—	.0007	.002	.003	M-2		60	—	.0007	.002	.004	C-2	
		.050	35	.0003	.0015	.002	.004	M-2		90	.0005	.002	.003	.005	C-2	
		.015	45	.0005	.002	.003	.005	M-2		115	.0007	.002	.004	.006	C-2	
13V-11Cr-3Al Solution Annealed	310-350	.250	30	—	.0007	.003	.004	M-2	Helix: 30° RR: 10° End Cl: 3° Per. Cl: 7° ECEA: 3° CA: 45° x .060" or Radius As Required	75	—	.0007	.003	.005	C-2	Helix: 15° RR: 0° End Cl: 12° Per. Cl: 12° ECEA: 3° CA: 45° x .040"
		.125	35	—	.001	.003	.004	M-2		90	—	.0015	.003	.005	C-2	
		.050	45	.0005	.002	.004	.005	M-2		115	.0005	.002	.005	.007	C-2	
		.015	55	.0007	.003	.005	.006	M-2		140	.001	.003	.006	.008	C-2	
13V-11Cr-3Al Solution Treated & Aged	375-440	.250	20	—	.0004	.001	.002	M-2	Helix: 30° RR: 10° End Cl: 3° Per. Cl: 7° ECEA: 3° CA: 45° x .060" or Radius As Required	50	—	.0004	.001	.002	C-2	Helix: 15° RR: 0° End Cl: 12° Per. Cl: 12° ECEA: 3° CA: 45° x .040"
		.125	25	—	.0007	.002	.003	M-2		60	—	.0007	.002	.004	C-2	
		.050	35	.0003	.0015	.002	.004	M-2		90	.0005	.002	.003	.005	C-2	
		.015	45	.0005	.002	.003	.005	M-2		115	.0007	.002	.004	.006	C-2	

(a) Nominal Tool Life to be Expected for the Recommended End Milling under Slotting Conditions with High Speed Steel and Carbide Cutters is 40 inches/tooth

Table courtesy of RMI Company

Table C.14 End milling: peripheral data (a)

MATERIAL	Brinell Hardness Number	Depth of Cut (Inches)	HIGH SPEED STEEL TOOL Speed (fpm)	Feed—Inches/Tooth Cutter Diameter—Inch ⅛	½	¾	1 To 2	Tool Matl.	Tool Geometry	CARBIDE TOOL Speed (fpm)	Feed—Inches/Tooth Cutter Diameter—Inch ⅛	½	¾	1 To 2	Tool Matl.	Tool Geometry
Grade 1	110-170	.250	100	—	.002	.005	.007	M-2	Helix: 30° RR: 10° End Cl: 3° Per. Cl: 7° ECEA: 3° CA: 45° x .060″ or Radius As Required	250	—	.002	.005	.007	C-2	Helix: 15° RR: 0° End Cl: 12° Per. Cl: 12° ECEA: 3° CA: 45° x .040″
Annealed		.125	135	—	.0035	.006	.008	M-2		335	—	.002	.005	.008	C-2	
		.050	175	.0015	.005	.007	.009	M-2		375	.001	.002	.006	.010	C-2	
		.015	200	.002	.006	.008	.010	M-2		400	.001	.002	.006	.010	C-2	
Grade 2	140-200	.250	90	—	.002	.005	.007	M-2	Helix: 30° RR: 10° End Cl: 3° Per. Cl: 7° ECEA: 3° CA: 45° x .060″ or Radius As Required	225	—	.002	.005	.007	C-2	Helix: 15° RR: 0° End Cl: 12° Per. Cl: 12° ECEA: 3° CA: 45° x .040″
Grade 3		.125	115	—	.0035	.006	.008	M-2		285	—	.002	.005	.008	C-2	
Ti-Pd		.050	150	.0015	.005	.007	.009	M-2		375	.001	.002	.006	.010	C-2	
Annealed		.015	180	.002	.006	.008	.010	M-2		400	.001	.002	.006	.010	C-2	
Grade 4	200-275	.250	60	—	.001	.004	.005	M-2	Helix: 30° RR: 10° End Cl: 3° Per. Cl: 7° ECEA: 3° CA: 45° x .060″ or Radius As Required	150	—	.001	.004	.006	C-2	Helix: 15° RR: 0° End Cl: 12° Per. Cl: 12° ECEA: 3° CA: 45° x .040″
Annealed		.125	70	—	.0015	.004	.005	M-2		175	—	.002	.004	.006	C-2	
		.050	85	.0008	.003	.005	.006	M-2		215	.0008	.003	.006	.007	C-2	
		.015	100	.001	.004	.006	.007	M-2		250	.001	.004	.007	.008	C-2	
5Al-2.5Sn	300-340	.250	55	—	.001	.004	.005	M-2	Helix: 30° RR: 10° End Cl: 3° Per. Cl: 7° ECEA: 3° CA: 45° x .060″ or Radius As Required	135	—	.001	.004	.006	C-2	Helix: 15° RR: 0° End Cl: 12° Per. Cl: 12° ECEA: 3° CA: 45° x .040″
5Al-2.5Sn ELI		.125	65	—	.0015	.004	.005	M-2		160	—	.002	.004	.006	C-2	
6Al-2Ca-1Ta-1Mo		.050	80	.0008	.003	.005	.006	M-2		200	.0008	.003	.006	.007	C-2	
4Al-3Mo-1V		.015	95	.001	.004	.006	.007	M-2		240	.001	.004	.007	.008	C-2	
Annealed																

(Continued)

Table C.14 (Continued)

MATERIAL	Brinell Hardness Number	HIGH SPEED STEEL TOOL								CARBIDE TOOL						
		Depth of Cut (Inches)	Speed (fpm)	Feed—Inches/Tooth Cutter Diameter—Inch				Tool Matl.	Tool Geometry	Speed (fpm)	Feed—Inches/Tooth Cutter Diameter—Inch				Tool Matl.	Tool Geometry
				⅛	⅜	¾	1 To 2				⅛	⅜	¾	1 To 2		
6Aʟ-4V 6Aʟ-4V ELI 8Mɴ Annealed	310-350	.250 .125 .050 .015	50 60 75 90	— — .0008 .001	.001 .0015 .003 .004	.004 .004 .005 .006	.005 .005 .006 .007	M-2 M-2 M-2 M-2	Helix: 30° RR: 10° End Cl: 3° Per. Cl: 7° ECEA: 3° CA: 45° x .060″ or Radius As Required	125 150 190 225	— — .0008 .001	.001 .002 .003 .004	.004 .004 .006 .007	.006 .006 .007 .008	C-2 C-2 C-2 C-2	Helix: 15° RR: 0° End Cl: 12° Per. Cl: 12° ECEA: 3° CA: 45° x .040″
7Aʟ-4Mo 8Aʟ-1Mo-1V 6Aʟ-6V-2Sɴ Annealed	320-370	.250 .125 .050 .015	50 55 70 85	— — .0008 .001	.001 .0015 .003 .004	.004 .004 .005 .006	.005 .005 .006 .007	M-2 M-2 M-2 M-2	Helix: 30° RR: 10° End Cl: 3° Per. Cl: 7° ECEA: 3° CA: 45° x .060″ or Radius As Required	125 135 175 215	— — .0008 .001	.001 .002 .003 .004	.004 .004 .006 .007	.006 .006 .007 .008	C-2 C-2 C-2 C-2	Helix: 15° RR: 0° End Cl: 12° Per. Cl: 12° ECEA: 3° CA: 45° x .040″
1Aʟ-8V-5Fᴇ Annealed	320-380	.250 .125 .050 .015	40 50 65 80	— — .0008 .001	.001 .0015 .003 .004	.004 .004 .005 .006	.005 .005 .006 .007	M-2 M-2 M-2 M-2	Helix: 30° RR: 10° End Cl: 3° Per. Cl: 7° ECEA: 3° CA: 45° x .060″ or Radius As Required	100 125 165 200	— — .0008 .001	.001 .002 .003 .004	.004 .004 .006 .007	.006 .006 .007 .008	C-2 C-2 C-2 C-2	Helix: 15° RR: 0° End Cl: 12° Per. Cl: 12° ECEA: 3° CA: 45° x .040″
6Aʟ-4V Solution Treated & Aged	350-400	.250 .125 .050 .015	40 50 65 80	— — .001 .0015	.0008 .0015 .003 .004	.003 .004 .004 .005	.004 .005 .005 .007	M-2 M-2 M-2 M-2	Helix: 30° RR: 10° End Cl: 3° Per. Cl: 7° ECEA: 3° CA: 45° x .060″ or Radius As Required	100 125 165 200	— — .001 .0015	.0008 .0015 .003 .004	.003 .004 .005 .006	.005 .006 .006 .008	C-2 C-2 C-2 C-2	Helix: 15° RR: 0° End Cl: 12° Per. Cl: 12° ECEA: 3° CA: 45° x .040″

(Continued)

Table C.14 (Continued)

MATERIAL	Brinell Hardness Number	Depth of Cut (Inches)	HIGH SPEED STEEL TOOL Speed (fpm)	HSS Feed ⅛	HSS Feed ⅜	HSS Feed ¾	HSS Feed 1 To 2	HSS Tool Matl.	HSS Tool Geometry	CARBIDE TOOL Speed (fpm)	Carb Feed ⅛	Carb Feed ⅜	Carb Feed ¾	Carb Feed 1 To 2	Carb Tool Matl.	Carb Tool Geometry
6Al-6V-2Sn 7Al-4Mo 4Al-3Mo-1V Solution Treated & Aged	375-420	.250 .125 .050 .015	40 50 65 80	— — .0008 .001	.0008 .0015 .002 .004	.0015 .003 .004 .005	.003 .004 .005 .007	M-2 M-2 M-2 M-2	Helix: 30° RR: 10° End Cl: 3° Per. Cl: 7° ECEA: 3° CA: 45° x .060″ or Radius As Required	100 125 165 200	— — .0008 .001	.0008 .0015 .002 .004	.003 .004 .005 .006	.005 .005 .006 .008	C-2 C-2 C-2 C-2	Helix: 15° RR: 0° End Cl: 12° Per. Cl: 12° ECEA: 3° CA: 45° x .040″
1Al-8V-5Fe Solution Treated & Aged	375-440	.250 .125 .050 .015	30 40 55 75	— — .0005 .0007	.0006 .001 .002 .003	.0015 .003 .003 .004	.003 .004 .005 .006	M-2 M-2 M-2 M-2	Helix: 30° RR: 10° End Cl: 3° Per. Cl: 7° ECEA: 3° CA: 45° x .060″ or Radius As Required	75 100 140 190	— — .0005 .0007	.0006 .001 .002 .003	.0015 .003 .004 .005	.003 .005 .006 .007	C-2 C-2 C-2 C-2	Helix: 15° RR: 0° End Cl: 12° Per. Cl: 12° ECEA: 3° CA: 45° x .040″
13V-11Cr-3Al Solution Annealed	310-350	.250 .125 .050 .015	50 55 70 85	— — .0007 .001	.001 .0015 .003 .004	.004 .004 .005 .006	.005 .005 .006 .007	M-2 M-2 M-2 M-2	Helix: 30° RR: 10° End Cl: 3° Per. Cl: 7° ECEA: 3° CA: 45° x .060″ or Radius As Required	125 140 175 215	— — .0007 .001	.001 .0015 .003 .004	.004 .005 .006 .007	.006 .006 .007 .008	C-2 C-2 C-2 C-2	Helix: 15° RR: 0° End Cl: 12° Per. Cl: 12° ECEA: 3° CA: 45° x .040″
13V-11Cr-3Al Solution Treated & Aged	375-440	.250 .125 .050 .015	30 40 55 75	— — .0005 .0007	.0006 .001 .002 .003	.0015 .003 .003 .004	.003 .004 .005 .006	M-2 M-2 M-2 M-2	Helix: 30° RR: 10° End Cl: 3° Per. Cl: 7° ECEA: 3° CA: 45° x .060″ or Radius As Required	75 100 140 190	— — .0005 .0007	.0006 .001 .002 .003	.0015 .003 .004 .005	.003 .005 .005 .007	C-2 C-2 C-2 C-2	Helix: 15° RR: 0° End Cl: 12° Per. Cl: 12° ECEA: 3° CA: 45° x .040″

(a) Nominal Tool Life to be Expected for the Recommended End Milling under Peripheral Conditions with High Speed Steel and Carbide Cutters is 40 inches/tooth

Table courtesy of RMI Company

SURFACE GRINDING

The proper combination of grinding fluid, abrasive wheel, and wheel speeds can expedite the shaping of titanium by means of surface grinding. The procedure recommended is to use considerably lower wheel speeds than in conventional grinding of steels.

Abrasives recommended by a leading manufacturer of grinding wheels are silicon carbide wheels for cutoff and portable grinding, and aluminum oxide wheels for cylindrical and surface grinding.

The following guidelines generally and typically apply to grinding operations:

- Use a sharply dressed wheel.

- Use the largest wheel diameter and thickness that are feasible.

- Tend to use harder wheels.

- Have ample power available at the spindle for grinding.

Reduced wheel speeds, as contrasted with those employed with steel, aid in grinding performance. In surface grinding, for example, 279 sm^2/min (3000 sft^2/min) has been found to cause minimum surface stresses and distortion. The basic reasons for using slower wheel speeds when grinding titanium are:

- High temperatures developed at the chip-grit interface

- Titanium's abrasive character

- Wheel loading

- Smearing

- Titanium's heat sensitivity

- Fire hazard

To minimize residual stresses in the ground surfaces of titanium parts, the following down feeds should be employed: 0.025 mm/pass to last 0.051 mm; then 0.0127 mm, 0.0102 mm, 0.0076 mm, 0.0051 mm, 0.0025 mm/pass (0.001 in./pass to last 0.002 in., then 0.0005 in., 0.0004 in., 0.0003 in., 0.0002 in., 0.0001 in./pass) no sparkout.

Nitride amine base fluids have been successfully used in a majority of operations. A water-sodium nitrite mixture gives excellent results as a coolant. For form grinding, straight grinding oil is recommended.

Because of the potential fire hazard when grinding titanium in the presence of a grinding oil, especially at increased speeds, the following precautions should be taken:

- Extra cutting fluid lines should be installed to quench the sparking as much as possible.

- Filters should be installed, whenever possible, to remove the fine titanium particles from the cutting fluid.

- External surfaces of the machines should be cleaned of titanium dust frequently.

- Oil is to be changed more often than is customary with steels.

- Material such as soapstone should be available in the vicinity of the machine to quench any fires.

THERMAL CUTTING

Oxygen cutting is a very suitable and economical process for fabrication operations. Oxyacetylene cutting of titanium can be performed using the same procedures as for steel and at speeds several times as fast. Line cutting, shape cutting, and severing of heavy sections is relatively easy. Preheating of the edge is necessary for starting the cut as with steel. If the metal is heavily scaled, scale removal by grinding may be necessary at the starting edge. Metal reaction with the cutting gases is limited to a few thousandths of an inch. The original chemical quality of the cut surface can be restored by light grinding to obtain a clean metal surface. The heat-affected zone, in which microstructure transformations occur, is generally not more than 0.254 mm (0.10 in.) deep, and is only slightly harder than the unaffected metal.

Gas tungsten-arc cutting is also suitable for cutting titanium, but the quality of the cut edge is generally inferior to that obtained with oxyacetylene cutting.

TOOL LIFE CHART REFERENCES

The following listing provides references which contain tool-life charts. These items may prove very helpful when making selections for milling titanium.

1. Norman Zlatin, John D. Christopher, and John T. Cammett, "Machining of New Materials," *USAF Technical Report AFML-TR-73-165*, Metcut Research Associates, Inc., Cincinnati, Ohio, July 1973.

2. Norman Zlatin, Michael Field, and William P. Koster, "Final Report on Machinability of Materials," *USAF Technical Report AFML-TR-65-444*, Metcut Research Associates, Inc., Cincinnati, Ohio, 1966.

3. Norman Zlatin, Michael Field, and William P. Koster, "Machining of New Materials," *USAF Technical Report AFML-TR-67-339*, Metcut Research Associates Inc., Cincinnati, Ohio, 1967.

4. Norman Zlatin and Michael Field, "Machinability Parameters on New and Selective Aerospace Materials," *USAF Technical Report AFML-TR-69-144*, Metcut Research Associates, Inc., Cincinnati, Ohio, 1969.

5. Norman Zlatin and Michael Field, "Machinability Parameters on New and Selective Aerospace Materials," *USAF Technical Report AFML-TR-71-95*, Metcut Research Associates, Inc., Cincinnati, Ohio, 1971.

6. Norman Zlatin, "Establishment of Production Machinability Data," *USAF Technical Report AFML-TR-75-120*, Metcut Research Associates, Inc., Cincinnati, Ohio, 1975.

7. *Machining Data Handbook*, 3rd edition, Metcut Research Associates, Inc., Machinability Data Center, Cincinnati, Ohio, 1980.

8. J. F. Kahles, ed., "Machining," *Metals Handbook; Desk Edition*, American Society for Metals, Metals Park, Ohio, 1984, pp. 27-48 to 27-51.

Appendix D

Filler Metals

AWS SPECIFICATION

The American Welding Society (AWS) issues the primary industry guidelines for filler metals in the United States. For titanium and its alloys, the specification is "Titanium and Titanium Alloy Bare Welding Rods and Electrodes" (A5.16). Items included are also listed across the top of Table D.1.

AMS SPECIFICATION

Frequently, as in Table D.2, references are made to the Aerospace Material Specification (AMS). This is published by the Society of Automotive Engineers. The relevant material is contained in the 4900 Series. (Refer to Appendix I at the AMS SPECIFICA-TIONS paragraph.)

FILLER METALS

The material included in Table D.1 will aid the user in identifying available types of filler metals designed for use with titanium and titanium alloys. The listing is neither an endorsement of a product nor a recommendation of a specific manufacturer. There is no guaranty that the listing is complete. (Consult your local representatives for the most current products and technical information.)

Table D.2 lists many of the more common titanium-related filler metals, as ordered by the Unified Numbering System (UNS) scheme. Other equivalent specifications, if any, are also cross-referenced. (Refer to Appendix I at the UNS NUMBERING SYSTEM paragraph.)

CHEMICAL COMPOSITIONS

The chemical compositions of the various filler metals are listed in Table 10.1 in Chapter 10.

Table D.1 Titanium, Titanium alloy bare welding rods, electrodes (a) (b)

Manufacturers	AWS Classification	ERTI-1	ERTI-2	ERTI-3	ERTI-4	ERTI-0.2Pd	ERTI-3Al-2.5V	ERTI-3Al-2.5V-1
Astrolite Alloys Corporation		Astrolite ERTi-1	Astrolite ERTi-2	Astrolite ERTi-3	Astrolite ERTi-4	Astrolite ERTi-0.2Pd	Astrolite ERTi-3Al-2.5V	Astrolite ERTi-3Al-2.5V1
Brite-Weld		Brite-Weld ERTi-1	Brite-Weld ERTi-2	Brite-Weld ERTi-3	Brite-Weld ERTi-4	—	—	—
Champion Welding Products, Inc.		Champion Ti-1	Champion Ti-2	Champion Ti-3	Champion Ti-4	Champion Ti-0.2P	Champion Ti-3Al-2.5V	Champion Ti-3Al-2.5V1
Cronatron Welding Systems, Inc.		Cronatron ERTi-1	—	—	—	—	—	—
J.W. Harris Co., Inc.		Harris ERTi-1 Titanium	Harris ERTi-2 Titanium	—	—	—	—	—
International Welding Products, Inc.		Inweld ERTi-1	Inweld ERTi-2	Inweld ERTi-3	Inweld ERTi-4	Inweld ERTi-0.2Pd	Inweld ERTi-3Al-2.5V	Inweld ERTi-3Al-2.5V-1
J.B. Alloy Company, Inc.		JB ERTi-1	JB ERTi-2	JB ERTi-3	JB ERTi-4	—	—	—
Johnston Stainless Welding Rods		Johnston CPTi-1	Johnston CPTi-2	Johnston CPTi-3	Johnston CPTi-4	Johnston 0.2Pd	Johnston 3Al-2.5V	Johnston 3Al-2.5V-1
Pac-Weld, Inc.		PAC-WELD 901-TA	PAC-WELD 902-TA	PAC-WELD 903-TA	PAC-WELD 904-TA	PAC-WELD 905-TA	PAC-WELD 906-TA	PAC-WELD 907-TA
Titanium Industries		TiWire ERTi-1	TiWire ERTi-2	TiWire ERTi-3	TiWire ERTi-4	TiWire ERTi-0.2Pd	TiWire ERTi-3Al-2.5V	TiWire ERTi-3Al-2.5V-1
United States Welding Corporation		U.S.W. 1525	U.S.W. 1526	U.S.W. 1527	U.S.W. 1528	—	U.S.W. 1521	U.S.W. 1521EW
Universal Welding Products		ERTi-1	ERTi-2	ERTi-3	ERTi-4	ERTi-7	ERTi-9	ERTi-9 ELI
WASAWELD		WASAWELD ERTi-1	WASAWELD ERTi-2	WASAWELD ERTi-3	WASAWELD ERTi-4	—	—	—

(a) Source: AWS-FMC-86 (formerly A5.0), "Filler Metals Comparison Charts, American Welding Society, 1986
(b) See AWS A5.16-70, "Specification for Titanium and Titanium Alloy Bare Rods and Electrodes.

Note: Table D.1 is reprinted with the permission of the **American Welding Society**, Miami, FL.

(Continued)

SUPPLIERS' DATA

A sampling of a manufacturer's typical welding material specifications is included in the following paragraphs as an indication of the type of data that is available to the designer.

The information included here is provided courtesy of **Universal Wire Works, Inc**.

AMS 4951 Wire

Commercially pure (CP) titanium is extremely versatile for applications requiring high-temperature resistance together with resistance to many chemical reagents. Commer-

Table D.1 (Continued)

Manufacturers	AWS Classification	ERTI-5Al-2.5Sn	ERTI-5Al-2.5Sn-1	ERTI-6Al-2Cb-1Ta-1Mo	ERTI-6Al-4V	ERTI-6Al-4V-1	ERTI-8Al-1Mo-1V	ERTI-13V-11Cr-3Al
Astrolite Alloys Corporation		Astrolite ERTi-5Al-2.5Sn	Astrolite ERTi-5Al-2.5Sn-1	Astrolite ERTi-6Al-2Cb-1Ta	Astrolite ERTi-6Al-4V	Astrolite ERTi-6Al-4V-1	—	—
Brite-Weld		—	—	—	—	—	—	—
Champion Welding Products, Inc.		Champion Ti-5Al-2.5Sn	Champion Ti-5Al-2.5Sn-1	Champion Ti-6Al-2Cb-1Ta-1Mo	Champion Ti-6Al-4V	Champion Ti-6Al-4V-ELI	—	—
Cronatron Welding Systems, Inc.		—	—	—	—	—	—	—
J.W. Harris Co., Inc.		—	—	—	Harris ER6Al-4V	Harris ERTi6Al4VELI	—	—
International Welding Products, Inc.		Inweld ERTi-5Al-2.5Sn	Inweld ERTi-5Al-2.5Sn-1	Inweld ERTi-6Al-2Cb-1Ta-1Mo	Inweld ERTi-6Al-4V	Inweld ERTi-6Al-4V-1	Inweld ERTi-8Al-1Mo-1V	Inweld ERTi-13-V 11Cr-3Al
J.B. Alloy Company, Inc.		—	—	—	JB 6-4	—	—	—
Johnston Stainless Welding Rods		Johnston 5Al-2.5Sn	Johnston 5Al-2.5Sn-ELI	Johnston 6-2-1-1	Johnston 6-4	Johnston 6-4 ELI	Johnston 8-1-1	Johnston 13-11-3
Pac-Weld, Inc.		PAC-WELD 908TA	PAC-WELD 909TA	PAC-WELD 910TA	PAC-WELD 911TA	PAC-WELD 913TA	PAC-WELD 914TA	PAC-WELD 915TA
Titanium Industries		ERTi-6	ERTi-6 ELI	ERTi-15	ERTi-5	ERTi-5 ELI	—	—
United States Welding Corporation		U.S.W. HQ4953	U.S.W. HQ4953ELI	U.S.W. MC1304	U.S.W. HQ4954	U.S.W. MC4956	U.S.W. HQ4955	—
Universal Welding Products		ERTi-6	ERTi-6 ELI	ERTi-15	ERTi-5	ERTi-5 ELI	—	—
WASAWELD		—	—	—	—	—	—	—

cially pure titanium has many unique properties providing industry with answers to a host of equipment problems. CP titanium is available in sheet, strip, plate, bar, billet, tubing, and wire. It is used in aircraft engine and airframe applications where tensile strength to weight ratios are important.

Mechanical Properties:

    Tensile Strength, ksi: 50
    Yield Strength, (0.2% offset) ksi: 25
    Elongation, %: 35

Chemical Composition, % (max except Ti):

    0.08C; 0.18O; 0.005H; 0.05N; 0.20Fe; 0.60 other; rem Ti

Sizes Available: .020, .025, .030, .035, .040, .045, 1/16, 3/32, 1/8″
            18″ lengths - 36″ lengths - 30# spools

AMS 4953 Wire

Ti-5Al-2.5Sn titanium alloy is easily welded with the gas tungsten arc, gas metal arc, or resistance welding process. A joint efficiency of 100% can be obtained, provided proper

Table D.2 Filler metal designations and equivalent specifications

AWS A5.16 (4/73) designations	Proposed revisions (as of 6/87 not accepted) (a)	UNS No. (equivalent) (b)	MIL. spec. (equivalent)
ERTi-1	Same	R50100	MIL-R-81586 (CP-E)
ERTi-2	Same	R50120	-
ERTi-3	Same	R50125	MIL-R-81588 (CP-F)
ERTi-4	Same	R50130	-
ERTi-6Al-4V	ERTi-5	R56400 (d)	-
ERTi-6Al-4V-1	ERTi-5-ELI	R56402	MIL-R-81588
ERTi-5Al-2.5Sn	ERTi-6	R54522	-
ERTi-5Al-2.5Sn-1	ERTi-6-ELI	R54523	MIL-R-81588
ERTi-0.2Pd	ERTi-7	R52401	-
ERTi-3Al-2.5V	ERTi-9	R56320 (d)	-
ERTi-3Al-2.5V-1	ERTi-9-ELI	R56321	-
- (c)	ERTi-12	R53400	-
ERTi-6Al-2Cb-1Ta-1Mo	ERTi-15	R56210 (d)	-
ERTi-8Al-1Mo-1V	Withdrawn	R54810 (d)	-
ERTi-13V-11Cr-3Al	Withdrawn	R58010 (d)	-

(a) As of the date of the publication of this guide, AWS's Executive Committee had not accepted these new designations, although various subcommittees has proposed the changes.
(b) Unless noted, numbers from fourth edition (1986) of Metals & Alloys in the Unified Numbering System. (See UNS Numbering System paragraph, below.)
(c) No 1973 equivalent; ERTi-12 is a new designation. See Appendix I.
(d) Proposed UNS numbers; not in fourth edition.

shielding procedures are used. Alloy 4953 is often employed for applications where weldability and oxidation resistance is of major interest. Good high-temperature strength and stability may be expected.

Annealed Mechanical Properties:

Tensile Strength, ksi: 120
Yield Strength, (0.2% offset) ksi: 115
Elongation, %: 10

Chemical Composition, % (max except Ti):

0.08C; 0.175O; 0.015H; 0.05N; 5.75Al; 3.0Sn; 0.50Fe; rem Ti

Sizes Available: .020, .025, .030, .035, .040, .045, 1/16, 3/32, 1/8″
18″ lengths - 36″ lengths - 30# spools

AMS 4954 Wire

Ti-6Al-4V titanium alloy is often welded with the gas tungsten arc or gas metal arc welding process, provided adequate inert gas shielding of the filler wire and deposited weld metal is used. As with all titanium alloys, the affinity of hot weld metal for oxygen, nitrogen, and hydrogen is high. Properly shielded deposits are sound and ductile and perform well in airframe and turbine engine parts (blades, discs, wheels, spacer rings), ordinance equipment, pressure vessels and rocket motor cases.

Mechanical Properties:

Tensile Strength, ksi: 130
Yield Strength, (0.2% offset) ksi: 120
Elongation, %: 10

Chemical Composition, % (max except Ti):

0.05C; 0.18O; 0.015H; 0.03N; 6.75Al; 4.50V; 0.30Fe; oth. 0.40

Sizes Available: .020, .025, .030, .035, .040, .045, 1/16, 3/32, 1/8″
18″ lengths - 36″ lengths - 30# spools

AMS 4955 Wire

Titanium alloy 8Al-1Mo-1V filler metal is for gas metal arc and gas tungsten arc welding of base metals of similar analysis.

Chemical Composition, % (max except Ti):

8.0Al; 1.0Mo; 1.0V; 0.30Fe; 0.12O; 0.08C; 0.05N; 0.01H; oth. 0.40; rem Ti

Sizes Available: .020, .025, .030, .035, .040, .045, 1/16, 3/32, 1/8″
18″ lengths - 36″ lengths - 30# spools

AMS 4956 Wire

Titanium alloy 6Al-4V-ELI is similar to type 6Al-4V in chemical analysis with the exception of extra low-interstitial gases (ELI). By controlling these contaminating gases, the fracture toughness is greatly increased. Alloy 4956 is used to weld liquid hydrogen tankage and airframe parts requiring a high fracture toughness.

Mechanical Properties:

Tensile Strength, ksi: 130 (-196°C/-320°F = 218)
Yield Strength, (0.2% offset) ksi: 120 (-196°C/-320°F = 202)
Elongation, %: 10 (-196°C/-320°F = 13.5)

Chemical Composition, %

0.03C; 0.08O; 0.005H; 0.012N; 6.75Al; 4.50V; 0.15Fe; 0.10Mn; oth. 0.10; rem Ti

Sizes Available: .020, .025, .030, .035, .040, .045, 1/16, 3/32, 1/8″
18″ lengths - 36″ lengths - 30# spools

Appendix E

Symbols

INTRODUCTION

This Appendix includes a listing of Greek symbols and their corresponding definitions. (Refer to Table E.1.) The symbols are commonly used in the more technical discussions of titanium. Understanding their meanings is especially helpful when reading graphs and figures in the general published literature involving titanium.

Table E.1 Greek symbols

Symbol	Description	Symbol	Description
α	Hexagonal-close-packed (hcp) crystal structure in Ti-base alloys	ν	Poisson's ratio
α'	Solute-lean, hcp, martensite	ρ	Electrical resistivity
α''	Solute-rich, orthorhombic, martensite	ρ_d	Density (mass)
β	Body-centered-cubic (bcc) crystal structure in Ti-base alloys	σ	Stress
β'	Solute-lean β-phase	σ_β	Fracture strength; in materials with limited ductility, $\sigma_\beta = Y_{ULT}$
β''	Solute-rich β-phase	σ_f	Flow stress
γ	Intermetallic compound (generic) in Ti-alloy binary phase diagrams	$\sigma_{0.01}$	Stress at 0.01% strain; the proportional limit $\sigma_{0.2}$
γ	Ordered face-centered-tetragonal compound of composition near TiAl in Ti-Al alloys	$\sigma_{0.2}$	Stress at 0.2% strain; the "0.2%-offset" yield stress
δ_β	Elongation of a tensile-test specimen just prior to fracture	ϕ	As superscript, this signifies the diameter of a wire, small sphere, electron beam, etc.
ϵ	Strain, % deformation	ϕ_c	Texturization parameter in magnetic texture measurement
θ	Angle of rotation	χ	Magnetic susceptibility
μm	Micrometer, micron, 10^{-6} m	ω	ω phase

E

Appendix F

Weights and Conversions

APPENDIX SUMMARY

The information contained in this Appendix assists the user on the most practical level—that of determining the overall weights of the various titanium grades in a variety of product forms. Additionally, various types of more general conversion tables are also provided.

For ease of reference, the following list indicates each table's contents:

F.1 - CP and Alloyed Titanium Weights

F.2 - Flat Titanium Bar Weights

F.3 - Rounds, Squares, Bar, Billet, Wire Weights

F.4 - Rectangular Bar Weights

F.5 - Sheet Weights

F.6 - Plate Weights

F.7 - Wire Weights

F.8 - Ti-6Al-4V Bar, Coil, Wire Weights

F.9 - Pipe Data, Weights

F.10 - Tube, Pipe Weights

F.11 - Mils/in./mm Conversions

F.12 - Mesh-Sieve Relations

A quick summary of conversion facts and formulas is contained in Figure F.1.

Note that the data contained in the following tables must be considered close approximations, not precise constants. Consult your supplier in instances when precision is required, since manufacturing processes affect weights directly.

- **Elemental titanium's density is nominally computed on the basis of 0.163 lb/in.^{3}**

- **Apply the following factors to determine the density of commercially pure Ti and various Ti alloys:**

CP	**x 1.019**
Ti-6Al-6V-2Sn	**x 1.025**
Ti-8Mo-8V-2Fe-3Al	**x 1.094**
Ti-13V-11Cr-3Al	**x 1.094**

  (See Table F.8 for Ti-6Al-4V.)

- **To determine the weight of a flat bar, multiply width by thickness by a density of 0.163/in.**

- **To determine the weight of a round bar, multiply diameter squared by a density of 1.536/ft.**

- **To determine tube weight per foot, use this formula: 6.145 x (OD - wall thickness) x wall thickness.**

- **To determine the theoretical average tube wall, divide the specified minimum wall by 0.9.**

- **Always consider the results of these formulas as approximations. Consult individual suppliers for more precise data.**

- **All computations and factors in this figure are based on English units.**

F

Figure F.1 - Summary of conversion facts and formulas

Table F.1 CP and Ti alloy weights (lb/sq ft) (a)

GAGE	Lbs. sq. ft.	Mtl. (b)	GAGE	Lbs. sq. ft.	Mtl. (b)	GAGE	Lbs. sq. ft.	Mtl. (b)	GAGE	Lbs. sq. ft.	Mtl. (b)
.008	.188	s	.060	1.408	s	.125	2.934	s	.625	14.670	p
.010	.235	s	.063	1.479	s	.130	3.051		.750	17.604	p
.012	.282	s	.071	1.667		.140	3.286		1.000	23.472	
.016	.376	s	.075	1.760		.150	3.521				
.020	.469	s	.080	1.878	s	.160	3.756				
.025	.587	s	.085	1.995		.170	3.990				
.030	.704		.090	2.112		.180	4.225				
.032	.751	s	.095	2.230		.187	4.389				
.035	.822	s	.100	2.347		.1875	4.401	p			
.040	.939	s	.106	2.488		.250	5.868				
.045	1.056		.112	2.629		.375	8.802				
.050	1.174	s	.118	2.770		.500	11.736	p			
.056	1.314	s									

(a) Table computations based on density of 0.063 lb/cu in.

(b) Mtl. = material; s = sheet; p = plate

Table courtesy of Supra Alloys, Inc.

Table F.2 Ti flat bar weights (lb/ft)

THICKNESS Inches	WIDTH—Inches								
	1/2	5/8	3/4	7/8	1	1-1/4	1-1/2	1-3/4	2
1/4	0.244	0.305	0.366	0.427	0.488	0.611	0.733	0.855	0.978
5/16	0.305	0.382	0.458	0.534	0.611	0.763	0.916	1.070	1.222
3/8	0.366	0.458	0.549	0.641	0.733	0.916	1.100	1.284	1.467
7/16	0.427	0.534	0.641	0.748	0.855	1.070	1.284	1.497	1.712
1/2	0.488	0.611	0.733	0.855	0.978	1.222	1.467	1.712	1.956
9/16	0.549	0.687	0.824	0.962	1.100	1.375	1.650	1.925	2.200
5/8	0.611	0.763	0.916	1.070	1.222	1.528	1.834	2.139	2.445
11/16	0.672	0.839	1.008	1.177	1.344	1.681	2.017	2.353	2.690
3/4	0.733	0.916	1.100	1.284	1.467	1.834	2.200	2.567	2.934
13/16	0.794	0.993	1.192	1.391	1.589	1.986	2.385	2.780	3.178
7/8	0.855	1.070	1.284	1.497	1.712	2.139	2.567	2.994	3.423
15/16	0.916	1.146	1.375	1.604	1.834	2.293	2.751	3.209	3.668
1	0.978	1.222	1.467	1.712	1.956	2.445	2.934	3.423	3.912
1-1/8	1.100	1.375	1.650	1.925	2.200	2.750	3.301	3.852	4.401
1-1/4	1.222	1.528	1.834	2.139	2.445	3.056	3.668	4.278	4.890
1-3/8	1.345	1.681	2.017	2.353	2.690	3.361	4.035	4.707	5.380
1-1/2	1.467	1.834	2.200	2.567	2.934	3.668	4.401	5.135	5.868
1-5/8	1.589	1.986	2.385	2.780	3.178	3.973	4.768	5.562	6.356
1-3/4	1.712	2.139	2.567	2.994	3.423	4.278	5.135	5.990	6.846
1-7/8	1.834	2.292	2.751	3.209	3.668	4.586	5.502	6.419	7.336
2	1.956	2.445	2.934	3.423	3.912	4.890	5.868	6.846	7.824

(Continued)

Table 11.2 (Continued)

THICKNESS Inches	2-1/4	2-1/2	2-3/4	3	3-1/4	3-1/2	3-3/4	4	THICKNESS Inches
				WIDTH—Inches					
1/4	1.100	1.222	1.344	1.467	1.589	1.712	1.833	1.956	1/4
5/16	1.375	1.528	1.681	1.834	1.986	2.139	2.292	2.444	5/16
3/8	1.650	1.834	2.017	2.200	2.385	2.567	2.751	2.934	3/8
7/16	1.925	2.139	2.353	2.567	2.780	2.994	3.209	3.424	7/16
1/2	2.200	2.445	2.690	2.934	3.178	3.423	3.668	3.912	1/2
9/16	2.476	2.751	3.027	3.301	3.576	3.852	4.125	4.400	9/16
5/8	2.751	3.056	3.361	3.668	3.973	4.278	4.584	4.890	5/8
11/16	3.026	3.361	3.699	4.035	4.370	4.707	5.043	5.380	11/16
3/4	3.301	3.668	4.035	4.401	4.768	5.135	5.501	5.868	3/4
13/16	3.576	3.973	4.370	4.768	5.165	5.562	5.958	6.356	13/16
7/8	3.852	4.278	4.707	5.135	5.562	5.990	6.417	6.846	7/8
15/16	4.126	4.585	5.043	5.502	5.959	6.419	6.877	7.336	15/16
1	4.401	4.890	5.380	5.868	6.356	6.846	7.335	7.824	1
1-1/8	4.957	5.501	6.051	6.602	7.151	7.701	8.253	8.802	1-1/8
1-1/4	5.501	6.112	6.725	7.335	7.946	8.558	9.168	9.780	1-1/4
1-3/8	6.051	6.725	7.397	8.069	8.740	9.413	10.087	10.760	1-3/8
1-1/2	6.602	7.335	8.069	8.802	9.536	10.270	11.003	11.736	1-1/2
1-5/8	7.151	7.948	8.741	9.536	10.330	11.124	11.918	12.712	1-5/8
1-3/4	7.701	8.558	9.413	10.270	11.124	11.981	12.836	13.692	1-3/4
1-7/8	8.252	9.170	10.086	11.004	11.920	12.838	13.755	14.672	1-7/8
2	8.802	9.780	10.758	11.730	12.712	13.692	14.670	15.648	2

Table courtesy of Supra Alloys, Inc.

Table F.3 Ti Rounds, squares, bar, billet, and wire weights (lb/ft)

SIZE (Inches)	ROUNDS (Lbs/Ft)	SQUARES (Lbs/Ft)	SIZE (Inches)	ROUNDS (Lbs/Ft)	SQUARES (Lbs/Ft)
1/16	0.006	0.008	2-5/16	8.215	10.460
1/8	0.024	0.031	2-3/8	8.665	11.034
3/16	0.054	0.069	2-7/16	9.128	11.621
1/4	0.096	0.123	2-1/2	9.601	12.225
5/16	0.150	0.192	2-9/16	10.087	12.844
3/8	0.216	0.276	2-5/8	10.585	13.478
7/16	0.294	0.375	2-11/16	11.096	14.127
1/2	0.384	0.489	2-3/4	11.618	14.792
9/16	0.487	0.619	2-13/16	12.152	15.47
5/8	0.600	0.764	2-7/8	12.698	16.17
11/16	0.726	0.924	2-15/16	13.255	16.88
3/4	0.864	1.100	3	13.826	17.60
13/16	1.015	1.291	3-1/8	15.00	19.10
7/8	1.177	1.497	3-1/4	16.226	20.66
15/16	1.351	1.719	3-3/8	17.50	22.28
1	1.536	1.956	3-1/2	18.82	23.96
1-1/16	1.734	2.207	3-5/8	20.19	25.70
1-1/8	1.944	2.476	3-3/4	21.60	27.51
1-3/16	2.166	2.759	3-7/8	23.07	29.37
1-1/4	2.400	3.056	4	24.58	31.30
1-5/16	2.645	3.369	4-1/8	26.14	33.28
1-3/8	2.904	3.699	4-1/4	27.75	35.33
1-7/16	3.175	4.042	4-3/8	29.40	37.44
1-1/2	3.456	4.401	4-1/2	31.11	39.61
1-9/16	3.750	4.775	4-5/8	32.86	41.84
1-5/8	4.056	5.165	4-3/4	34.66	44.13
1-11/16	4.374	5.570	4-7/8	36.51	46.49
1-3/4	4.705	5.990	5	38.41	48.90
1-13/16	5.047	6.425	5-1/8	40.35	51.38
1-7/8	5.400	6.876	5-1/4	42.34	53.91
1-15/16	5.766	7.342	5-3/8	44.38	56.51
2	6.144	7.824	5-1/2	46.47	59.17
2-2/16	6.534	8.320	5-5/8	48.61	61.89
2-1/8	6.937	8.832	5-3/4	50.79	64.67
2-3/16	7.350	9.359	5-7/8	53.03	67.51
2-1/4	7.777	9.902	6	55.30	70.42

(Continued)

Table 11.3 (Continued)

SIZE (Inches)	ROUNDS (Lbs/Ft)	SQUARES (Lbs/Ft)
6-1/8	57.63	73.38
6-1/4	60.01	76.41
6-3/8	62.43	79.49
6-1/2	64.91	82.64
6-5/8	67.43	85.85
6-3/4	70.00	89.12
6-7/8	72.61	92.45
7	75.28	95.84
7-1/4	80.75	102.81
7-1/2	86.41	110.03
7-3/4	92.27	117.48
8	98.32	125.18
8-1/2	111.0	141.3
9	124.4	158.4
9-1/2	138.6	176.5
10	153.6	195.6
10-1/2	169.4	215.6
11	186.0	237.0
11-1/2	203.0	259.0
12	221.0	282.0
12-1/2	240.0	306.0
13	260.0	331.0
14	301.0	383.0
15	347.0	440.0
16	393.0	500.0
17	444.0	565.0
18	498.0	634.0
19	555.0	
20	614.0	
21	677.0	
22	744.0	
23	813.0	
24	885.0	

Table F.4 Ti rectangular bar weights (lb/linear ft)

width	gauge			
	3/16"	**1/4"**	**3/8"**	**1/2"**
1 inch	.367	.489	.733	.978
1¼"	.458	.611	.917	1.22
1½"	.550	.733	1.10	1.47
1¾"	.640	.856	1.28	1.71
2 inch	.732	.978	1.47	1.96

Table courtesy of Industrial Titanium Corp.

Table F.5 Ti sheet weights (lb/sq ft)

Thickness (Gauge)	Approx. Wgt. (lbs/sq ft)	Weight (lbs/36 x 96" sheet)
.012"	0.282	6.760
.016"	0.376	9.013
.020"	0.469	11.267
.025"	0.587	14.083
.032"	0.751	18.026
.040"	0.939	22.533
.050"	1.174	28.166
.063"	1.477	35.490
.070"	1.643	39.432
.075"	1.760	42.250
.080"	1.878	45.066
.090"	2.112	50.700
.093"	2.143	52.390
.110"	2.582	61.967
.125"	2.934	70.416

Table F.6 Ti plate weights (lb/sq ft)

Thickness (Gauge)	Approx. Wgt. (lbs/sq ft)
3/16″ (.187)	4.401
1/4″ (.250)	5.868
3/8″ (.375)	8.802
1/2″ (.500)	11.736
5/8″ (.625)	14.674
3/4″ (.750)	17.582
1″ (1.000)	23.443

Table courtesy of Industrial Titanium Corp.

Table F.7 Ti wire weights

SIZE	FT/lb.	Lbs/100 ft.
1/32″ (.032)	635	0.15748
.050	260	0.38462
1/16″ (.063)	169	0.60976
3/32″ (.093)	75	1.31579
1/8″ (.125)	42	2.40053
5/32″ (.156)	27	3.73860
3/16″ (.187)	19	5.26316
1/4″ (.250)	10.5	9.60150
5/16″ (.312)	6.7	15.00234

Table courtesy of Industrial Titanium Corp.

Table F.8 Ti-6Al-4V bar, coil, and wire weights (lb/100 ft; ft/lb)

Diameter, Inches	Lbs per 100 Ft	Ft per Pound	Diameter, Inches	Lbs per 100 Ft	Ft per Pound	Size, Inches	Rounds, Lbs/Ft	Square Lbs/Ft
0.010	0.015	6,630	0.100	1.508	66	0.3125	0.147	0.188
0.012	0.022	4,605	0.105	1.663	60	0.375	0.212	0.270
0.014	0.030	3,380	0.110	1.825	54	0.4375	0.289	0.368
0.016	0.039	2,590	0.115	1.994	50	0.50	0.377	0.480
0.018	0.049	2,045	0.120	2.172	46	0.5625	0.477	0.608
0.020	0.060	1,655	0.125	2.356	42	0.625	0.589	0.750
0.022	0.073	1,370	0.130	2.549	39	0.6875	0.713	0.908
0.024	0.087	1,150	0.135	2.748	36	0.750	0.848	1.08
0.026	0.102	980	0.140	2.956	33	0.8125	0.995	1.27
0.028	0.118	845	0.145	3.170	31	0.875	1.15	1.47
0.030	0.136	735	0.150	3.393	29	0.9375	1.33	1.69
0.032	0.154	645	0.155	3.623	27	1.0	1.51	1.92
0.034	0.174	570	0.160	3.860	25	1.0625	1.70	2.17
0.036	0.195	510	0.165	4.105	24	1.125	1.91	2.43
0.038	0.218	455	0.170	4.358	22	1.1875	2.13	2.71
0.040	0.241	414	0.175	4.618	21	1.25	2.36	3.00
0.042	0.266	375	0.180	4.886	20	1.3125	2.60	3.31
0.044	0.292	342	0.185	5.161	19	1.375	2.85	3.63
0.046	0.319	313	0.190	5.444	18	1.4375	3.12	3.97
0.048	0.347	287	0.195	5.734	17	1.50	3.39	4.32
0.050	0.377	265	0.200	6.032	16	1.5675	3.68	4.69
0.052	0.408	245	0.205	6.337	15	1.625	3.98	5.07
0.054	0.440	227	0.210	6.650	15	1.6875	4.29	5.47
0.056	0.473	211	0.215	6.971	14	1.75	4.62	5.88
0.058	0.507	197	0.220	7.299	13	1.8125	4.95	6.31
0.060	0.543	184	0.225	7.634	13	1.875	5.30	6.75
0.062	0.580	172	0.230	7.977	12	1.9375	5.66	7.21
0.064	0.618	161	0.235	8.328	12	2.0	6.03	7.68
0.068	0.697	143	0.240	8.686	11	2.0625	6.41	8.17
0.070	0.739	135	0.245	9.052	11	2.125	6.81	8.67
0.072	0.782	127	0.250	9.423	10	2.1875	7.22	9.19
0.074	0.826	121	0.255	9.806	10	2.25	7.63	9.72
0.076	0.871	114	0.260	10.194	9	2.3125	8.06	10.27
0.078	0.917	108	0.265	10.590	9	2.375	8.51	10.83
0.080	0.965	103	0.270	10.993	9	2.4375	8.96	11.41
0.082	1.014	98	0.275	11.404	8	2.50	9.43	12.00
0.084	1.064	93	0.280	11.822	8	2.5625	9.90	12.61
0.086	1.115	89	0.285	12.248	8	2.625	10.39	13.23
0.088	1.168	85	0.290	12.682	7	2.6875	10.89	13.87
0.090	1.221	81	0.295	13.123	7	2.75	11.40	14.52
0.092	1.276	78	0.300	13.572	7	2.8125	11.93	15.19
0.094	1.332	75	0.305	14.028	7	2.875	12.46	15.87
0.096	1.390	71	0.310	14.492	7	2.9375	13.01	16.57
0.098	1.448	69	0.315	14.963	6	3.0	13.57	17.28
						3.0625	14.14	18.01

Weights are based on a density of .160 lbs/in³. Apply following factors for other grades:

		3.125	14.73	18.75
Ti-6Al-6V-2Sn	x 1.025	3.1875	15.32	19.51
C.P.	x 1.019	3.25	15.93	20.28
Ti-13V-11Cr-3Al	x 1.094	3.3125	16.55	21.07
Ti-8Mo-8V-2Fe-3Al	x 1.094	3.375	17.18	21.87
		3.4375	17.82	22.69
		3.50	18.47	23.52

Table courtesy of Dynamet, Inc.

Table F.9 Ti pipe weights (wt/ft by lb)

PIPE SIZE	Outside Diameter (inches)	Schedule 10 Light Weight		Schedule 40 Standard Weight		Schedule 80 Extra Heavy Weight	
		Wall (inches)	Wt/ft (lbs)	Wall (inches)	Wt/ft (lbs)	Wall (inches)	Wt/ft (lbs)
⅛ "	0.405	0.049	0.107	0.068	0.141	0.095	0.181
¼	0.540	0.065	0.182	0.088	0.244	0.119	0.308
⅜	0.675	0.065	0.244	0.091	0.326	0.126	0.425
½	0.840	0.083	0.386	0.109	0.489	0.147	0.626
¾	1.050	0.083	0.493	0.113	0.650	0.154	0.848
1	1.315	0.109	0.807	0.133	0.965	0.179	1.249
1¼ "	1.660	0.109	1.038	0.140	1.307	0.191	1.723
1½	1.900	0.109	1.199	0.145	1.563	0.200	2.088
2	2.375	0.109	1.517	0.154	2.100	0.218	2.888
2½	2.875	0.120	2.030	0.203	3.331	0.276	4.405
3	3.500	0.120	2.491	0.216	4.356	0.300	5.894
3½	4.000	0.120	2.849	0.226	5.238	0.318	7.188
4	4.500	0.120	3.227	0.237	6.204	0.337	8.614

Table courtesy of Industrial Titanium Corp.

Table F.10 Ti tube and pipe weights (lb/linear ft)

Wall	1/8"	3/16	1/4	5/16	3/8	7/16	1/2	9/16	5/8	3/4	7/8	1	1 1/4	1 3/8	1 1/2	1 5/8	1 3/4	1 7/8	2	2 1/4	2 3/8	2 1/2	2 3/4	3
.020"	.013	.020	.028	.035	.043	.051	.058	.066	.073	.089	.104	.119	.151											
.025	.015	.025	.034	.044	.053	.063	.072	.081	.091	.110	.129	.148	.186	.207	.226									
.028	.016	.027	.038	.048	.059	.069	.080	.091	.101	.123	.144	.165	.208	.229	.253	.274								
.032	.018	.030	.042	.055	.067	.079	.091	.103	.115	.140	.164	.188	.237	.261	.285	.313	.337							
.035	.019	.032	.046	.060	.073	.086	.100	.113	.126	.154	.180	.205	.259	.287	.312	.341	.368	.395						
.042	.021	.037	.053	.069	.086	.101	.118	.133	.151	.182	.215	.244	.309	.340	.373	.405	.437	.472	.505					
.049		.041	.060	.079	.098	.116	.135	.153	.173	.211	.248	.286	.361	.395	.436	.469	.512	.549	.587	.662				
.058			.068	.090	.113	.134	.157	.178	.202	.246	.291	.336	.424	.465	.513	.558	.603	.647	.686	.780	.825	.870	.959	1.04
.065			.073	.098	.123	.147	.173	.196	.223	.273	.323	.373	.473	.520	.573	.622	.673	.722	.773	.872	.922	.972	1.07	1.17
.072				.106	.134	.160	.189	.215	.244	.300	.355	.411	.521	.571	.631	.686	.742	.797	.853	.963	1.01	1.07	1.18	1.29
.083					.148	.179	.212	.242	.276	.340	.404	.468	.595	.651	.722	.786	.850	.913	.977	1.10	1.16	1.23	1.35	1.48
.095					.163	.198	.236	.270	.309	.382	.455	.528	.674	.741	.820	.892	.966	1.03	1.11	1.25	1.33	1.40	1.54	1.69
.109						.220	.261	.301	.345	.429	.513	.586	.764	.836	.932	1.01	1.09	1.18	1.26	1.43	1.51	1.60	1.76	1.93
.120							.280	.327	.372	.464	.556	.649	.833	.916	1.01	1.10	1.20	1.29	1.38	1.57	1.66	1.75	1.93	2.12

Outside diameter (in.)

Table courtesy of Industrial Titanium Corp.

Table F.11 Mills/in./mm conversions

Mils	Inches	Mm
0.01	0.00001	0.000254
0.05	0.00005	0.00127
0.1	0.0001	0.00254
0.5	0.0005	0.0127
1	0.001	0.0254
2	0.002	0.0508
3	0.003	0.0762
4	0.004	0.1016
5	0.005	0.1270
6	0.006	0.1524
7	0.007	0.1778
8	0.008	0.2032
9	0.009	0.2286
10	0.010	0.2540
15	0.015	0.3810
20	0.020	0.5080
25	0.025	0.6350
30	0.030	0.7620
35	0.035	0.8890
40	0.040	1.016
45	0.045	1.143
50	0.050	1.270
60	0.060	1.524
70	0.070	1.778
80	0.080	2.032
90	0.090	2.286
100	0.100	2.540
125	0.125	3.175
150	0.150	3.810
200	0.20	5.080
250	0.25	6.350
300	0.30	7.620
400	0.40	10.160
500	0.50	12.700
750	0.75	19.050
1,000	1.0	25.40
	2	50.80
	3	76.20
	4	101.60
	5	127.00
	6	152.40
	7	177.80
	8	203.20
	9	228.60
	10	254.00
	11	279.40
	12	304.80

Table F.12 Mesh-sieve relations

"Mesh" Sieve Designation	Sieve Opening inches	mm.
1	1.00	25.4
3/4	0.750	19.0
5/8	0.625	16.0
1/2	0.500	12.7
3/8	0.375	9.51
5/16	0.312	8.00
1/4	0.250	6.35
3-1/2	0.223	5.66
4	0.187	4.76
5	0.157	4.00
6	0.132	3.36
7	0.111	2.83
8	0.0937	2.38
10	0.0787	2.00
12	0.0661	1.68
14	0.0555	1.41
16	0.0469	1.19
18	0.0394	1.00
		Microns
20	0.0331	841
25	0.0278	707
30	0.0234	595
35	0.0197	500
40	0.0165	420
45	0.0139	354
50	0.0117	297
60	0.0098	250
70	0.0083	210
80	0.0070	177
100	0.0059	149
120	0.0049	125
140	0.0041	105
170	0.0035	88
200	0.0029	74
230	0.0025	63
270	0.0021	53
325	0.0017	44
400	0.0015	37

A – mesh designation denotes all particles are smaller than the indicated opening.
A + mesh designation denotes all particles are larger than the indicated opening.
Example: – 3/8 + 10 mesh, particle sizes range from 2.00–9.51 mm.

Appendix G

General Corrosion Rates

GENERAL

Although titanium alloys are still vital to the aerospace industry, recognition of the excellent resistance of titanium to many highly corrosive environments, particularly oxidizing and chloride-containing process streams, has led to widespread nonaerospace (industrial) applications. Because of decreasing cost and the increasing availability of titanium alloy products, of them have become standard engineering materials for a host of common industrial applications. In fact, a growing trend involves the use of high-strength, aerospace-founded titanium alloys for industrial service in which the combination of strength-to-density and also corrosion-resistance properties is critical and desirable.

The excellent corrosion resistance of titanium alloys results from the formation of very stable, continuous, highly adherent, and protective oxide films on metal surfaces. Because titanium metal itself is highly reactive and has an extremely high affinity for oxygen, these beneficial surface oxide films form spontaneously and instantly when fresh metal surfaces are exposed to air and/or moisture. In fact, a damaged oxide film can generally reheal itself instantaneously if at least traces (that is, parts per million) of oxygen or water (moisture) are present in the environment. However, anhydrous conditions in the absence of a source of oxygen may result in titanium corrosion, because the protective film may not be regenerated if damaged.

The nature, composition, and thickness of the protective surface oxides that form on titanium alloys depend on environmental conditions. In most aqueous environments, the oxide is typically TiO_2, but may consist of mixtures of other titanium oxides, including TiO_2, Ti_2O_3, and TiO. High-temperature oxidation tends to promote the formation of the chemically resistant, highly crystalline form if TiO_2 known as rutile, whereas lower temperatures often generate the more amorphous form of TiO_2, anatase, or a mixture of rutile and anatase. Although these naturally formed films are typically less than 10 nm thick and are invisible to the eye, the TiO_2 oxide is highly chemically resistant and is attacked by very few substances, including hot, concentrated HCl, H_2SO_4, NaOH, and (most notably) HF. This thin surface oxide is also a highly effective barrier to hydrogen.

Furthermore, the TiO_2 film, being an n-type semiconductor, possesses electronic conductivity. As a cathode, titanium permits electrochemical reduction of ions in an aqueous electrolyte. On the other hand, very high resistance to anodic current flow through the passive oxide film can be expected in most aqueous solutions. Because the passivity of titanium stems from the formation of a stable oxide film, an understanding of the corrosion behavior of titanium is obtained by recognizing the conditions under which this oxide is thermodynamically stable.

Thus, successful use of titanium alloys can be expected in mildly reducing-to-highly oxidizing environments in which protective TiO_2 and Ti_2O_3 films form spontaneously and remain stable. On the other hand, uninhibited, strongly reducing acidic environments may attack titanium, particularly as temperature increases. However, shifting the alloy potential in the noble (positive) direction by various means can induce stable oxide film formation, often overcoming the corrosion resistance limitations of titanium alloys in normally aggressive reducing media.

CORROSION RATES

This Appendix is a compilation of general corrosion rate values for the two broad groupings of titanium-based alloys:

- Commercially pure, ASTM Grades 1 thru 4 (Table G.1)

- Other titanium alloys (Table G.2)

These values were derived from various published sources, and, also, from in-house laboratory tests.

Table G.3 supports the user by breaking out into more readable forms the abbreviated designations used in the other two tables.

This data should be used only as a starting point or initial guideline for titanium's corrosion performance. Actual rates may vary, depending on changes in media chemistry, temperature, length of exposure, and various other factors. Also, note that the suitability of a specific CP grade or alloy for an application cannot necessarily be inferred from these values alone, since other modes of corrosion, such as localized attack, may be limiting. In complex, variable and/or dynamic environments, *in situ* testing may provide more reliable data.

Note: tables in this appendix were contributed by **Ronald W. Schutz**, Supervisor, Industrial Product Development and Corrosion Research, TIMET Corp., Henderson, NV.

Table G.1 Commercially pure titanium corrosion rates (a)

Medium	Concentration, %	Temperature, °C	Corrosion rate, mm/yr
Acetaldehyde	75	149	0.001
	100	149	nil
Acetate, *n*-propyl	...	87	nil
Acetic acid	5–99.7	124	nil
	33–vapor	Boiling	nil
	99	Boiling	0.003
	65	121	0.003
	58	130	0.381
	99.7	124	0.003
Acetic acid + 3% acetic anhydride	Glacial	204	1.02
Acetic acid + 1.5% acetic anhydride	Glacial	204	0.005
Acetic acid + 109 ppm Cl	31.2	Boiling	0.259

(Continued)

Table G.1 (Continued)

Medium	Concentration, %	Temperature, °C	Corrosion rate, mm/yr
Acetic acid + 106 ppm Cl	62.0	Boiling	0.272
Acetic acid + 5% formic acid	58	Boiling	0.457
Acetic anhydride	100	21	0.025
	100	150	0.005
	99.5	Boiling	0.013
Adipic acid + 15–20% glutaric + 2% acetic acid	25	199	nil
Adipic acid	67	240	nil
Adipylchloride and chlorobenzene solution	. . .	. . .	nil
Adiponitrile	Vapor	371	0.008
Aluminum chloride, aerated	10	100	0.002
	25	100	3.15
Aluminum chloride	10	100	0.002
	10	150	0.03
	25	60	nil
	25	100	6.55
Aluminum	Molten	677	164.6
Aluminum fluoride	Saturated	Room	nil
Aluminum nitrate	Saturated	Room	nil
Aluminum sulfate	Saturated	Room	nil
	10	80	0.05
	10	Boiling	0.12
Aluminum sulfate + 1% H_2SO_4	Saturated	Room	nil
Ammonium acid phosphate	10	Room	nil
Ammonium aluminum chloride	Molten	350–380	Very rapid attack
Ammonia, anhydrous	100	40	<0.127
Ammonia, steam, water	. . .	222	11.2
Ammonium acetate	10	Room	nil
Ammonium bicarbonate	50	100	nil
Ammonium bisulfite, pH 2.05	Spent pulping liquor	71	0.015
Ammonium carbamate	50	100	nil
Ammonium chloride	Saturated	100	<0.013
Ammonium chlorate	300 g/L	50	0.003
Ammonium fluoride	10	Room	0.102
Ammonium hydroxide	28	Room	0.003
	28	100	nil
Ammonium nitrate	28	Boiling	nil
Ammonium nitrate + 1% nitric acid	28	Boiling	nil
Ammonium oxalate	Saturated	Room	nil
Ammonium perchlorate	20	88	nil
Ammonium sulfate	10	100	nil
Ammonium sulfate + 1% H_2SO_4	Saturated	Room	0.010
Aniline	100	Room	nil
Aniline + 2% $AlCl_3$	98	158	>1.27
Aniline hydrochloride	5	100	nil
	20	100	nil
Antimony trichloride	27	Room	nil
Aqua regia	3:1	Room	nil
	3:1	80	0.86
	3:1	Boiling	1.12
Arsenous oxide	Saturated	Room	nil
Barium carbonate	Saturated	Room	nil
Barium chloride	5	100	nil
	20	100	nil
	25	100	nil
Barium hydroxide	Saturated	Room	nil
Barium nitrate	10	Room	nil
Barium fluoride	Saturated	Room	nil
Benzaldehyde	100	Room	nil
Benzene (traces of HCl)	Vapor and liquid	80	0.005
	Liquid	50	0.025

(Continued)

Table G.1 (Continued)

Medium	Concentration, %	Temperature, °C	Corrosion rate, mm/yr
Benzene.............................	Liquid	Room	nil
Benzoric acid	Saturated	Room	nil
Bismuth	Molten	816	High
Bismuth/lead.........................	Molten	300	Good resistance
Boric acid	Saturated	Room	nil
	10	Boiling	nil
Bromine.............................	Liquid	30	Rapid attack
Bromine, moist........................	Vapor	30	<0.003
Bromine gas, dry	· · ·	21	Dissolves rapidly
Bromine-water solution	· · ·	Room	nil
Bromine in methyl alcohol..............	0.05	60	0.03 (cracking possible)
N-butyric acid........................	Undiluted	Room	nil
Calcium bisulfite......................	Cooking liquor	26	0.001
Calcium carbonate.....................	Saturated	Boiling	nil
Calcium chloride	5	100	0.005
	10	100	0.007
	20	100	0.015
	55	104	0.001
	60	149	<0.003
	62	154	0.406
	73	175	0.80
Calcium hydroxide.....................	Saturated	Room	nil
	Saturated	Boiling	nil
Calcium hypochlorite	2	100	0.001
	6	100	0.001
	18	21	nil
	Saturated	21	nil
Carbon dioxide........................	100	. . .	Excellent
Carbon tetrachloride	99	Boiling	0.005
	Liquid	Boiling	nil
	Vapor	Boiling	nil
Carbon tetrachloride + 50% H_2O........	50	25	0.005
Chlorine gas, wet......................	>0.7 H_2O	Room	nil
	>0.95 H_2O	140	nil
	>1.5 H_2O	200	nil
Chlorine saturated water	Saturated	97	nil
Chlorine gas, dry	<0.5 H_2O	Room	May react
Chlorine dioxide.......................	5	82	<0.003
Chlorine dioxide + HOCl, H_2O + Cl_2	15	43	nil
Chlorine dioxide in steam	5	99	nil
Chlorine dioxide.......................	10	70	0.03
Chlorine monoxide (moist)..............	Up to 15	43	nil
Chlorine trifluoride	100	30	Vigorous reaction
Chloracetic acid.......................	30	82	<0.127
	100	Boiling	<0.127
Chlorosulfonic acid	100	Room	0.312
Chloroform	Vapor and liquid	Boiling	0.000
Chloroform + 50% H_2O..............	50	25	0.000
Chloropicrin	100	95	0.003
Chromic acid.........................	10	Boiling	0.003
	15	24	0.006
	15	82	0.015
	50	24	0.013
	50	82	0.028
Chromic acid + 5% nitric acid	5	21	<0.003
Citric acid	10	100	0.009
	25	100	0.001
	50	60	0.000

Table G.1 (Continued)

Medium	Concentration, %	Temperature, °C	Corrosion rate, mm/yr
	50	Boiling	0.127–1.27
	672	149	Corroded
Citric acid (aerated).....................	50	100	<0.127
Copper nitrate........................	Saturated	Room	nil
Copper sulfate	50	Boiling	nil
Copper sulfate + 2% H_2SO_4	Saturated	Room	0.018
Cupric carbonate + cupric hydroxide.....	Saturated	Ambient	nil
Cupric chloride........................	20	Boiling	nil
	40	Boiling	0.005
	55	118	0.003
Cupric cyanide	Saturated	Room	nil
Cuprous chloride	50	90	<0.003
Cyclohexylamine	100	Room	nil
Cyclohexane (plus traces of formic acid).........................	· · ·	150	0.003
Dichloroacetic acid	100	Boiling	0.007
Dichlorobenzene + 4–5% HCl	· · ·	179	0.102
Diethylene triamine	100	Room	nil
Ethyl alcohol.........................	95	Boiling	0.013
	100	Room	nil
Ethylene dichloride	100	Boiling	0.005–0.127
Ethylene dichloride + 50% water.........	50	25	0.005
Ethylene diamine......................	100	Room	nil
Ferric chloride	10–20	Room	nil
	1–30	100	0.004
	10–40	Boiling	nil
	1–30	Boiling	nil
	50	150	0.003
Ferric chloride	10	Boiling	0.00
Ferric sulfate.........................	10	Room	nil
Ferrous chloride + 0.5% HCl	30	79	0.006
Ferrous sulfate	Saturated	Room	nil
Fluoboric acid........................	5–20	Elevated	Rapid attack
Fluorine, commercial	Gas–liquid	Gas-109	0.864
Fluorine, HF free......................	Liquid	-196	0.011
	Gas	-196	0.011
Fluorosilicic acid	10	Room	47.5
Formaldehyde.........................	37	Boiling	nil
Formamide vapor......................	· · ·	300	nil
Formic acid, aerated	10	100	0.005
	25	100	0.001
	50	100	0.001
	90	100	0.001
Formic acid, nonaerated................	10	100	nil
	25	100	2.44
	50	Boiling	3.20
	90	100	3.00
Formic acid...........................	9	50	<0.127
Furfural	100	Room	nil
Gluconic acid	50	Room	nil
Glycerin.............................	· · ·	Room	nil
Hydrogen chloride, gas.................	Air mixture	25–100	nil
Hydrochloric acid, aerated	1	60	0.004
	2	60	0.016
	5	60	1.07
	1	100	0.46
	5	35	0.01
	10	35	1.02
	20	35	4.45
Hydrochloric acid	0.1	Boiling	0.10
	1	Boiling	1.8
Hydrochloric acid + 4% $FeCl_3$ + 4% $MgCl_2$	19	82	0.51
Hydrochloric acid + 4% $FeCl_3$ + 4% $MgCl_2$ + Cl_2 saturated	19	82	0.46

(Continued)

Table G.1 (Continued)

Medium	Concentration, %	Temperature, °C	Corrosion rate, mm/yr
Hydrochloric acid, chlorine saturated	5	190	<0.025
	10	190	28.5
Hydrochloric acid, + 200 ppm Cl₂	36	25	0.432
Hydrochloric acid			
+1% HNO₃	5	40	nil
+1% HNO₃	5	95	0.091
+5% HNO₃	5	40	0.025
+5% HNO₃	5	95	0.030
+10% HNO₃	5	40	nil
+10% HNO₃	5	95	0.183
+3% HNO₃	8.5	80	0.051
+5% HNO₃	1	Boiling	0.074
Hydrochloric acid			
+2.5% NaClO₃	10.2	80	0.009
+5.0% NaClO₃	10.2	80	0.006
Hydrochloric acid			
+0.5% CrO₃	5	38	nil
+0.5% CrO₃	5	95	0.031
+1% CrO₃	5	38	0.018
+1% CrO₃	5	95	0.031
Hydrochloric acid			
+0.05% CuSO₄	5	38	0.040
+0.05% CuSO₄	5	93	0.091
+0.5% CuSO₄	5	38	0.091
+0.5% CuSO₄	5	93	0.061
+1% CuSO₄	5	38	0.031
+1% CuSO₄	5	93	0.091
+5% CuSO₄	5	38	0.020
+5% CuSO₄	5	93	0.061
+0.05% CuSO₄	5	Boiling	0.064
+0.5% CuSO₄	5	Boiling	0.084
Hydrochloric acid			
+0.05% CuSO₄	10	66	0.025
+0.20% CuSO₄	10	66	nil
+0.5% CuSO₄	10	66	0.023
+1% CuSO₄	10	66	0.023
+0.05% CuSO₄	10	Boiling	0.295
+0.5% CuSO₄	10	Boiling	0.290
Hydrochloric acid + 0.1% FeCl₃	5	Boiling	0.01
Hydrochloric acid + 1 g/L Ti⁴⁺	10	Boiling	0.000
Hydrochloric acid + 5.8 g/L Ti⁴⁺	20	Boiling	0.000
Hydrochloric acid + 18% H₃PO₄ +			
5% HNO₃	18	77	0.000
Hydrofluoric acid	1	26	127
Hydrofluoric acid, anhydrous	100	Room	0.127–1.27
Hydrofluoric-nitric acid 5 vol%			
HF-35 vol% HNO₃	. . .	25	452
Hydrofluoric-nitric acid 5 vol%			
HF-35 vol% HNO₃	. . .	35	571
Hydrogen peroxide	3	Room	<0.127
	6	Room	<0.127
	30	Room	<0.305
Hydrogen peroxide			
+ 2% NaOH	1	60	55.9
Hydrogen peroxide			
pH 4	5	66	0.061
pH 1	5	66	0.152
pH 1	20	66	0.69
pH 11	0.08	70	0.42
Hydrogen sulfide (water saturated)	. . .	21	<0.003
Hydrogen sulfide, steam,			
and 0.077% mercaptans	7.65	93–110	nil
Hydroxy-acetic acid	. . .	40	0.003
Hypochlorous acid + ClO			
and Cl₂ gases	17	38	0.000

(Continued)

Table G.1 (Continued)

Medium	Concentration, %	Temperature, °C	Corrosion rate, mm/yr
Iodine, dry or moist gas.................	· · ·	25	0.1
Iodine in water + potassium iodide.......	· · ·	Room	nil
Iodine in alcohol	Saturated	Room	Pitted
Lactic acid..........................	10–85	100	<0.127
	10	Boiling	<0.127
Lead................................	· · ·	816	Attacked
	· · ·	324–593	Good
Lead acetate	Saturated	Room	nil
Linseed oil, boiled....................	· · ·	Room	nil
Lithium, molten	· · ·	316–482	nil
Lithium chloride......................	50	149	nil
Magnesium	Molten	760	Limited resistance
Magnesium chloride	5–20	100	<0.010
	5–40	Boiling	0.005
Magnesium hydroxide.................	Saturated	Room	nil
Magnesium sulfate....................	Saturated	Room	nil
Manganous chloride	5–20	100	nil
Maleic acid	18–20	35	0.002
Mercuric chloride.....................	1	100	0.000
	5	100	0.011
	10	100	0.001
	Saturated	100	0.001
Mercuric cyanide.....................	Saturated	Room	nil
Mercury.............................	100	Up to 38	Satisfactory
	100	Room	nil
	· · ·	371	3.03
Methyl alcohol	91	35	nil
	95	100	<0.01
Mercury + iron	· · ·	371	0.079
Mercury + copper.....................	· · ·	371	0.063
Mercury + zirconium	· · ·	371	0.033
Mercury + magnesium	· · ·	371	0.083
Monochloracetic acid	30	80	0.02
	100	Boiling	0.013
Nickel chloride.......................	5	100	0.004
	20	100	0.003
Nickel nitrate	50	Room	nil
Nitric acid, aerated	10	Room	0.005
	30	Room	0.004
	40	Room	0.002
	50	Room	0.002
	60	Room	0.001
	70	Room	0.005
	10	40	0.003
	20	40	0.005
	30	50	0.015
	40	50	0.016
	50	60	0.037
	60	60	0.040
	70	70	0.040
	40	200	0.610
	70	270	1.22
	20	290	0.305
Nitric acid	35	80	0.051–0.102
	70	80	0.025–0.076
	17	Boiling	0.076–0.102
	35	Boiling	0.127–0.508
	70	Boiling	0.064–0.900
Nitric acid, not refreshed..............	5–60	35	0.002–0.007
	5–60	60	0.01–0.02
	30–50	100	0.10–0.18
	5–20	100	0.02
	30–60	190	1.5–2.8
	70	270	1.2
	20	290	0.4

(Continued)

Table G.1 (Continued)

Medium	Concentration, %	Temperature, °C	Corrosion rate, mm/yr
	70	290	1.1
Nitric acid, white fuming	Liquid or vapor	Room	nil
	...	82	0.152
	...	122	<0.127
	...	160	<0.127
Nitric acid, red fuming	<About 2% H₂O	Room	Ignition sensitive
	>About 2% H₂O	Room	Not ignition sensitive
Nitric acid	40	Boiling	0.63
+0.01% K₂Cr₂O₇	40	Boiling	0.01
+0.01% CrO₃	40	Boiling	0.01
+0.01% FeCl₃	40	Boiling	0.68
+1% FeCl₃	40	Boiling	0.14
+1% NaClO₃	40	Boiling	0.31
+1% NaClO₃	40	Boiling	0.02
+1% Ce(SO₄)₂	40	Boiling	0.10
+0.1% K₂Cr₂O₇	40	Boiling	0.016
Nitric acid, saturated with zirconyl nitrate	33–45	118	nil
Nitric acid + 15% zirconyl nitrate	65	127	nil
Nitric acid + 179 g/L NaNO₃ and 32 g/L NaCl	20.8	Boiling	0.127–0.295
Nitric acid + 170 g/L NaNO₃ and 2.9 g/L NaCl	27.4	Boiling	0.483–2.92
Oxalic acid	1	35	0.03
	5	35	0.13
	1	Boiling	107
	25	60	11.9
	Saturated	Room	0.508
Perchloroethylene + 50% H₂O	50	25	nil
Perchloryl fluoride + liquid ClO₃	100	30	0.002
Perchloryl fluoride + 1% H₂O	99	30	Liquid 0.290
	...	...	Vapor 0.003
Phenol	Saturated solution	25	0.102
Phosphoric acid	10–30	Room	0.020–0.051
	30–80	Room	0.051–0.762
	5.0	66	0.005
	6.0	66	0.117
	0.5	Boiling	0.094
	1.0	Boiling	0.266
	12	25	0.005
	20	25	0.076
	50	25	0.19
	9	52	0.03
	10	52	0.38
	5	Boiling	3.5
	10	80	1.83
Phosphoric acid + 3% nitric acid	81	88	0.381
Phosphorus oxychloride	100	Room	0.004
Phosphorus trichloride	Saturated	Room	nil
Photographic emulsions	...	...	<0.127
Phthalic acid	Saturated	Room	nil
Potassium bromide	Saturated	Room	nil
Potassium chloride	Saturated	Room	nil
	Saturated	60	nil
Potassium dichromate	Saturated	Room	nil
Potassium ethyl xanthate	10	Room	nil
Potassium ferricyanide	Saturated	Room	nil
Potassium hydroxide + 13% potassium chloride	13	29	nil
Potassium hydroxide	50	29	0.010
	10	Boiling	<0.127
	25	Boiling	0.305

(Continued)

Table G.1 (Continued)

Medium	Concentration, %	Temperature, °C	Corrosion rate, mm/yr
Potassium hydroxide	50	Boiling	2.74
	50 anhydrous	241–377	1.02–1.52
Potassium iodide	Saturated	Room	nil
Potassium permanganate	Saturated	Room	nil
Potassium perchlorate	20	Room	0.003
	0–30	50	0.003
Potassium sulfate	10	Room	nil
Potassium thiosulfate	1	Room	nil
Propionic acid	Vapor	190	Rapid attack
Pyrogallic acid	355 g/L	Room	nil
Salicylic acid	Saturated	Room	nil
Seawater	...	24	nil
Seawater, 4¹/₂-year test	...	Ambient	nil
Sebacic acid	...	240	0.008
Silver nitrate	50	Room	nil
Sodium	100	To 1100 (593)	Good
Sodium acetate	Saturated	Room	nil
Sodium aluminate	25	Boiling	0.091
Sodium bifluoride	Saturated	Room	Rapid
Sodium bisulfate	Saturated	Room	nil
	10	66	1.83
Sodium bisulfite	10	Boiling	nil
	25	Boiling	nil
Sodium carbonate	25	Boiling	nil
Sodium chlorate	Saturated	Room	nil
Sodium chlorate			
+ NaCl 80–250 g/L	0–721 g/L	40	0.003
Sodium chloride	Saturated	Room	nil
pH 7	23	Boiling	nil
pH 1.5	23	Boiling	nil
pH 1.2	23	Boiling	0.71
pH 1.2, some dissolved chlorine	23	Boiling	nil
Sodium citrate	Saturated	Room	nil
Sodium cyanide	Saturated	Room	nil
Sodium dichromate	Saturated	Room	nil
Sodium fluoride	Saturated	Room	0.008
pH 7	1	Boiling	0.001
pH 10	1	Boiling	0.001
pH 7	1	204	0.000
Sodium hydrosulfide +			
sodium sulfide and polysulfides	5–12	110	<0.003
Sodium hydroxide	5–10	21	0.001
	10	Boiling	0.021
	28	Room	0.003
	40	80	0.127
	50	57	0.013
	50	Boiling	0.051
	73	129	0.178
	50–73	188	>1.09
	50	38	0.023
Sodium hypochlorite	6	Room	nil
Sodium hypochlorite + 15% NaCl +			
1% NaOH	1.5–4	66–93	0.030
Sodium nitrate	Saturated	Room	nil
Sodium perchlorate	900 g/L	50	0.003
Sodium phosphate	Saturated	Room	nil
Sodium silicate	25	Boiling	nil
Sodium sulfate	10–20	Boiling	nil
	Saturated	Room	nil
Sodium sulfide	10	Boiling	0.027
	Saturated	Room	nil
Sodium sulfite	Saturated	Boiling	nil
Sodium thiosulfate	25	Boiling	nil
Sodium thiosulfate + 20% acetic acid	20	Room	nil
Soils, corrosive	...	Ambient	nil
Stannic chloride	5	100	0.003

(Continued)

Table G.1 (Continued)

Medium	Concentration, %	Temperature, °C	Corrosion rate, mm/yr
	24	Boiling	0.045
Stannic chloride, molten...............	100	66	nil
Stannic chloride.......................	100	35	nil
	Saturated	Room	nil
Steam + air...........................	...	82	nil
Steam + 7.65% hydrogen sulfide	...	93–110	nil
Stearic acid, molten...................	100	180	0.003
Succinic acid.........................	100	185	nil
	Saturated	Room	nil
Sulfanilic acid........................	Saturated	Room	nil
Sulfamic acid	3.75 g/L	Boiling	nil
	7.5 g/L	Boiling	2.74
Sulfamic acid + 0.375 g/L FeCl₃	7.5 g/L	Boiling	0.030
Sulfur, molten........................	100	240	nil
Sulfur monochloride	...	202	>1.09
Sulfur dioxide, dry.....................	...	21	nil
Sulfur dioxide, water saturated..........	Near 100	Room	0.003
Sulfur dioxide gas + small amount SO₃ and approximately 3% O₂	18	316	0.006
Sulfuric acid, aerated	1	60	0.008
	3	60	0.013
	5	60	4.83
	10	35	1.27
	40	35	8.64
	75	35	1.07
	75	Room	10.8
	1	100	0.005
	3	100	23.4
	Concentrated	Room	1.57
	Concentrated	Boiling	5.38
	1	100	7.16
	3	100	21.1
Sulfuric acid	1	Boiling	17.8
	5	Boiling	25.4
Sulfuric acid + 0.25% CuSO₄	5	95	nil
	30	38	0.061
	30	95	0.088
Sulfuric acid + 0.5% CuSO₄	30	38	0.067
	30	95	0.823
Sulfuric acid + 1.0% CuSO₄	30	38	0.020
	30	95	0.884
Sulfuric acid + 0.5% CrO₃...............	5	95	nil
	30	95	nil
Sulfuric acid + 1.0% CuSO₄	30	Boiling	1.65
Sulfuric acid vapors....................	96	38	nil
	96	66	nil
	96	200–300	0.013
Sulfuric acid + 10% HNO₃	90	Room	0.457
Sulfuric acid + 50% HNO₃	50	Room	0.635
Sulfuric acid + 70% HNO₃	30	Room	0.102
Sulfuric acid + 90% HNO₃	10	Room	nil
Sulfuric acid + 90% HNO₃	10	60	0.011
Sulfuric acid + 95% HNO₃	5	60	0.005
Sulfuric acid + 50% HNO₃	50	60	0.399
Sulfuric acid + 20% HNO₃	80	60	1.59
Sulfuric acid saturated with chlorine......	45	24	0.003
	62	16	0.002
	5, 10	190	<0.025
	82	50	>1.19
Sulfuric acid + 4 g/L Ti⁴⁺	40	100	nil
Sulfurous acid........................	6	Room	nil
Tannic acid	25	100	nil
Tartaric acid	10–50	100	<0.127
	10	60	0.003
	25	60	0.003

(Continued)

Table G.1 (Continued)

Medium	Concentration, %	Temperature, °C	Corrosion rate, mm/yr
	50	60	0.001
	10	100	0.003
	25	100	nil
	50	100	0.0121
Terephthalic acid	77	218	nil
Tetrachloroethane, liquid and vapor	100	Boiling	0.001
Tetrachloroethylene + H_2O	· · ·	Boiling	0.127
Tetrachloroethylene.	100	Boiling	nil
Tetrachloroethylene, liquid and vapor.	100	Boiling	0.001
Titanium tetrachloride.	99.8	300	1.57
Trichloroacetic acid.	100	Boiling	14.6
Trichloroethylene.	99	Boiling	0.003–0.127
Trichloroethylene + 50% H_2O	50	25	0.001
Uranium chloride.	Saturated	21–90	nil
Uranyl ammonium phosphate filtrate + 25% chloride + 0.5% fluoride + 1.4% ammonia + 2.4% uranium	20.9	165	<0.003
Uranyl nitrate containing 25.3 g/L Fe^{3+}, 6.9 g/L Cr^{3+}, 2.8 g/L Ni^{2+}, 4.0 M HNO_3 + 1.0 M Cl	120 g/L	Boiling	nil
Uranyl sulfate + 3.1 M Li_2SO_4 + 100–200 ppm O_2	3.1 M	250	<0.020
Uranyl sulfate + 3.6 M Li_2SO_4, 50 psi oxygen	3.8 M	350	0.006–0.432

(a) Table G.1 data apply to unalloyed titanium, ASTM Grades 1 thru 4.
(See Appendix I, ASTM Specification B 265.)
(b) Room temperature assumed to be 25°C (77°F).

Table G.2 Titanium alloy corrosion rates (a)

Medium	Alloy	Concentration, %	Temperature, °C	Corrosion rate, mm/yr
Acetic acid	Grade 9	99.7	Boiling	nil
Acetic acid + 5% formic acid	Grade 12	58	Boiling	nil
Ammonium hydroxide	Grade 12	30	Boiling	nil
Aluminum chloride	Grade 12	10	Boiling	nil
	Grade 7	10	100	<0.025
	Grade 7	25	100	0.025
Ammonium chloride	Grade 12	10	Boiling	nil
Ammonium hydroxide	Grade 9	8, 28	150	nil
Aqua regia.	Grade 7	3:1	Boiling	1.12
	Grade 12	3:1	Boiling	0.61
	Grade 9	3:1	Boiling	1.29
	Grade 9	3:1	25	0.015
Calcium chloride	Grade 7	62	150	nil
	Grade 7	73	177	nil
Chlorine, wet	Grade 7	. . .	25	nil
Chromic acid.	Grade 7	10	Boiling	nil
	Grade 9	10	Boiling	0.008
	Grade 9	30	Boiling	0.053
	Grade 9	50	Boiling	0.26
Citric acid	Grade 7	50	Boiling	0.025
	Grade 12	50	Boiling	0.013
	Grade 9	50	Boiling	0.38
Ferric chloride	Grade 7	10	Boiling	nil
	Grade 12	10	Boiling	nil
	Ti-5Ta	10	Boiling	nil
	Grade 7	30	Boiling	nil
	Ti-6-4	10	Boiling	nil

(Continued)

Table G.2 (Continued)

Medium	Alloy	Concentration, %	Temperature, °C	Corrosion rate, mm/yr
	Ti-3-8-6-4-4	10	Boiling	nil
	Ti-10-2-3	10	Boiling	nil
	Ti-6-2-4-6	10	Boiling	0.06
	Transage 207	10	Boiling	0.19
	Ti-550	10	Boiling	nil
	Grade 9	10	Boiling	nil
	Ti-6-2-1-.8	10	Boiling	nil
Formic acid	Grade 9	25	88	<0.13
Formic acid, nitrogen-sparged	Grade 9	25	35	<0.13
Formic acid	Grade 9	50	Boiling	5.08
	Grade 7	45	Boiling	nil
	Grade 12	45, 50	Boiling	nil
	Grade 7	50	Boiling	0.01
	Ti-6-4	50	Boiling	7.92
	Transage 207	50	Boiling	0.90
	Ti-6-2-4-6	50	Boiling	0.62
	Ti-3-8-6-4-4	50	Boiling	0.98
	Ti-5Ta	50	Boiling	3.16
	Ti-550	50	Boiling	0.02
	Grade 12	90	Boiling	0.56
	Grade 7	90	Boiling	0.056
Hydrochloric acid	Ti-550	0.5	Boiling	0.056
	Ti-550	1.0	Boiling	0.64
	Transage 207	0.5	Boiling	0.005
	Transage 207	1.0	Boiling	0.025
	Ti-6-2-4-6	0.5	Boiling	nil
	Ti-6-2-4-6	1.0	Boiling	0.03
Hydrochloric acid, aerated	Ti-6-2-4-6	pH 1	Boiling	0.01
Hydrochloric acid	Ti-10-2-3	0.5	Boiling	1.10
	Ti-3-8-6-4-4	0.5	Boiling	0.003
	Ti-3-8-6-4-4	1.0	Boiling	0.058
	Ti-3-8-6-4-4	1.5	Boiling	0.26
Hydrochloric acid, aerated	Ti-3-8-6-4-4	pH 1	Boiling	nil
Hydrochloric acid	Ti-5Ta	0.5	Boiling	0.013
	Ti-5Ta	1.5	Boiling	2.10
	Ti-6-4	1.0	Boiling	2.52
Hydrochloric acid, aerated	Ti-6-4	pH 1	Boiling	0.60
Hydrochloric acid	Grade 9	0.5	Boiling	1.08
	Grade 9	1	88	0.009
	Grade 9	3	88	3.10
Hydrochloric acid, deaerated	Grade 7	3	82	0.013
	Grade 7	5	82	0.051
	Grade 7	10	82	0.419
Hydrochloric acid	Grade 9	1	Boiling	2.79
Hydrochloric acid, aerated	Grade 9	5	35	0.001
Hydrochloric acid, nitrogen saturated	Grade 9	5	35	0.185
Hydrochloric acid	Ti-6-2-1-.8	0.5	Boiling	0.020
	Ti-6-2-1-.8	1.0	Boiling	1.07
	Grade 7	0.5	Boiling	nil
	Grade 7	1.0	Boiling	0.008
	Grade 7	1.5	Boiling	0.03
	Grade 7	5.0	Boiling	0.23
	Grade 12	0.5	Boiling	nil
	Grade 12	1.0	Boiling	0.04
	Grade 12	1.5	Boiling	0.25
Hydrochloric acid, hydrogen saturated	Grade 7	1–15	25	<0.025
	Grade 7	20	25	0.102
	Grade 7	5	70	0.076
	Grade 7	10	70	0.178
	Grade 7	15	70	0.33
	Grade 7	3	190	0.025
	Grade 7	5	190	0.102

(Continued)

Table G.2 (Continued)

Medium	Alloy	Concentration, %	Temperature, °C	Corrosion rate, mm/yr
	Grade 7	10	190	8.9
Hydrochloric acid,				
oxygen saturated . . . . Grade 7		3, 5	190	0.127
	Grade 7	10	190	9.3
Hydrochloric acid,				
chlorine saturated . . . . Grade 7		3, 5	190	<0.03
	Grade 7	10	190	29.0
Hydrochloric acid, aerated . . . . Grade 7		1, 5	70	<0.03
	Grade 7	10	70	0.05
	Grade 7	15	70	0.15
Hydrochloric acid +				
4% $FeCl_3$ + 4% $MgCl_2$ . . . . . . Grade 7		19	82	0.49
Hydrochloric acid +				
4% $FeCl_3$ + 4% $MgCl_2$,				
chlorine saturated . . . . Grade 7		19	82	0.46
Hydrochloric acid				
+5 g/L $FeCl_3$ . . . . Grade 7		10	Boiling	0.279
+16 g/L $FeCl_3$ . . . . Grade 7		10	Boiling	0.076
+16 g/L $CuCl_2$ . . . . Grade 7		10	Boiling	0.127
Hydrochloric acid				
+2 g/L $FeCl_3$ . . . . Grade 12		4.2	91	0.058
+0.2% $FeCl_3$ . . . . Grade 9		1	Boiling	0.005
+0.2% $FeCl_3$ . . . . Grade 9		5	Boiling	0.033
+0.2% $FeCl_3$ . . . . Grade 9		10	Boiling	0.305
+0.1% $FeCl_3$ . . . . Grade 9		5	Boiling	0.008
+0.1% $FeCl_3$ Ti-550		5	Boiling	0.393
+0.1% $FeCl_3$ . . . . Transage 207		5	Boiling	0.048
+0.1% $FeCl_3$ Ti-6-2-4-6		5	Boiling	0.068
+0.1% $FeCl_3$ Ti-10-2-3		5	Boiling	0.008
+0.1% $FeCl_3$ Ti-3-8-6-4-4		5	Boiling	0.018
+0.1% $FeCl_3$ Ti-5Ta		5	Boiling	0.020
+0.1% $FeCl_3$ Ti-6-4		5	Boiling	0.015
+0.1% $FeCl_3$ Ti-6-2-1.-.8		5	Boiling	0.051
+0.1% $FeCl_3$ . . . . Grade 7		5	Boiling	0.013
+0.1% $FeCl_3$ . . . . Grade 12		5	Boiling	0.020
Hydrochloric acid +				
18% H_3PO_4 + 5% HNO_3 . . . . Grade 7		18	77	nil
Hydrogen peroxide				
pH 1 . . . . Grade 7		5	23	0.062
pH 4 . . . . Grade 7		5	23	0.010
pH 1 . . . . Grade 7		5	66	0.127
pH 4 . . . . Grade 7		5	66	0.046
+500 ppm Ca^{2+}, pH 1 . . . . . . Grade 7		5	66	nil
+500 ppm Ca^{2+}, pH 1 . . . . . . Grade 7		20	66	0.76
Hydrogen peroxide,				
pH 1 + 5% NaCl . . . . Grade 7		20	66	0.008
Magnesium chloride . . . . Grade 7		Saturated	Boiling	nil
Methyl alcohol . . . . Grade 9		99	Boiling	nil
Oxalic acid . . . . Grade 7		1	Boiling	1.14
Nitric acid . . . . Grade 9		10	Boiling	0.084
	Grade 9	30	Boiling	0.497
Phosphoric acid,				
naturally aerated . . . . Grade 12		25	25	0.019
	Grade 12	30	25	0.056
	Grade 12	45	25	0.157
	Grade 12	8	52	0.02
	Grade 12	13	52	0.066
	Grade 12	15	52	0.52
	Grade 12	5	66	0.038
	Grade 12	7	66	0.15
	Grade 12	0.5	Boiling	0.071
	Grade 12	1.0	Boiling	0.14
	Grade 7	40	25	0.008
	Grade 7	60	25	0.07
	Grade 7	15	52	0.036
	Grade 7	23	52	0.15

(Continued)

Table G.2 (Continued)

Medium	Alloy	Concentration, %	Temperature, °C	Corrosion rate, mm/yr
	Grade 7	8	66	0.076
	Grade 7	15	66	0.104
	Grade 7	0.5	Boiling	0.050
	Grade 7	1.0	Boiling	0.107
	Grade 7	5.0	Boiling	0.228
Potassium hydroxide . . . . Grade 9		50	150	9.21
Seawater . . . . Grade 9		. . .	Boiling	nil
Sodium chloride, pH 1 . . . . Grade 9		Saturated	93	nil
Sodium fluoride				
pH 7 . . . . Grade 12		1	Boiling	0.001
pH 7 . . . . Grade 7		1	Boiling	0.002
Sodium hydroxide . . . . Grade 9		50	150	0.49
Sodium sulfate, pH 1 . . . . Grade 7		10	Boiling	nil
Sulfamic acid . . . . Grade 12		10	Boiling	11.6
	Grade 7	10	Boiling	0.37
Sulfuric acid,				
naturally aerated	Grade 12	9	24	0.003
	Grade 12	9.5	24	0.006
	Grade 12	10	24	0.38
	Grade 12	3.5	52	0.013
	Grade 12	3.75	52	1.73
	Grade 12	2.75	66	0.015
	Grade 12	3.0	66	1.65
	Grade 12	0.75	Boiling	0.003
	Grade 12	1.0	Boiling	0.91
	Grade 7	1.0	204	0.005
	Grade 7	2.0	204	nil
	Grade 12	1.0	204	0.91
	Grade 9	0.5	Boiling	8.48
Sulfuric acid,				
nitrogen saturated . . . . Grade 7		5	70	0.15
	Grade 7	10	70	0.25
	Grade 7	1, 5	190	0.13
	Grade 7	10	190	1.50
Sulfuric acid,				
oxygen saturated . . . . Grade 7		1–10	190	0.13
Sulfuric acid,				
chlorine saturated . . . . Grade 7		10	190	0.051
	Grade 7	20	190	0.38
Sulfuric acid,				
nitrogen saturated . . . . Grade 7		10	25	0.025
	Grade 7	40	25	0.23
Sulfuric acid, aerated . . . . Grade 9		5	35	0.025
Sulfuric acid,				
nitrogen saturated . . . . Grade 9		5	35	0.405
Sulfuric acid,				
naturally aerated Ti-3-8-6-4-4		1	Boiling	nil
	Ti-3-8-6-4-4	5	Boiling	1.85
Sulfuric acid, aerated . . . . Grade 7		10	70	0.10
	Grade 7	40	70	0.94
Sulfuric acid +				
5 g/L Fe$_2$(SO$_4$)$_3$ . . . . Grade 7		10	Boiling	0.178
Sulfuric acid +				
16 g/L Fe$_2$(SO$_4$)$_3$ . . . . Grade 7		10	Boiling	<0.03
Sulfuric acid +				
16 g/L Fe$_2$(SO$_4$)$_3$ . . . . Grade 7		20	Boiling	0.15
Sulfuric acid +				
15% CuSO$_4$ . . . . Grade 7		15	Boiling	0.64
Sulfuric acid +				
3% Fe$_2$(SO$_4$)$_3$ Ti-3-8-6-4-4		50	Boiling	<0.03
Sulfuric acid +				
1 g/L FeCl$_3$ Ti-3-8-6-4-4		10	Boiling	0.15
Sulfuric acid +				
50 g/L FeCl$_3$ Ti-3-8-6-4-4		10	Boiling	0.05

(Continued)

Table G.2 (Continued)

Medium	Alloy	Concentration, %	Temperature, °C	Corrosion rate, mm/yr
Sulfuric acid + 1% CuSO$_4$ .... Grade 7		30	Boiling	1.75
Sulfuric acid + 100 ppm Cu^{2+} + 1% thiourea (deaerated) ..... Grade 7		1	100	nil
Sulfuric acid + 100 ppm Cu^{2+} + 1% thiourea (deaerated) ..... Grade 12		1	100	0.23
Sulfuric acid + 1000 ppm Cl$^-$ .... Grade 7		15	49	0.015

(a) Gr = ASTM grade (See Appendix I, ASTM Specification B 265.)
(b) For unabbreviated designations, see Table G.3.

Table G.3 Abbreviation Explanation

Abbreviation	Alloy
Gr. 7	Ti-0.2Pd
Gr. 9	Ti-3Al-2.5V
Gr. 12	Ti-0.3Mo-0.8Ni
Ti-6-4	Ti-6Al-4V
Ti-3-8-6-4-4	Ti-3Al-8V-6Cr-4Mo-4Zr
Ti-10-2-3	Ti-10V-2Fe-3Al
Ti-6-2-4-6	Ti-6Al-2Sn-4Zr-6Mo
Ti-6-2-1-0.8	Ti-6Al-2Nb-1Ta-0.8Mo
Ti-10-2-3	Ti-10V-2Fe-3Al
Ti-550	Ti-4Al-4Mo-2Sn
Transage 207	Ti-2.5Al-2Sn-9Zr-8Mo (a)

(a) Lockheed Missile & Space Company's developmental alloy.

Appendix H

Listing of Manufacturers, Suppliers, Services

INTRODUCTION

The following is a selected listing of manufacturers, suppliers, and services related to titanium and its alloys. Suppliers are listed by their primary product or service line. Complete names and mailing addresses can also be found in these alphabetical listings; however, after the first complete reference, only the company name is indicated.

This appendix is neither an endorsement of a product nor a recommendation of a specific manufacturer. Regarding product forms, there is no guaranty that the listing is complete nor that the individual company currently makes that particular form available.

The listing that makes up this appendix does not include other services that the various companies frequently offer. Typical of these additional functions are:

- R&D
- Alloy development
- Applications assistance
- Custom melting
- Analytical and testing
- Heat treating
- Sawing
- Shearing
- Plasma cutting

- Machining
- Chemical milling
- Sand, grit blasting
- Laser drilling, cutting
- Cold, hot working operations
- Custom welding
- HIPing and atomizing for P/M
- Cold, hot isostatic pressing

For more information about these services, refer to either of the following two publications:

- *Thomas Register* (latest year), "Products and Services" volume, titanium entry. Available at many libraries.

• Titanium Development Association's "Buyer's Guide" (latest year). Available at:

11 W. Monument Ave.
Dayton, OH 45401

ALLOYS

Aluminum Mill Supply Corp.
3400 Lawson Blvd.
Oceanside, NY 11572

Amerimet Corp.
17 Metacom Dr.
Simbury, CT 06070

Cezus
Tour Manhatten
Cedex 21
92087 Paris-La-Défense
France

Chemalloy Co., Inc.
P. O. Box 350
Bryn Mawr, PA 19010

Dynamet, Inc.
195 Museum Rd.
Washington, PA 15301

Goldman Titanium Company
P. O. Box 246
Buffalo, NY 14240

Howmet Turbine Components Corp.
Titanium Ingot Div.
555 Benston Rd.
Whitehall, MI 49461

IMI Titanium
5400 S. Delaware St.
Littleton, CO 80120

A. Johnson Metals Corp.
215 Welsh Pool Rd.
Lionville, PA 19353

Kennametal Inc.
P. O. Box 231
Latrobe, PA 15650

Kolon Trading Co., Inc.
145 W. 67th St.
New York, NY 10023

Milward Alloys, Inc.
500 Mill St.
Lockport, NY 14094

OREMET Titanium
(Oregon Metallurgical Corp.)
P. O. Box 580
Albany, OR 97321

President Titanium & Steel Co., Inc.
243 Franklin St
Hanson, MA 02341

Quanex/Viking Metallurgical Corp.
#1 Erik Circle
Verdi, NV 89439

Reading Alloys, Inc.
P. O. Box 53
Robesonia, PA 19551

RMI Co.
1000 Warren Ave.
Niles, OH 44446

Supra Alloys, Inc.
1185 Calle Suerte
Camarillo, CA 93010

Teledyne Allvac
P. O. Box 5030
Monroe, NC 28110

Teledyne Allvac
P. O. Box E-915
New Bedford, MA 02742

Teledyne Wah Chang Albany
P. O. Box 460
Albany, OR 97321

TIMET Corp.
P. O. Box 2824
Pittsburgh, PA 15230

Wyman-Gordon Co.
244 Worcester St.
North Grafton, MA 01536

BAR, ROD

Aluminum Mill Supply
Amerimet Corp.
Dynamet, Inc.
Goldman Titanium Co.
Howmet Turbine Components Corp.
IMI Titanium

Industrial Titanium Corp.
3041 Commercial Ave.
Northbrook, IL 60062

Kobe Steel, Ltd., Titanium Metals Div.
299 Park Ave.
New York, NY 10171

OREMET Titanium
President Titanium & Steel Co., Inc.
Quanex/Viking Metallurgical Corp.
RMI Co.
Supra Alloys, Inc.
Teledyne Allvac
Teledyne Wah Chang Albany
TIMET Corp.

Titanium Industries
110 Lehigh Dr.
Fairfield, NJ 07006

Tico Titanium, Inc.
24581 Crestview Court
Framington, MI 48018

Wyman-Gordon Co.

BILLETS

Amerimet Corp.
IMI Titanium
Howmet Turbine Components Corp.
Industrial Titanium Corp.
Kobe Steel, Ltd., Titanium Metals Div.
OREMET Titanium
President Titanium & Steel Co., Inc.
Quanex/Viking Metallurgical Corp.
RMI Co.
Supra Alioys, Inc.
Teledyne Allvac
Teldyne Wah Chang Albany
TIMET Corp.
Titanium Industries
Wyman-Gordon Co.

CASTINGS

Amerimet Corp.

Howmet Turbine Components Corp.
Hampton Casting Div.
P. O. Box 9365
Hampton, VA 23670

Howmet Turbine Components Corp.
Ti Cast Div.
1600 S. Warner Rd.
Whitehall, MI 49461

OREMET Titanium

Precision Cast Parts Corp.
S.E. Harney Dr.
Portland, OR 97206

TiTech International, Inc.
4000 W. Valley Blvd.
Pomona, CA 91769

COIL

Dynamet, Inc.
Kobe Steel, Ltd., Titanium Metals Div.
President Titanium & Steel Co., Inc.
Teledyne Wah Chang Albany

EXTRUSIONS

Al Tech Specialty Steel Corp.
Spring Street
Watervliet, NY 12189

FABRICATORS

Astro Metallurgical Div.
(Harsco Corp.)
3225 Lincoln Way West
Wooster, OH 44691
    (Butt-weld fittings; welding wire;
    ball valves; pressure vessels;
    tanks; piping systems)

Jet Die Div.
(Barnes Group, Inc.)
P. O. Box 25066
Lansing, MI 48909

Tico Titanium, Inc.

Titanium Industries
110 Lehigh Dr.
Fairfield, NJ 07006
  (Pipe spools; tanks; heat
  exchangers; structures;
  especially for offshore
  applications)

Western Titanium
(Ellett Group of Companies)
1575 Kingsway Ave.
Port Coquitiam, BC
Canada, V3C 4E5

FOIL, STRIP

Amerimet Corp.

Arnold Engineering Co.
300 NOrth West St.
Marengo, IL 60152

Chemalloy Co., Inc.
IMI Titanium
Kobe Steel, Ltd., Titanium Metals Div.
President Titanium & Steel Co., Inc.
RMI Co.
Supra Alloys, Inc.

Teledyne Rodney Metals
1357 East Rodney French Blvd.
New Bedford, MA 02742

Teledyne Wah Chang Albany
Tico Titanium, Inc.
TIMET Corp.
Titanium Industries

FORGINGS

Aluminum Co. of America
Forging Div.
1600 Harvard Ave.
Cleveland, OH 44105

Aluminum Mill Supply Corp.
Amerimet Corp.

Cameron Iron Works, Inc.
P. O. Box 1212
Houston, TX 77251

Dynamet, Inc.
Jet Die & Engineering
Kobe Steel Ltd., Titanium Metals Div.

Ladish Co., Inc.
5481 S. Packard Ave.
Cudahy, WI 53110

President Titanium & Steel Co., Inc.
Quanex/Viking Metallurgical Corp.
Teledyne Wah Chang Albany
TIMET Corp.
Wyman Gordon Co.

METAL

Aluminum Mill Supply Corp.
Amerimet Corp.
Chemalloy Co., Inc.
IMI Titanium
Goldman Titanium Corp.
A. Johnson Metals Corp.
Kobe Steel, Ltd.
Kolon Trading Co., Inc.
OREMET Titanium
President Titanium & Steel Co., Inc.
Quanex/Viking Metallurgical Corp.
RMI Co.

Specialloy, Inc.
4025 S. Keeler Ave.
Chicago, IL 60632

Sumitomo Metal Industries Ltd.
1-1-3 Ohtema-chi, Chiyoda-ku
Tokyo, Japan

Teledyne Allvac
Teledyne Rodney Metals
Teledyne Wah Chang Albany
TIMET Corp.

V/O Techsnabexport
c/o Kolon Trading Co., Inc.
145 W. 67th St.
New York, NY 10023

ORES

Amerimet Corp.

PLATE

Aluminum Mill Supply Corp.
Amerimet Corp.
Haynes International, Inc.
   (Cabot Wrought Products Div.)
IMI Titanium
Industrial Titanium Corp.
Kobe Steel, Ltd., Titanium Metals Div.
President Titanium & Steel Co., Inc.
Quanex/Viking Metallurgical
RMI Co.
Teledyne Wah Chang Albany
Tico Titanium, Inc.
TIMET Corp.
Titanium Industries

POWDER

Amerimet Corp.
Chemalloy Co., Inc.

Nuclear Metals, Inc.
2229 Main St.
Concord, MA 01742

Teledyne Wah Chang Albany
Timet Corp.
Titanium Industries

SHEET

Aluminum Mill Supply Corp.
Amerimet Corp.
Haynes International, Inc.
   (Cabot Wrought Products Div.)
Goldman Titanium Co.
IMI Titanium
Industrial Titanium Corp.
Kobe Steel, Ltd., Titanium Metals Div.
President Titanium & Steel Co., Inc.
Quanex/Viking Metallurgical
RMI Co.
Teledyne Rodney Metals
Teledyne Wah Chang Albany
Tico Titanium, Inc.
TIMET Corp.
Titanium Industries

SPONGE

Amerimet Corp.
Chemalloy Co., Inc.
Goldman Titanium Corp.

International Titanium Inc.
1320 Wheeler Rd.
Moses Lake, WA 98837

Mitsui & Co., Ltd.
1-2-1 Otemachi, Chiyoda-ku
Tokyo, Japan

OREMET Titanium
RMI Co.
TIMET Corp.
V/O Techsnabexport

TUBE

Aluminum Mill Supply Corp.
Amerimet Corp.

Haynes International, Inc.
   (Cabot Wrought Products Div.)
1020 W. Park Ave.
Kokomo, IN 46901

High Performance Tube, Inc.
1435 Morris Ave.
Union, NJ 07083

IMI Titanium
Industrial Titanium Corp.
Kobe Steel, Ltd., Titanium Metals Div.

Nikko Wolverine, Inc.
2525 Beech-Daly Rd.
Dearborne Hts., MI 48125

President Titanium & Steel Co., Inc.
RMI Co.
Teledyne Allvac
Tico Titanium, Inc.
TIMET Corp.
Titanium Industries

WIRE

Aluminum Mill Supply Corp.

Amerimet Corp.
Dynamet, Inc.
IMI Titanium
Kobe Steel, Ltd., Titanium Metals Div.
President Titanium & Steel Co., Inc.
Teledyne Wah Chang Albany
Tico Titanium, Inc.
Titanium Industries

Universal Wire Works, Inc.
7111 Long Dr.
Houston, TX 77087

BUYERS OF SCRAP

Amerimet Corp.
Haynes International, Inc.
   (Cabot Wrought Products Div.)

Chemalloy Co., Inc. Goldman Titanium
   Corp.
A. Johnson & Co.
Kolon Trading Co., Inc.
Milward Alloys, Inc.
OREMET Titanium
President Titanium & Steel Co., Inc.
Quanex/Viking Metallurgical Corp.

Schiavone-Bonomo Corp.
Foot of Jersey Ave.
Jersey City, NJ 07302

Suisman Titanium Corp.
500 Flatbush Ave.
Hartford, CT 06106

Teledyne Allvac
Teledyne Wah Chang Albany
TIMET Corp.

Appendix I

Standards and Specifications

INTRODUCTION

This Appendix presents the widest possible range of international standards, specifications, and designations, as related to titanium and titanium alloys. It reflects the evolving international character of trade and commerce. The countries and organizations included here are representative of the industrial European and Western Hemisphere geographic groupings, along with the Pacific Basin area. Although not every possible country is included, it is currently perhaps the most complete, consolidated, up-to-date listing of its type.

In instances when a specific country or standards-issuing organization does not have titanium-related specifications, a comment to that effect is included. In cases where there is no national standard, an authoritative correspondent's comment is frequently noted to give an indication of exactly which standards are used by way of substitution.

The standards and specifications contained in this appendix are divided as follows:

- MIL Spec (U.S. military)

- DOD (U.S. Department of Defense)

- AMS (Aerospace Material Specifications)

- ASTM (American Society for Testing Materials)

- AIA (Aerospace Industries Association)

- SAE (Society of Automotive Engineers)

- UNS (Unified Numbering System)

- ANSI (American National Standards Institute)

- ASME (American Society of Mechanical Engineers)

- NBS (U. S. National Bureau of Standards)

349

- International cross-reference listings
- Austria
- Czechoslovakia
- Europe/AECMA
- Federal Republic of Germany
- Finland
- France
- India
- International/ISO
- Israel
- Japan

- Korea
- Netherlands
- Pan America
- People's Republic of China
- Portugal
- Republic of China (Taiwan)
- Republic of South Africa
- Spain
- Turkey
- United Kingdom
- USSR

The organizational arrangement for the United States is based on a frequency of usage. Countries other than the United States are alphabetized for quick reference.

Refer to each of these divisions for a listing of tables contained therein. Addresses of each organization are also included.

Two other organizations deserve mention here. They are:

American National Standards Institute (ANSI)
1430 Broadway
New York, NY 10018

American Welding Society (AWS)
550 N.W. LeJeune Rd.
Miami, FL 33135

Refer to Appendix D, "Filler Metals," for AWS specifications.

MIL SPECIFICATIONS

Military specifications are issued by the Department of Defense to define materials, products, and/or services used only, or predominantly, by military entities.

U.S. Military Specifications
Engineering Specifications and
  Standards Division
Naval Air Engineering Center
Philadelphia, PA 19112

All military specifications begin with the upper case letters MIL. The actual specification that follows begins with an upper case code letter representing the first letter of the title for the item specified. (Thus T is used for titanium and F for forgings.) A serial number follows.

Although not included in this listing, there are also military standards which provide procedures for design, manufacturing, and testing, rather than providing a particular material's description. A standard is formatted as MIL-STD-, followed by a serial number.

Individual specifications and standards may have serial numbers which include a revision letter suffix: MIL-T-9046J.

The revision levels of titanium specifications are very important since so many changes have occurred over the years. Such modifications involve methods of citing a "designation," or an individual alloy's identification; nominal compositions and physical property maximum percents and/or range limits; and line items added or deleted. Manufacturers and product literature continue to reference earlier, superseded revisions, perhaps to aid the customer or perhaps because an individual product may actually be produced according to the earlier specification.

At times amendments are issued to correct, not supersede, a specification's revision level. It is important to determine if amendments exist, since they do much to clarify the basic specification.

In the following tables, the latest revision levels, as of the publication date of this volume, are represented and detailed. No attempt, however, is made to specify superseded revisions, with a single exception: earlier product designations, when they exist, are related to current designations to aid the reader.

The titanium and titanium alloy MIL specs included here are:

  Table I.1 - MIL-T-9046J (1/83): Sheet, strip, plate

  Table I.2 - MIL-T-9047G (12/78): Bars, reforging stock
      MIL-T-9047G, Amendment 2 (8/86)

  Table I.3 - MIL-T-81556A (1/83): Titanium and titanium alloys,
      extruded bars and shapes, aircraft quality

Table I.4 - MIL-T-46035A (10/66): High-strength, wrought
        alloy for critical components
        MIL-T-46035A, Amendment 1 (10/72)

Table I.5 - MIL-T-81915 (3/73): Investment castings

Table I.6 - MIL-T-83142A (12/69): Forgings, premium quality

Note: The following tables reflect the typical, nominal chemical compositions and mechanical properties of a given titanium designation. However the individual military specifications should be consulted for precise, qualifying type information and for alternative mechanical properties resulting from various types of conditions. The specifications noted in the tables are not complete in these respects.

Military specifications may be ordered directly from the Naval Publications & Forms Center, noted below at the "DOD Specifications" paragraph. However they may be more conveniently accessed at major public libraries by means of microfiche. One such reference service is the Information Handling Services' *U.S. Government Specifications Service; Numeric Index* (latest issue). If necessary, contact:

Information Handling Services
P. O. Box 1154
Englewood, CO 80150

Other MIL specs not covered in this book are:

• MIL-H-81200, "Heat Treatment of Titanium and Titanium Alloys"

• MIL-STD-412, "Alloy Designation System for Titanium."
  (This document was formally cancelled.)

• MIL-HDBK 697A, *Titanium and Titanium Alloys Handbook.*
  (This document was published in 1974 and is somewhat out of date, although it does contain much valuable information.)

• MIL-T-13405E, "Titanium, Pyrotechnic Powder."
  (This concerns pyrotechnic mixtures and M36 bomb clusters.)

• MIL-T-24585A, "Titanium Alloy Rod, Discs, and Upset Forgings Composition Ti-6Al-6V-2.5Sn." (A U.S. Navy controlled-distribution document.)

For a comprehensive treatment of MIL standards related to titanium and titanium alloys, refer to: Chapter 5, "Titanium," in *Military Standardization Handbook; Metallic Materials and Elements for Aerospace Vehicle Structures*, MIL-HDBK-5D (1 Jun 83), Vol 2 of 2, Pub. No. FSC 1560.

(Available from Naval Publications and Forms Center, noted above.)

Table I.1 Mil-T-9046J (1/83): Sheet, strip, plate; classification by composition

Grade/ alloy	Rev. J code desig- nation	Rev. H code desig- nation (a)	Rev. H. common designation	Chemical composition (%) max. unless noted	Mechanical properties, hardness values (b)
Commer- cially pure Ti	CP-1	Type I, Comp. B	70 ksi min YS	0.5Fe; 0.05N; 0.08C; 0.015H; 0.4O; 0.3 other; remain. Ti.	Condition (c) Thickness: 25.4 mm (1.0 in.) and under Tensile strength: 551 MPa (80) ksi min Yield strength (d): 482-655 MPa (70-95 ksi) Elongation: 15% in 50.8 mm (2 in.)
	CP-2	Type I, Comp. C.	55 ksi min YS	0.03Fe; 0.05N; 0.08C; 0.015H; 0.3O; 0.3 other; re- main. Ti.	Condition: (c) Thick.: 25.4 mm (1.0 in.) and under TS: 448 MPa (65 ksi) main YS: (d): 379-551 MPa (55-80 ksi) EL: 18% in 50.8 mm (2 in.)
	CP-3	Type I, Comp. A	40 ksi min YS	0.03Fe; 0.05N; 0.08C; 0.015H; 0.2O; 0.3 other; re- main. Ti.	Condition: (c) Thick.: 25.4 mm (1.0 in.) and under TS: 344 MPa (50 ksi) min YS: (d) 275-448 MPa (40-65 ksi) EL: 20% in 50.8 mm (2 in.)
	CP-4	-	25 ksi min YS	0.2Fe; 0.05N; 0.08C; 0.015H; 0.15O; 0.3 other; remain. Ti.	Condition (c) Thick.: 25.4 mm (1.0 in.) and under TS: 241 MPa (35 ksi) min YS: (d): 172-310 MPa (25-45 ksi) EL: 24% in 50.8 mm (2 in.)
Alpha alloys	A-1	Type II, Comp. B	Ti-5Al- 2.5Sn	0.05Fe; 0.05N; 0.08C; 0.020H; 0.2O; 4.5-5.5Al; 2.0-3.0 Sn; 0.005Y; 0.4 other; remain. Ti.	Condition: (c) Thick.: 38 mm (1.5 in) and under TS: 827 MPa (120 ksi) min YS: (d) 770 MPa (113 ksi) EL: 10% in 50.8 mm (2.0 in.)
	A-2	Type II, Comp. B	Ti-5Al- 2.5Sn- ELI	0.25Fe; 0.035N; 0.05C; 0.0125H; 0.12O; 4.5-5.75Al; 2.0-3.0Sn; 0.005Y; 0.3 other; remain. Ti.	Condition (c) Thick.: 0.2-0.4 mm (0.008-0.014 in.) (e) TS: 689 MPa (100 ksi) min YS: (d): 655 MPa (95 ksi) EL: 6% in 50.8 mm (2.0 in.)

(Continued)

Table I.1 (Continued)

Grade/alloy	Rev. J code desig-nation	Rev. H code desig-nation (a)	Rev. H. common designation	Chemical composition (%) max. unless noted	Mechanical properties, hardness values (b)
	A-3	Type II, Comp. G	Ti-6Al-2Cb -1Ta-0.8Mo	0.25 Fe; 0.03N; 0.05C; 0.0125H; 0.1O; 5.5-6Al 1.5-2.5Cb; 0.5-1.5Ta; 0.5-1.0Mo; 0.005Y; 0.4other; remain. Ti.	Condition: (c) Thick.: 4.76-100 mm (0.1875-4.0 in.) TS: 710 MPa (103 ksi) min YS: (d) 655 MPa (95 ksi) El: 10% in 50.8 mm (2.0 in.)
	A-4	Type II, Comp F	Ti-8Al-1Mo -1V	0.3Fe; 0.05N; 0.08C; 0.015H; 0.15O; 7.35-8.35Al; 0.75-1.25Mo; 0.75-1.25; 0.005Y; 0.4 other, remain. Ti	Condition: (c) Thick.: 0.2-0.4 mm (0.008-0.014 in.) (e) TS: 999 MPa (145 ksi) min YS: (d): 930 MPa (135 ksi) El: 6% in 50.8 mm (2.0 in.)
Alpha-beta	AB-1	Type III, Comp. C	Ti-6Al-4V	0.3Fe; 0.05N; 0.08C; 0.0125H; 0.2O; 5.5-5-6.75Al; 3.5-4.5V; 0.005Y; 0.4 other; remain. Ti	Condition: (c) (f) Thick.: 1.57 mm (0.062 in.) and under TS: 923 MPa (134 ksi) min YS: (d) 868 mm (126 ksi) EL: 8% in 50.8 mm (2.0 in.)
	AB-2	Type III, Comp. D	Ti-6Al-4V-ELI	0.25Fe; 0.05N; 0.08C; 0.012H; 0.13O; 5.5-6.5Al; 3.5-4.5V; 0.005Y; 0.3 other; remain Ti	Condition: (c) Thick.: 0.2-0.4 mm (0.008-0.014 in.) (e) TS: 896 Pa (130 ksi) min YS: (d): 827 MPa (120 ksi) EL: 6% in 50.8 mm (2.0 in.)
	AB-3	Type III, Comp. E	Ti-6Al-6V-2Sn	0.35-1.0Fe; 0.04N; 0.05C; 0.015H; 0.2O; 5.0-6.0Alp 1.5-2.5Sn; 5.0-6.0V; 0.35-1.0Cu; 0.005Y; 0.3 other, remain. Ti	Condition: (c) (f) Thick.: 6.0 mm (0.024 in.) and under TS: 1069 MPa (155 ksi) min YS: (d) 999-1172 MPa (145-170 ksi) EL: 8% in 50.8 mm (2.0 in.) transverse
	AB-4	Type III, Comp. G	Ti-6Al-2Sn-4Zr-2Mo	0.25Fe 0.04N; 0.05C; 0.015H; 0.15O; 5.5-6.5Al; 1.8-2.2Sn 1.8-2.2Mo; 3.6-4.4Zr; 0.13Si; 0.005Y; 0.3 other; remain. Ti	Condition: Duplex an-nealed (f) Thick.: 1.57 mm (0.062 in.) and under TS: 930 MPa (135 ksi) min YS (d): 861 MPa (125 ksi) EL: 8% in 50.8 mm (2.0 in.)

(Continued)

Table I.1 (Continued)

Grade/alloy	Rev. J code desig- nation	Rev. H code desig- nation (a)	Rev. H. common designation	Chemical composition (%) max. unless noted	Mechanical properties, hardness values (b)
	AB-5	–	Ti-3Al-2.5V	0.3Fe; 0.02N; 0.05C; 0.015H; 0.12O; 2.5-3.5Al; 2.0-3.00V; 0.005Y; 0.4 other; remain. Ti	Condition: (c) Thick.: 25.4 mm (1.0 in.) and under TS: 620 MPa (90 ksi) min YS (d): 517 MPa (75 ksi) El: 15% in 50.8 mm (2.0 in.)
	AB-6	–	Ti-8Mn	0.5Fe; 0.05N; 0.08C; 0.015H; 0.2O; 6.5-9.0Mn; 0.005Y; 0.4 other; remain. Ti	Condition: (c) Thick.: 4.76 mm (0.1875 in.) and under TS: 861 MPa (125 ksi) min YS (d): 758-965 mm (110-140 ksi) El: 10% in 50.8 mm (2.0 in.)
Beta alloys	B-1	Type IV, Comp A.	Ti-13V-11Cr-3Al	0.15-0.35Fe; 0.05N; 0.05C; 0.025H; 0.17O; 2.5-3.5Al; 12.5-14.5V; 10.0-12.0Cr; 0.004Y; 0.4 other; remain. Ti	Condition: Solution treated (f) Thick.: 1.24 mm (0.049 in.) and under TS: 910 MPa (132 ksi) min YS (d): 868 MPa (126 ksi) El: 8% in 50.8 mm (2.0 in.)
	B-2	Type IV, Comp. B	Ti-11.5Mo-6Zr-4.5Sn	0.35Fe; 0.05N; 0.1C; 0.020H; 0.18O; 3.75-5.25Sn; 4.5-7.5Zr; 10.0-13.0Mo; 0.005Y; 0.4 other; remain. Ti	Condition: Solution treated (f) Thick.: 6.0 mm (0.024 in.) and under Ts: 689 MPa (100 ksi) min YS (d): 620 MPa (90 ksi) El: 10% in 50.8 mm (2.0 in.)
	B-3	Type IV, Comp. C	Ti-3Al-8V-6Cr-4Mo-4Zr	0.3Fe; 0.03N; 0.05C; 0.02H; 0.12O; 3.0-4.0Al; 3.5-4.5Mo; 5.5-6.5Cr; 3.5-4.5Zr; 7.5-8.5V; 0.005Y; 0.4 other; remain. Ti	Condition: Solution treated (f) Thick.: 0.73 mm (0.029 in.) and under TS: 861 Mpa (125 ksi) min YS (d): 827 MPa (120 ksi) max El: 6% in 50.8 mm (2.0 in.)

(Continued)

Table I.1 (Continued)

Grade/ alloy	Rev. J code desig- nation	Rev. H code desig- nation (a)	Rev. H. common designation	Chemical composition (%) max. unless noted	Mechanical properties, hardness values (b)
	B-4	Type IV, Comp. D	Ti-8Mo-8V- 2Fe-3Al	1.6-2.4Fe; 0.05N; 0.05C; 0.015H; 0.16O; 2.6-3.4Al; 7.5-8.5Mo; 7.5-8.5V; 0.005Y; 0.4 other; remain. Ti.	Condition: Solution treated (f) Thick.: 4.76 mm (0.1875 in.) and under TS: 861 MPa (125 ksi) min YS (d): 827 MPa (120 ksi) max El: 18% in 50.8 mm (2.0 in.)

(a) Note that code designations in Rev. J vary greatly from the earlier Rev. H. Note also that in many instances nominal compositions changed from revision to revision. Only the latest revision's composition and properties are indicated in this table.
(b) MIL spec are issued in English units. Metric units are conversions that have been conservatively rounded.
(c) Unless noted, properties apply to annealed condition.
(d) Yield strength at 0.2% offset.
(d) Other thickness(es) specified, which may affect following properties.
(f) Other conditions specified, which may affect other properties.

Table I.2 MIL-T-9047G (12/78): Bars, reforging stock; classification by composition (a)

Grade/alloy	Rev. G. code designation	Rev. E., F code designation (b)	Chemical composition (%) max. unless noted	Mechanical properties, hardness values
Commercially pure Ti	Ti-CP-70	Comp. 1	0.5Fe; 0.05N; 0.08C; 0.0125H; 0.4O; 0.30 other remain. Ti	Condition: N/A Diam. (e): 101.6 mm (4.0 in.) and under Tensile strength: 551 MPa (80 ksi) Yield strength: 482 MPa (70 ksi) Elongation: 15% Reduction in area: 30%
Alpha alloys	Ti-5Al-2.5Sn	Comp. 2	0.5Fe; 0.05N; 0.08C; 0.02H; 0.2O; 4.5-5.75Al; 2.0-3.0Sn; 0.005Y; 0.4 other; remain. Ti	Condition: N/A Diam. (e): 101.6 mm (4.0 in.) and under TS: 792 MPa (115 ksi) YS: 758 MPa (110 ksi) El: 10% Red.: 25%
	Ti-5Al-2Sn-ELI	Comp. 3	0.25Fe ;0.035N; 0.05C; 0.0125H; 0.12O; 4.5-5.75Al; 2.0-3.0Sn; 0.005Y; 0.3 other; remain. Ti	Condition: N/A Diam. (e): 76.2 mm (3.0 in.) and under TS: 689 MPa (100 ksi) YS: 620 MPa (90 ksi) El: 10% Red.: 25%
	Ti-6Al-2Cb-1Ta-0.8Mo	-	0.25Fe; 0.03N; 0.05C; 0.0125H; 0.1O; 5.5-6.5Al; 1.5-2.5Cb; 0.005Y; 0.5-1.5 Ta; 0.5-1.0Mo; 0.4 other; remain Ti	Condition: N/A Diam. (e): 101.6 mm (4.0 in.) and under TS: 710 MPa (103 ksi) YS: 655 MPa (95 ksi) El: 10% Red.: 20%
	Ti-8Al-1Mo-1V	Comp. 5	0.3Fe; 0.05N; 0.08C; 0.015H; 0.15O; 7.35-8.35Al; 0.75-1.25Mo; 0.75-1.25V; 0.005Y; 0.4 other; remain. Ti	Condition: N/A; Duplex annealed also specified Diam. (e); 63.5 mm (2.5 in.) and under TS: 896 MPa (130 in.) or under YS: 827 MPa (120 in.) El: 10% Red.: 20%
Alpha-beta alloys	Ti-3Al-2.5V	-	0.3Fe; 0.02N; 0.05C; 0.15H; 0.12O; 2.5-3.5Al; 2.0-3.0V; 0.005Y; 0.04 other; remain. Ti	Condition: (c) (d) Diam. (e): 25.4 mm (1.0 in.) and under TS: 620 MPa (90 ksi) YS: 517 MPa (75 ksi) El: 15% Red.: 30%
	Ti-6Al-4V	Comp. 6	0.3Fe; 0.05N; 0.08C; 0.015H; 0.2O; 5.5-6.75Al; 3.5-4.5V; 0.005Y; 0.4 other; remain. Ti	Conditon: (c) (d) (f) Diam. (e)(g): 101.5 mm (4.0 in.) and under TS: 896 MPa (130 ksi) YS: 827 MPa (120 ksi) El: 10% Red.: 25%

(Continued)

Table I.2 (Continued)

	Ti-6Al-4V-ELI	Comp. 7	0.25Fe; 0.05N; 0.08C; 0.0125H; 0.13O; 5.5-6.5Al; 3.5-4.5V; 0.005Y; 0.3 other; remain. Ti	Condition: (c) (d) Diam. (e)(g): 38.1 mm (1.5 in.) and under TS: 896 MPa (130 ksi) YS: 827 MPa (120 ksi) El: 10% Red.: 25%
	Ti-6Al-6V-2Sn	–	0.35-1.0Fe; 0.04N; 0.05C; 0.015H; 0.2O; 5.0-6.0Al; 5.0-6.0V; 1.5-2.5Sn; 0.35-1.0Cu; 0.005Y; 0.3 other; remain. Ti	Condition: (c) (d) (f) Diam. (e)(g): 38.1 mm (1.5 in.) and under TS: 1020 MPa (148 ksi) YS: 944 MPa (137 ksi) El: 10% Red.:20%
	Ti-6Al-2Sn-4Zr-2Mo	Comp. 11	0.25Fe; 0.04N; 0.05C: 0.015H; 0.15O; 5.5-6.5Al; 1.8-2.2Sn; 3.6-4.4Zr; 1.0-2.2Mo; 0.13Si; 0.005Y; 0.3 other; remain. Ti	Condition: duplex annealed (d) Diam. (e): 76.2 mm (3 in.) and under TS: 896 MPa (130 ksi) YS: 827 MPa (120 ksi) El: 10% Red.: 25%
	Ti-6Al-2Sn-4Zr-6Mo	Comp. 14	0.15Fe; 0.04N; 0.04C; 0.0125H; 0.15O; 5.5-6.5Al; 1.75-2.25Sn; 3.6-4.4Zr; 5.5-6.5Mo; 0.005Y; 0.4 other; remain. Ti	Condition: duplex annealed (d) Diam. (e)(g): 50.8 mm (2.0 in.) and under TS: 1103 MPa (160 ksi) YS: 1034 MPa (150 ksi) El: 10% Red.: 25%
	Ti-7Al-4Mo	Comp. 9	0.3Fe; 0.05N; 0.1C; 0.013H; 0.2O; 6.5-7.3Al; 3.5-4.5Mo; 0.005Y; 0.4 other; remain. Ti	Condition: (c) (d) (f) Diam. (e) (g): 25.4 mm (1.0 in.) and under TS: 999 MPa (145 ksi) YS: 930 MPa (135 ksi) El: 10% Red.: 20%
Beta alloys	Ti-8Mo-8V-2Fe-3Al	–	1.6-2.4Fe; 0.05N; 0.05C; 0.015H; 0.16O; 2.6-3.4Al; 7.5-8.5Mo; 7.5-8.5V; 0.005Y; 0.4 other; remain. Ti	Condition: solution treated (d) (f) Diam. (e) (g): 50.8 mm (2.0 in.) and under TS: 896 MPa (130 ksi) YS: 827 MPa (120 ksi) El: 10% Red.: 24%
	Ti-11.5Mo-6Zr-4.5Sn	Comp. 13	0.35Fe; 0.05N; 0.1C; 0.02H; 0.18O; 10.0-13.0 Mo; 3.75-5.25Sn; 4.5-7.5Zr; 0.005Y; 0.4 other; remain. Ti	Condition: solution treated (d)(f) Diam. (e)(g): 41.27 mm (1.625 in.) and under TS: 756 MPa (110 ksi) YS: 620 MPa (90 ksi) El: 15% Red.: 50%

(Continued)

Table I.2 (Continued)

Ti-3Al-8V- 6Cr-4Mo- 4Zr	-	0.3Fe; 0.03N; 0.05C; 0.02H; 0.12O; 3.0-4.0Al; 3.5-4.5Mo; 7.5-8.5V; 5.5-6.5Cr; 3.5-4.5Zr; 0.005Y; 0.4 other; remain. Ti	Condition: solution treated (d)(f) Diam. (e)(g): 38.1 mm (1.5 in.) and under TS: 861-MPa (125 ksi) YS: 827 MPa (120 ksi) El: 10% Red.: 30%
Ti-13V-11Cr -3Al	Comp. 12	0.35Fe; 0.05N; 0.05C; 0.025H; 0.17O; 2.5-3.5Al; 12.5-14.5V; 10.0-12.0 Cr; 0.4 other; 0.005Y; re- main. Ti	Condition: solution treated (d)(f) Diam. (e)(g): 177.8 mm (7.0 in.) and under TS: 861 MPa (125 ksi) YS: 827 MPa (120 ksi) El: 10% Red.: 25%

(a) Bars: rolled or forged. Only general specifications given. See the specification for complete details, including all conditions, widths, and reductions.

(b) Note that code designations in Rev. G very greatly from the earlier and superseded Rev. E and F. The much earlier and superseded Rev. D., which used "Composition" plus an alpha letter, is not included here. Note also that in many instances nominal compositions changed from revision to revision. Only the latest revision's composition and properties are indicated in this table.

(c) Properties apply to annealed condition, unless noted.

(d) Conditions. Bars shall be hot-worked and supplied in one of following conditions: A = annealed; DA = duplex annealed; ST = solution treated; STA = solution treated and aged. Reforging stock shall be furnished in condition orderd by the forging manufacturer.

(e) Thickness, diameter or distance between flats.

(f) Other conditions specified but not noted here. These may affect properties stated here.

(g) Other thickness(es), diameter, etc. specified, but not noted here. These may affect properties stated here.

Table I.3 MIL-T-81556A (1.83): Extruded bars and shapes, aircraft quality

Grade/alloy	Code designation	Common designation	Chemical composition (%) max. unless noted	Mechanical properties, hardness values (a)
Commercially pure Ti	CP-1	70 ksi min YS (483 MPa)	0.5Fe; 0.08C; 0.05N; 0.015H; 0.4O; 0.3 total other; remain. Ti	Condition: Annealed Thickness (b): 4.78-25.4 mm (0.188-1.0 in.) Tensile strength: 552 MPa (80 ksi) min Yield strength: (c): 483 MPa (70 ksi) min Elongation (d): 15% min in 50 mm (2 in.)
	CP-2	55 ksi min YS (379 MPa)	0.3Fe; 0.08C; 0.05N; 0.015H; 0.3O; 0.3 total other; remain. Ti	Condition: Annealed Thick. (b): 4.78-25.4 mm (0.188-1.0 in.) TS: 448 MPa (65 ksi) min YS (c): 379 MPa (55 ksi) min El (d): 18% min in 50 mm (2 in.)
	CP-3	40 ksi min YS (276 MPa)	0.3Fe; 0.08C; 0.05N; 0.015H; 0.2O; 0.3 total other; remain. Ti	Condition: Annealed Thick. (b): 4.78-25.4 mm (0.188-1.0 in.) TS: 345 MPa (50 ksi) min YS (c): 276 MPa (40 ksi) min El (d): 20% min in 50 mm (2 in.)
	CP-4	30 ksi min YS (207 MPa)	0.2Fe; 0.08C; 0.05N; 0.015H; 0.15O; 0.3 total other; remain. Ti	Condition: Annealed Thick. (b): 4.78-25.4 mm (0.188-1.0 in.) TS: 276 MPa (40 ksi) min YS (c): 207 MPa (30 ksi) min El (d); 25% min in 50 mm (2 in.)
Alpha alloys	A-1	Ti-5Al-2.5Sn	4.5-5.75Al; 2.0-2.0Sn; 0.5Fe; 0.08C; 0.05N; 0.02H; 0.2O; 0.005Y; 0.4 total other; remain. Ti	Condition: Annealed Thick. (b): 4.78-25.4 mm (0.188-1.0 in.) TS: 827 MPa (120 ksi) min YS (c): 793 MPa (115 ksi) min El (d): 10% min in 50 mm (2 in.)
	A-2 (e)	Ti-5Al-2.5Sn-ELI	4.5-5.75Al; 2.0-3.0Sn; 0.25Fe; 0.05C; 0.035N; 0.0125H; 0.12O; 0.005Y; 0.3 total other; remain. Ti (f) (g)	Condition: Annealed Thick.: 4.78-25.4 mm (0.188-1.0 in.) TS: 690 MPa (100 ksi) min YS (c): 655 MPa (95 ksi) min El (d): 10% min in 50 mm (2 in.)

(Continued)

Table I.3 (Continued)

	A-4	Ti-8Al-1Mo-1V	7.35-8.35Al; 0.75-1.25Mo; 0.75-1.25V; 0.3Fe; 0.08C; 0.05N; 0.015H; 0.15O; 0.005Y; 0.4 total other; remain. Ti	Condition: Annealed Thick.: (b): 4.78-12.7 mm (0.188-0.5 in.) TS: 1000 MPa (145 ksi) min YS (c): 931 MPa (135 ksi) min El (d): 10% min in 50 mm (2 in.) Condition: Duplex annealed Thick. (b): 4.78-25.4 mm (0.188-1.0 in.) TS: 896 MPa (130 ksi) min YS (c): 827 MPa (120 ksi) min El (d): 10% min in 50 mm (2 in.)
Alpha-beta	AB-1	Ti-6Al-4V	5.5-6.75Al; 3.5-4.5V; 0.3Fe; 0.08C; 0.05N; 0.0125H; 0.2O; 0.005Y; 0.4 total other; remain. Ti	Condition: Annealed Thick. (b): 4.78-31.73 mm (0.188-1.249 in.) TS: 896 MPa (130 ksi) min YS (c): 827-1000 MPa (120-145 ksi) El (d): 10% min in 50 mm (2 in.) longitudinal; 8% min for transverse Reduction in area: 20% min longitudinal; 5% min transverse Condition: Solution treated and aged Thick. (b): 4.78-12.7 mm (0.188-0.5 in.) TS: 1103 MPa (160 ksi) min YS (c) 1034 MPa (150 ksi) min El (d): 8% min in 50 mm (2 in.) Red.: 15% min
	AB-2	Ti-6l-4V-ELI	5.5-6.5Al; 3.5-4.5V; 0.25Fe; 0.08C; 0.05N; 0.0125H; 0.13O; 0.005Y; 0.3 total other; remain. Ti (g)	Condition: Annealed Thick. (b): 4.78-15.4 mm (0.188-1.0 in.) TS: 862 MPa (125 ksi) min YS (c): 793-1000 MPa (115-145 ksi) El (d): 10% min in 50 mm (2 in.) Red.: 20% min
	AB-3	Ti-6Al-6V-2Sn	5.5-6.5Al; 1.5-2.5Sn; 5.0-6.0V; 0.35-1.0Cu; 0.35-1.0Fe; 0.05C; 0.04N; 0.015H; 0.2O; 0.005Y; 0.3 total other; remain. Ti	Condition: Annealed Thick. (b): 4.78-38.1 mm (0.188-1.5 in.) TS: 1034 MPa (150 ksi) min YS (c): 965-1138 MPa (140-165 ksi) El (d): 10% min in 50 mm (2 in.) longitudinal; 5% transverse Red.: 20% min longitudinal; 5% min transverse

(Continued)

Table I.3 (Continued)

				Condition: Solution treated Thick. (b): **4.78-12.7 mm (0.188-0.5 in.)** TS: 1172 MPa (170 ksi) YS (c) 1103 MPa (160 ksi) min El (d): 8% min in 50 mm (2 in.) longitudinal; 5% min transverse Red.: 15% min longitudinal; 5% min transverse
AB-4	**Ti-6Al-2Sn-4Zr-2Mo+Si**	5.5-6.5Al; 1.8-2.2Sn; 3.6-4.4Zr; 0.06-0.1Si; 1.8-2.2Mo; 0.25Fe; 0.05C; 0.04N; 0.015H; 0.15O; 0.005Y; 0.3 total other; remain. Ti		Condition: Annealed Thick.: **4.78-102.0 mm (0.188-4.0 in.)** TS: 896 MPa (130 ksi) YS (c): 827-1000 MPa (120-145 ksi) El (d): 10% min in 50 mm (2 in.) longitudinal; 8% min transverse Red.: 20% min longitudinal; 15% min transverse
				Condition: Solution treated and aged Thick. (b): **4.78-12.7 mm (0.188-1.5 in.)** TS: 1034 MPa (150 ksi) YS (c): 965 MPa (140 ksi) min El (d): 10% min in 50 mm (2 in.) longitudinal; 8% min transverse Red.: 20% min longitudinal; 15% min transverse

(a)   MIL spec are issued in English units. Metric units here are conversions that have been conservatively rounded.
(b)   Other thickness(es) specified, which may affect following properties. Here "thickness" also includes diameter or distance between flats. Thickness is defined by the specification as the cross section from which the specimen is to be taken to test the tensile strength.
(c)   Yield strength at 0.2% offset.
(d)   Elongation in mm or 4 diameters.
(e)   A-3 does not apply to this MIL spec.
(f)   Iron plus oxygen shall not exceed 0.32%.
(g)   Other elements here each shall not exceed 0.05% max.

Table I.4 MIL-T-46035A (10/66): High-strength, wrought alloy for critical components; by chemical composition (a)

Code description	Designation	Chemical composition (% max) (b)	Mechanical properties
EL 1	Extra-low impurity	0.02N; 0.01C; 0.0125H; 0.10	(c)
L 1	Low impurity	0.03N; 0.1C; 0.0125H; 0.12O	(c)
N 1	Normal impurity	0.04N; 0.1C; 0.0125H; 0.18O	(c)

(a) Annealed and heat-treated shapes having a critial section thickness of 1/4 to 2-1/2 in., such as tubes, chambers, and nozzles, but nor armor.
(b) Carbon content corrected as per Amendment 1 (10/72).
(c) Mechanical properties are varied and, at times, subject to negotiation between the parties. Refer to the MIL spec for details.

Table I.5 MIL-T-81915 (3/73): Investment castings; classification by composition (a)

Grade/alloy	Code designation	Composition	Chemical composition (% by wgt) (b)	Mechancal properties, hardness values (c)
Commercially pure Ti	Type I, Comp. A	CP	0.2Fe; 0.05N, 0.08C; 0.015H; 0.2O; 0.6 total other; remain. Ti	Condition: Annealed Tensile strength: 241 MPa (35 ksi) min Yield strength (d): 172 MPa (25 ksi) min Elongation: 24% in 50 mm (2 in.) min
Alpha-titanium	Type II, Comp. A	Ti-5Al-2.5Sn	0.5Fe; 0.05N; 0.08C; 0.02H; 0.02O; 4.5-5.7Al; 2.0-3.0Sn; 0.4 total other; remain. Ti	Condition: Annealed TS: 758 MPa (110 ksi) min YS (d): 724 MPa (105 ksi) min El: 10% in 50 mm (2 in.) min Reduction in area: 20% min
Alpha-beta titanium	Type III, Comp. A	Ti-6Al-4V	0.3Fe;, 0.05N; 0.08C; 0.015H; 0.2O; 5.5-6.5Al; 3.5-4.5V; 0.4 total other; remain. Ti	Condition: Annealed TS: 861 MPa (125 ksi) min YS (d): 792 MPa (115 ksi) min El: 8% in 50 mm (2 in.) min Red.: 16% min
	Comp. B	Ti-6Al-2Sn-4Zr-2Mo	0.35Fe; 0.05N; 0.08C; 0.015H; 0.12O; 5.5-6.5Al; 3.5-4.5V; 0.4 total other; remain. Ti	Condition: Annealed TS: 861 MPa (125 ksi) min YS (d): 792 MPa (115 ksi) min El: 8% in 50 mm (2 in.) min Red.: 20% min

(a) Castings specified as Grades A, B, or C, which are defined in paragraph 6.3 of this MIL spec.
(b) Max unless otherwise noted.
(c) MIL spec are issued in English units. Metric units are conversions that have been conservatively rounded.
(d) Yield strength: 0.2% offset.

Table I.6 MIL-F-83142A (12/69): Forgings (premium quality); classification by composition (a) (b) (c)

Grade/alloy	Code designation	Chemical composition (d)	Mechanical properties
Commercially pure Ti	Comp. 1	Unalloyed	Tensile strength: 80 ksi min Yield strength (e): 70 ksi min Elongation (f): 15% min Reduction in area: 30% min
Alpha	Comp. 2	Ti-5Al-2.5Sn	TS: 115 ksi min YS (e): 110 ksi min El (f): 12% min Red.: 30% min
	Comp. 3	Ti-5Al-2.5Sn-ELI	TS: 100 ksi min YS (e) 90 ksi min El (f): 10% min Red.: 25% min
	Comp. 4	Ti-5Al-5Zr-5Sn	<u>Up to 2 in.:</u> TS: 120 ksi min YS (e): 100 ksi min El (f): 10% min Red.: 20% min <u>Up to 4 in.:</u> TS: 110 ksi min YS (e): 100 ksi min El (f): 10% min Red.: 20% min
	Comp. 5	Ti-8Al-1Mo-1V	TS: 130 ksi min YS (e): 120 ksi min El (f): 10% min Red.: 25% min
Alpha-beta alloy	Comp. 6	Ti-6Al-4V	TS: 130 ksi min YS (e): 120 ksi min El (f): 10% min Red.: 25% min
	Comp. 7	Ti-6Al-4V-ELI	<u>Up to 1.75 in.:</u> TS: 125 ksi min YS (e): 115 ksi min El (f): 10% min Red.: 25% min <u>1.75 to 4 in.:</u> TS: 120 ksi min YS (e): 110 ksi min El (f): 10% min Red.: 27% min
	Comp. 8	Ti-6Al-6V-2Sn	TS: 140 ksi min YS (e): 130 ksi min El (f): 8% min Red.: 20% min
	Comp. 9	Ti-7Al-4Mo	TS: 145 ksi min YS (e): 135 ksi min El (f): 10% min Red.: 20% min
	Comp. 10	Ti-11Sn-5Zr-3Al-1Mo	TS: 135 ksi min YS (e): 125 ksi min El (f): 11% min Red.: 25% min
	Comp. 11	Ti-6Al-2Sn-4Zr-2Mo	TS: 130 ksi min YS (e): 120 ksi min El (f): 10% min Red.: 25% min

(Continued)

Table I.6 (Continued)

Grade/alloy	Code designation	Chemical composition (d)	Mechanical properties
Beta alloy	Comp. 12	Ti-13V-11Cr-3Al	TS: 130 ksi min YS (e): 120 ksi min El (f): 10% min Red.: 25% min
	Comp. 13	Ti-11.5Mo-6Zr-4.5 Sn	TS: 130 ksi min YS (e): 120 ksi min El (f): 10% min Red.: 25% min

(a)	Forging stock in the form of as-cast ingots, billets or bars may be used assuming certain restrictions including an assurance that no signficant change in size, shape, form or grain flow orientation of the forging without notification and retests.
(b)	Billets and bar stock composition, structure and properties shall conform to the requirement of MIL-T-9047 for the respective alloy classification in the annealed condition.
(c)	When cast ingots are used as forging stock, certain restrictions must be met. Composition of castings shall conform to MIL-T-9047, as specified. Reforging bars and billets shall be procured to MIL-T-9047.
(d)	Design documentation includes forging drawings that indicate a number of items including the alloy composition to be used. See this MIL spec's Table III for composition correlations among MIL-T-9047C, D, E and -83142, A.
(e)	YS 0.2% offset.
(f)	In 4 diameters.

DOD SPECIFICATIONS, STANDARDS

Table I.7 is a listing of the United States Department of Defense's specifications for titanium and titanium alloys. The sources are *Index of Specifications and Standards; Part I (1 July 1986)* and *Supplement; Part I (1 Jan. 1987)*.

Use of the *Index of Specifications and Standards* is mandatory on all military activities. It requires that the Federal and military specifications, standards, and related standardization documents be considered in identifying items for procurement actions.

The document number column indicates the issuing agency of the individual standard or specification.

Copies of the DOD *Index . . .* may be obtained from:

Naval Publications & Forms Center
5801 Tabor Ave.
Philadelphia, PA 19120

Use DD Form 1425.

Table I.7 DOD specifications listing

Title	Document Number	Date
Ti & Ti alloys, chemical check analysis limits	AMS 2249C-85	01 Nov 85
Ti alloy rod, discs, & upset forgings composition Ti-6Al-6V-2Sn. Issue controlled; reqsts by other than DOD actys must be submitted via Naval Sea Systems Command (SEA 0982).	MIL-T-24585A	19 Sep 86
Ti and Ti alloy, sheet, strip and plate	MIL-T-9046J	11 Jan 83
Ti and Ti alloy bars (rolled or forged) and reforging stock, aircraft quality	MIL-T-9047G	30 Sep 86
Ti & Ti alloys, extruded bars & shapes, aircraft quality	MIL-T-81556A	31 Jan 83
Ti & Ti alloys, chemical analysis of Ti & Ti alloy surfaces, descaling & cleaning	ASTM E120-83 ASTM B600-74	22 Apr 85 18 Nov 81
Ti alloy, high-strength, wrought (for critical components)	MIL-T-46035A	05 Oct 72
Ti alloys, ultrasonic inspection of	AMS 2631-72	21 Mar 83
Ti and Ti alloys, chemical check analysis limits	AMS 2249B-85	01 Nov 85
Ti and Ti alloy, sheet, strip and plate	MIL-T-9046J INT AMD 1	11 Jan 83
Ti and Ti alloys	MIL-HDBK 697A	01 Jun 74
Ti and Ti alloy bars (rolled or forged) and reforging stock, aircraft quality	MIL-T-9047G INT AMD 1	15 Dec 78
Ti and Ti alloy castings, investment	MIL-T-81915	16 Mar 73
Ti, technical, power	MIL-T-13405E	23 Feb 84
Heat treatment of Ti and Ti alloys	MIL-H-81200A	24 Mar 69

AMS SPECIFICATIONS

Aerospace Material Specifications (AMS) are issued by Society of Automotive Engineers (SAE). They focus on materials intended for aerospace applications. These standards are frequently used in worldwide industry and government procurements. (See Table I.8.)

(At times the U.S. Department of Defense adopts AMS specifications. AMS standards have been accepted by ANSI and also as American National Standards. In some instances non-United States agencies adopt them.)

References to "AMS Number" indicate a document where the specifications may be found. Individual documents or complete sets may be ordered from:

Society of Automotive Engineers
Customer Service Department
Publications Group
400 Commonwealth Drive
Warrendale, PA 15096

There are sales outlets in England (for Europe), India, and Australia.

Table I.8   AMS specifications for titanium and titanium alloys

AMS No. (a)	Form	Condition	Grade/alloy (b)	Similar MIL specification (c)	UNS No.
4900	Plate, sheet, strip	Annealed	CP; 55 ksi YS	MIL-T-9046	R50550
4901	Plate, sheet, strip	Annealed	CP; 70 ksi YS	MIL-T-9046	R50700
4902	Plate, sheet, strip	Annealed	CP; 40 ksi YS	MIL-T-9046	R50400
4905	Plate (damage tolerant grade)	Annealed	Ti-6Al-4V	-	R56400
4906	Noncurrent, 1982	-			
4907	Plate, sheet, strip	Annealed	Ti-6Al-4V-ELI	MIL-T-9046	R56401
4908	Sheet, strip	Annealed	Ti-8M; 110 ksi YS	MIL-T-9046	R56080
4909	Plate, sheet, strip	Annealed	Ti-5Al-2.5Sn-ELI	MIL-T-9046	R54521
4910	Plate, sheet, strip	Annealed	Ti-5Al-2.5Sn	MIL-T-9046	R54520
4911	Plate, sheet, strip	Annealed	Ti-6Al-4V	MIL-T-9046	R56400
4912	Noncurrent, 1981	-	-	-	-
4913	Noncurrent, 1981	-	-	-	-
4914	Sheet, strip	Solution heat treated	Ti-15V-3Cr-3Sn-3Al	-	-
4915	Plate, sheet, strip	Single annealed	Ti-8Al-1Mo-1V	MIL-T-9046	R54810
4916	Plate, sheet, strip	Duplex annealed	Ti-8Al-1Mo-1V	MIL-T-9046	R54810
4917	Plate, sheet, strip	Solution treated	Ti-13V-11Cr-3Al	MIL-T-9046	R58010
4918	Plate, sheet, strip	Annealed	Ti-6Al-6V-2Sn	MIL-T-9046	R56620
4919	Plate, sheet, strip	Annealed	Ti-6Al-2Sn-4Zr-2Mo	MIL-T-9046	R54620
4920	Forging; alpha-beta or beta processed	Annealed	Ti-6Al-4V	-	R56400
4921	Bar, forging, ring	Annealed	CP; 70 ksi YS	MIL-T-9047	R50700
4923	Cancelled	-	-	-	-
4924	Bar, forging, ring	Annealed	Ti-5Al-2.5Sn-ELI; 90 ksi YS	MIL-T-9047	R54521
4925	Cancelled	-	-	-	-
4926	Bar, ring	Annealed	Ti-5Al-2.5Sn; 110 ksi YS	MIL-T-9047	R54520
4927	Cancelled	-	-	-	-
4928	Bar, forging, ring	Annealed	Ti-6Al-4V; 120 ksi YS	MIL-T-9047	R56400
4929	Cancelled	-	-	-	-
4930	Bar, forging, ring	Annealed	Ti-4Al-4V-ELI	MIL-T-9047	R56401
4933	Extrusion, flash-welded ring	Solution heat treated & stablized	Ti-8Al-1Mo-1V	-	R54810
4934	Extrusion, flash-welded ring	Solution heat treated & aged	Ti-6Al-4V	-	R56400
4935	Extrusion, flash-welded ring; beta processed	Annealed	Ti-6Al-4V	-	R56400

(Continued)

Table I.8 (Continued)

AMS No. (a)	Form	Condition	Grade/alloy (b)	Similar MIL specification (c)	UNS No.
4936	Extrusion, flash-welded ring; beta	Extruded plus annealed	Ti-6Al-6V-2Sn	MIL-T-81556	R56620
4941	Welded tubing	Cold drawn, annealed	CP; 40 ksi YS	-	R50400
4942	Seamless tubing	Cold reduced, annealed	CP; 40 ksi YS	-	R50400
4943	Seamless tubing; hydraulic	Annealed	Ti-3Al-2.5V	-	R56320
4944	Seamless tubing; hydraulic	Cold worked; stress relieved	Ti-3Al-2.5V	-	R56320
4951	Welding wire	As drawn	CP	-	R50550
4953	Welding wire	-	Ti-5Al-2.5Sn	AWS A5.16 (d)	R54520
4954	Welding wire	-	Ti-6Al-4V	-	R56400
4955	Welding wire	-	Ti-8Al-1Mo-1V	-	R54810
4956	Welding wire	Environmentally controlled packaging	Ti-6Al-4V-ELI	-	R56401
4959	Wire	Spring temper	Ti-13.5V-11Cr-3Al	-	R58010
4965	Bar, forging, ring	Solution & precip. heat treated	Ti-6Al-4V	-	R56400
4966	Forging	Annealed; for stock; as ordered	Ti-5Al-2.5Sn; 110 ksi YS	MIL-F-83142	R54520
4967	Bar, forging, ring	Annealed; heat treatable	Ti-6Al-4V	MIL-T-9047	R56400
4968	Cancelled	-	-	-	-
4969	Cancelled	-	-	-	-
4970	Bar, forging	Solution & precip. heat treated	Ti-7Al-4Mo	MIL-T-9047; MIL-F-83142	R56740
4971	Bar, forging, ring	Annealed; heat treatable	Ti-6Al-6V-2Sn	MIL-T-9047; MIL-F-83142	R56620
4972	Bar, ring	Solution heat treated, stablized	Ti-8Al-1Mo-1V	MIL-T-9047	R54810
4973	Forging	Solution heat treated, stablized	Ti-8Al-1Mo-1V	-	R54810
4974	Bar, forging	Solution & precip. heat treated	Ti-11Sn-5Zr-2.3 Al-1Mo-0.21Si	-	R54790

(Continued)

Table I.8 (Continued)

AMS No. (a)	Form	Condition	Grade/alloy (b)	Similar MIL specification (c)	UNS No.
4975	Bar, ring	Solution & precip. heat treated	Ti-6Al-2Sn-4Zr-2Mo	MIL-T-9047	R54620
4976	Forging	Solution & precip. heat treated	Ti-6Al-2Sn-4Zr-2Mo	MIL-T-9047	R54620
4977	Bar	Solution & recip. heat treated	Ti-11.5Mo-6Zr-4.5Sn 690°-730°C (1275°C -1350°F)	-	R58030
4978	Bar, forging, ring	Annealed Solution & precip. heat treated	Ti-6Al-6V-2Sn; 140 ksi YS	MIL-T-9047; MIL-F-83142	R56620
4979	Bar, forging, ring	Solution heat treated	Ti-6Al-6V-2Sn	MIL-T-9047; MIL-F-83142	R56620
4980	Bar	Solution & precip. heat treated	Ti-11.5Mo-6Zr-4.5Sn 745°C (1375°F)	-	R58030
4981	Bar, forging		Ti-6Al-2Sn-4Zr-6Mo	MIL-T-9047	R56260
4982	Wire	-	Ti-44.5Cb	-	R58450
4983	Forging	Solution heat treated & aged	Ti-10V-2Fe-3Al	-	-
4985	Investment or rammed graphite	Annealed	Ti-6Al-4V	-	R56400
4991	Investment casting	Annealed	Ti-6Al-4V	-	R56400
4993	B-nut compacts	Blended powder	Ti-6Al-4V	-	-
4995	Billet, platform; for P/M	Premium quality; as ordered	Ti-5Al-2Sn-2Zr-4Cr-4Mo-0.1O	-	R58650
4996	Billet, platform; for P/M	Premium quality	Ti-6Al-4V	-	R56400
4997	Powder	Premium quality; as manufactured	Ti-5Al-2Sn-2Zr-4Cr-0.1O	-	R58650
7460	Bolt, screws	Heat treated; roll threaded	Ti-6Al-4V (AMS 4967)	-	-

ASTM STANDARDS, SPECIFICATIONS

The American Society for Testing Materials (ASTM) is an industry-wide association which publishes the *Annual Book of ASTM Standards*. References to titanium and titanium alloys are contained in Volume 02.04, *Nonferrous Metals*, Section 2, "Nonferrous Metal Products."

American Society for Testing and Materials
1916 Race Street
Philadelphia, PA 19103

What ASTM terms "standard specifications," or documents, are listed in Table I.9. The designations and chemical compositions for titanium sponge appear in Table I.10. The ASTM "Grade" for various product forms, along with some mechanical properties, are indicated in Table I.11. Selected mechanical properties for surgical implants are listed in Table I.12. The specifications for commercial, wrought, nonferrous nuts, bolts, screws, and studs are noted in Table I.13.

Table I.9 ASTM standard specifications for Ti, Ti alloy

No.	Title
A 845	Titanium Scrap for Use in Deoxidation and Alloying of Steel; A Specification
B 265	Titanium and Titanium Alloy Strip, Sheet, Plate; A Specification
B 299	Standard Specification for Titanium Sponge
B 337	Seamless and Welded Titanium and Titanium Alloy Pipe; A Specification
B 338	Seamless and Welded Titanium and Titanium Alloy Tubes for Condensers and Heat Exchangers
B 348	Titanium and Titanium Alloy Bars, Billets; A Specification
B 363	Seamless and Welded Unalloyed Titanium for Welded Fittings
B 367	Titanium and Titanium Alloy Castings; A Specification
B 381	Titanium an Titanium Alloy Forgings: A Specification
B 481	Standard Practice for Preparation of Titanium and Titanium Alloys for Electroplating
B 600	Standard Recommended Practice for Descaling and Cleaning Titanium and Titanium Alloy Surfaces
E 120	Methods of Chemical Analysis of Titanium and Titanium Alloys
E 539	Methods for X-Ray Emission Spectrometric Analysis of 6Al-4V Titanium Alloy
F 136	Wrought Titanium 6Al-4V-ELI Alloys for Surgical Implant Applications
F 620	Titanium 6Al-4V-ELI Alloy Forgings for Surgical Implants; A Specification
G 41	Determining Cracking Suspectibility of Titanium Alloys Exposed Under Stress to a Hot Salt Environment; A Recommended Practice

Table I.10   ASTM B 299: titanium sponge chemical composition (%)

Designation	Fe	N	C	H	O	Si	Mg	Na	Cl	Impurities	Ti (rem.)
MD-120 (magnesium-reduced, distillation-finished)	0.12	0.015	0.02	0.01	0.01	0.04	0.08	–	0.12	0.05	99.3
ML-120 (magnesium-reduced, distillation-finished)	0.15	0.015	0.025	0.03	0.1	0.04	0.5	–	0.2	0.05	99.1
SL-120 Sodium-reduced, leaching-finished	0.05	0.015	0.02	0.05	0.1	0.4	–	0.19	0.2	0.05	99.3
GP-1 general purpose (a)	0.25	0.02	0.025	0.03	0.15	0.04	(b)	(b)	0.2	0.05	rem.

(a) Magnesium- or sodium-reduced; finished by leaching or by inert gas sweep.
(b) Sodium or magnesium: max. 0.5%

Table I.11   ASTM "specification standards" for Ti and Ti alloys

Specification	Grade	Alloy	Min 0.2% yield strength (a) MPa	Min 0.2% yield strength (a) ksi	Similar AMS specifications (b)
B 265	1	Unalloyed	170	25	....
(1979)	2	Unalloyed	275	40	4902
	3	Unalloyed	380	55	4900
	4	Unalloyed	485	70	4901
	5	Ti-6Al-4V	830	120	4911
	6	Ti-5Al-2.5Sn	795	115	4910
	7	Ti-0.2Pd	275	40	....
	19	Ti-4.5Sn-11.5Mo-6Zr	620	90	4977
	11	Ti-0.2Pd	170	25	....
	12	Ti-0.3Mo-0.8Ni	345	50	....
Seamless, welded pipe					
B 337	1	Unalloyed	170	25	....
(1983)	2	Unalloyed	275	40	4941 (c) 4942 (d)
	3	Unalloyed	380	55	....
	7	Ti-0.2Pd	275	40	....
	9	Ti-3Al-2.5V	485	70 (e)	4943
	9	Ti-3Al-2.5V	725 (f)	105 (f)	4943
	10	Ti-11.5Mo-6Zr-4.5Sn	620 (g)	90 (g)	4977, 4980
	11	Ti-0.2Pd	170	25	....
	12	Ti-0.3Mo-0.8Ni	345	50	....
Seamless, welded tube for condensers, heat exchangers					
B 338	1	Unalloyed	170	25	....
(1983)	2	Unalloyed	275	40	....
	3	Unalloyed	380	55	....
	7	Ti-0.2Pd	275	40	....
	9	Ti-3Al-2.5V	725	105	4943, 4944
	10	Ti-11.5Mo-6Zr-4.5Sn	620	90	4977, 4980
	11	Ti-0.2Pd	170	25	....
	12	Ti-0.3Mo-0.8Ni	345	50	....
Bar, billet					
B 348	1	Unalloyed	170	25	....
(1983)	2	Unalloyed	275	40	....
	3	Unalloyed	380	55	....
	4	Unalloyed	483	70	4921 (h)
	5	Ti-6Al-4V	825	120	4928 (h)
	6	Ti-5Al-2.5Sn	795	115	4926 (h)
	7	Ti-0.2Pd	275	40	....
	9	Ti-3Al-2.5V	483	70	....
	10	Ti-4.5Sn-11.5Mo-6Zr	620	90	4977, 4980
	11	Ti-0.2Pd	170	25	....
	12	Ti-3Mo-0.8Ni	345	50	....
Castings					
B 367	C-1	(Withdrawn, 1983)			
(1983)	C-2	Unalloyed	275	40	....
	C-3	Unalloyed	380	55	....
	C-4	(Withdrawn, 1983)			
	C-5	Ti-6Al-4V	825	120	....
	C-6	Ti-5Al-2.5Sn	725	105	....
	C-7A	(Withdrawn, 1983)			
	C-7B	Ti-0.2Pd	275	40	....
	C-8A	Ti-0.2Pd	380	55	....
	C-8B	(Withdrawn, 1983)			

(Continued)

Table I.11 (Continued)

Specification	Grade	Alloy	Min 0.2% yield strength (a) MPa	ksi	Similar AMS specifications (b)
		Forgings			
B 381	F1	Unalloyed	170	25	....
(1983)	F2	Unalloyed	275	40	....
	F3	Unalloyed	380	55	....
	F4	Unalloyed	483	70	4921
	F5	Ti-6Al-4V	825	120	4928
	F6	Ti-5Al-2.5Sn	795	115	4966
	F7	Ti-0.2Pd	275	40	....
	F9	Ti-3Al-2.5V	483	70	....
	F11	Ti-0.2Pd	170	25	....
	F12	Ti-0.3Mo-0.8Ni	345	50	....

(a) Annealed
(b) Interstitial and impurity levels, and mechanical property requirements may show minor differences compared with ASTM specifications.
(c) Welded tubing
(d) Seamless tubing
(e) Annealed
(f) Cold worked and stress relieved
(g) Solution treated
(h) AMS specifications cover bar and forgings but not billet.

Table I.12   ASTM surgical implant specifications

Specification	Grade	Alloy	Min 0.2% yield strength (a) MPa	ksi	Similar AMS specification(s)
		Surgical implants (bar, billet, or flat product)			
F 67	1	(a)	170	25	N/A
(1983)	2	(a)	275	40	N/A
	3	(a)	380	55	N/A
	4	(a)	485	70	N/A
		Wrought Ti-6Al-4V-ELI for surgical implants			
F 136	N/A	Ti-6Al-4V-ELI	827	120 (b)	N/A
(1984)			795	115 (b)	N/A

(a) See ASTM Standard Specification F 67 for chemical requirements.
(b) Varies with thickness or diameter.

Table I.13 ASTM specifications for nuts, bolts, screws, studs (a)

ASTM specifications	Item	ASTM grade	Chemical composition(%)(b)(c)	Nuts: mech. properties, hardness values	Bolts, screws, studs mechanical properties, hardness values (e)
F 467	Nuts	1	0.2Fe; 0.05N; 0.1C; 0.0125H; 0.18O; remain. Ti	Diam. 0.25 thru 1.5 in., as agreed. HV: 140, min. Proof stress: 40 ksi, min.	HV: 140 to 160. YS: 30 ksi, min. TS: 40 to 70 ksi.
F 468 (d)	Bolts, screws, studs	2	0.3Fe; 0.05N; 0.1C; 0.0125H; 0.25O; remain. Ti	Diam. 0.25 thru 1.5 in., as agreed. HV: 150, min. Proof stress: 55 ksi, min.	HV: 160 to 180. YS: 45 ksi, min. TS: 55 to 85 ksi.
		4	0.5Fe; 0.07N; 0.1C; 0.0125H; 0.4O; remain. Ti	Diam. 0.25 thru 1.5 in., as agreed. HV: 200, min. Proof stress: 85 ksi, min.	HV: 200 to 220. YS: 75 ksi, min. TS: 85 to 115 ksi.
		5	5.5-6.75Al; 0.4Fe; 0.05N; 0.1C; 0.0125H; 0.2O; 3.5 to 4.5V; remain Ti	Diam. 0.25 thru 1.5 in., as agreed. HRC: 30 ksi, min. Proof stress: 135 ksi, min.	HRC: 30-36. YS: 125 ksi, min. TS: 135 to 165 ksi.
		7	0.3Fe; 0.05N; 0.1C; 0.0125H; 0.25O; 0.12 to 0.25Pd; remain. Ti	Diam. 0.25 thru 1.5 in., as agreed. HV: 160, min. Proof stress: 55 ksi, min.	HV: 160-180. YS: 45, ksi, min. TS: 55 to 85 ksi.

(a) 1974 issue
(b) C, Fe, H, N and O are all max %.
(c) Chemical composition for F 467 and F 468 are identical.
(d) Metric versions of these specifications are available: F 467M and F 468M.
(e) Diameters same as nuts.

AIA STANDARDS

The Aerospace Industries Association, Inc. (AIA) is a trade association comprised of companies engaged in aerospace product research, development, and manufacturing. It is intended to serve as a vehicle for making public the consensus of member companies on non-competitive matters related to business prospects and operations.

AIA's Aerospace Technical Council, Technical Specifications Division, directs the National Aerospace Standards Committee. This group issues various aerospace-related standards which are proposed guidelines for the aerospace industry. They are designed to eliminate misunderstandings between manufacturers and purchasers; they may also be used to assist a purchaser with selection of a product for a specific application.

Note that the AIA standards are printed by the National Standards Association, Inc. which is only a publishing company, not a standards-issuing group. A listing of AIA standards may be found in the numerical *Index to National Aerospace Standards*, published by the National Standards Association, Inc. The listing in Table I.14 is current as of the April, 1987 (issue 87-2) edition. (Always consult the latest edition.)

To obtain a copy of the *Index to National Aerospace Standards,* order from:

National Standards Association, Inc.
5161 River Road
Bethesda, MD 20816

Other publishers also make the standards available on microfilm. These include:

Information Handling Services (IHS)
15 Inverness Way East
Englewood, CO 80150

Aircraft Technical Publishers
101 South Hill Drive
Brisbane, CA 94005

Table I.14 AIA standards for titanium parts

Document No. (a)	Revision No. (latest date) (b)	Standard title
NAS2406-2412	Rev. 3 (5/81)	Bolt: Lock, Shear, Protruding Head, Standard and Over-size, Pull-Type, Titanium Alloy
NAS2605-2621	Rev. 4 (12/81)	Bolt: Lock, Shear, Protuding Head, Stump-Type, Titanium Alloy
NAS2506-2512	Rev. 3 (5/81)	Bolt: Lock, Shear, 100° Head, Standard and Oversize, Pull-Type, Titanium Alloy
NAS2705-2712	Rev. 4 (12/81)	Bolt: Lock, Shear, 100° Head, Stump-Type, Titanium Alloy
NAS2005-2012	Rev. 4 (4/81)	Bolt: Lock, Tension, Protruding Head, Standard and Oversize, Pull-Tupe, Titanium Alloy
NAS2205-2212	Rev. 3 (12/81)	Bolt: Lock, Tension, Protruding Head, Stump-Type, Titanium Alloy
NAS2125-2132	-	Bolt: Lock, Tension, 100°C Crown Head, Standard and Oversize, Pull-Type, Titanium Alloy
NAS2115-2122	-	Bolt: Lock, Tension, 100°C Head (MS20426), Standard and Oversize, Pull-Type, Titanium Alloy
NAS2315-2322	Rev. 1 (9/81)	Bolt: Lock, Tension, 100°C Head (MS20426), Stump-Type, Titanium Alloy
NAS2105-2112	Rev. 4 (4/81)	Bolt: Lock, Tension 100°C Head (MS24694) Pull-Type, Titanium Alloy
NAS2306-2312	Rev. 3 (12/81)	Bolt: Lock, Tension, 100°C Head (MS24694) Stump-Type, Titanium Alloy
NAS1266-1270	Rev. 3 (6/80)	Bolt: Hex Head, Close Tolerance, Titanium Alloy
NAS1261-1265	Rev. 2 (6/80)	Bolt: Hex Head, Close Tolerance, Short Thread, Titanium Alloy
NAS6803-6820	- (2/81)	Bolt: Hex Head, Close Tolerance, 6Al-4V Titanium Alloy, Long Thread, Self-Locking and Nonlocking
NAS6804-6820	Rev. 2 (2/72)	Bolt: Hex Head, Close Tolerance, 6Al-4V Titanium Alloy, Long Thread, Self-Locking and Nonlocking
NAS6403-6420	Rev. 4 (11/80)	Bolt: Hex Head, Close Tolerance, 6Al-4V Titanium Alloy, Long Thread, Self-Locking and Nonlocking
NAS5200-5206	Rev. 5 (8/85)	Bolt: Pan Head, Close Tolerance, Short Thread, Tri-Wing Recess, Titanium Alloy 6Al-4V, Self-Locking and Nonlocking
NAS1083-1088	Rev. 4 (3/77)	Bolt: 100°C Flush Head Hi-Torque Recess Titanium Alloy (Short Thread)
NAS663-668	Rev. 8 (11/83)	Bolt: 100°C Flush Head Hi-Torque Recess Titanium Alloy (Long Thread)
NAS7600-7616	Rev. 2 (5/84)	Bolt: 100°C Head Phillips Recess, Close Tolerance, 6Al-4V Titanium Alloy, Short Thread, Self-Locking and Nonlocking
NAS7303-7316	Rev. 3 (4/84)	Bolt: 100°C Head Phillips Recess, Close Tolerance, 6Al-4V Titanium Alloy, Short Thread, Self-Locking and Nonlocking
NAS4304-4316	Rev. 3 (8/85)	Bolt: 100°C Tri-Wing, Recess, Close Tolerances, 6Al-4V Titanium Alloy, Short Thread, Self-Locking and Nonlocking
NAS4600-4616	Rev. 4 (8/85)	Bolt: 100°C Tri-Wing, Recess, Close Tolerances, 6Al-4V Titanium Alloy, Short Thread, Self-Locking and Nonlocking
NAS8802-8816	- (4/82)	Bolt: 100°C Reduced Head, Phillips Recess, Close Tolerance, 6Al-4V Titanium Alloy, Short Thread
Fasteners		
NAS4004	Rev. 5 (8/85)	Fasteners: 6Al-4V Titanium Alloy Externally Threaded

(Continued)

Table I.14 (Continued)

Document No. (a)	Revision No. (latest date) (b)	Standard title
		Rivet
NAS1806-1816	Rev. 1 (4/78)	Rivet: Hi-Shear, Flat Head, Interference Fit, Titanium Alloy
NAS1906-1916	Rev. 1 (4/78)	Rivet: Hi-Shear, 100°C Head, Interference Fit, Titanium Alloy
NAS1097	Rev. 7 (8/85)	Rivet: Solid 100°C Flush Shear Head, Aluminum Alloy, Titanium Columbian Alloy
		Screws
NAS9201-9206	– (8/84)	Screw: Hex Head, ACR Ribbed Recess, Torq-Set Full Thread, Titanium Alloy 6Al-4V, 160 ksi Tensile
NAS6100-6103	Rev. 4 (8/85)	Screw: Hex Head, Tri Wing, Recess, 6Al-4V Titanium Alloys, Full Thread, Nonlocking
NAS5500-5506	Rev. 5 (8/85)	Screw: Machine, Flat-Fillister Head, Full Thread, Tri-Wing Recess, Titanium Alloy 6Al-4V, Self-Locking and Nonlocking
NAS8200-8206	Rev. 1 (5/83)	Screw: Pan Head, Phillips Recess, Titanium Alloy 6Al-4V, Full Thread, Self-Locking and Nonlocking
NAS5800-5806	Rev. 4 (8/85)	Screw: 100°C Head, Tri-Wing Recess, 6Al-4V Titanium Alloy, Full Thread, Self-Locking and Nonlocking
		Miscellaneous
NAS821	–	Inspection Requirements for Titanium Alloy Sheet
NAS941	Rev. 1 (12/59)	Identification Marking of Titanium Sheet

(a) Not in numerical order. Items grouped according to subject.
(b) Revision number, if one exists, indicated. For convenience, in this table, date represents the later of the original issue date or a standard's "reaffirmation" date.

SAE STANDARDS

The Society of Automotive Engineers (SAE) is a professional engineering association that issues approved standards and recommended practices which are advisory in nature only. Compliance is entirely voluntary. These standards are used primarily by designers, manufacturers, and maintenance personnel in the automotive and aerospace industries.

SAE uses alphabetical letter prefixes to indicate various categories of materials. Among them is AMS, or Aerospace Materials Specifications. Table I.8, above, under the AMS specifications paragraph, lists many of the AMS numbers between 4900 and 7461, since these all relate to titanium wrought products. Table I.15, which follows, is more general in nature and tends to involve procedures, practices, or finished products.

References to SAE No. in Table I.15, column head, indicate the document where the standard may be found. Individual documents or complete sets may be ordered from:

Society of Automotive Engineers
Customer Service Department
Publications Group
400 Commonwealth Drive
Warrendale, PA 15096

Table I.15 SAE general type Ti standards

SAE No.	Description
ARP 1333 (a)	Nondestructive Testing of Electron Beam Welded Joints in Titanium-base Alloys
AS 1577 (a)	Tube end, Welding, Hydraulic
AS 1580	Ring, Tube Weld, 30 ksi, Hydraulic, Titanium
AS 1814	Terminology for Titanium Microstructures
ARP 982B	Minimizing Stress Corrosion Cracking in Wrought Titanium Alloy Products
ARP 1795A	Stress Corrosion of Titanium Alloys; Effect of Cleaning Agents on Aircraft Engine Materials
ARP 1843	Surface Preparation for Structural Adhesive Bonding Titanium Alloy Parts
ARP 1932	Anodize Treatment of Titanium and Titanium Alloys (pH 12.4 max)
AMS 2241J MAM 2241 (a) (b)	Tolerances: Corrosion and Heat Resistant Steel, Iron Alloy, Titanium and Titanium Alloy Bars and Wire
AMS 2242E MAM 2242	Tolerances: Corrosion and Heat Resistant Steel, Iron Alloy, Titanium and Titanium Alloy Sheet, Strip, Plate
AMS 2244A MAM 2244	Tolerances: Titanium and Titanium Alloy Tubing
AMS 2245A MAM 2245	Tolerances: Titanium and Titanium Alloy Extruded Bars, Rods, Shapes
AMS 2249C	Chemical Check Analysis Limits: Titanium and Titanium Alloys (DOD adopted.)
AMS 2269	Fusion Welding Titanium
AMS 2380A	Approval and Control of Premium Quality Titanium Alloys
AMS 2486B	Conversion Coating of Titanium Alloys: Fluoride-Phosphate Type
AMS 2631A	Ultrasonic Inspection of Titanium and Titanium Alloy Bar and Billet
AMS 2642	Structural Examination of Titanium Alloys; Etch-anodize Inspection Procedure
AMS 2643	Structural Examination of Titanium Alloys; Chemical-etch Inspection Procedure
AMS 2689	Fusion Welding Titanium and Titanium Alloys
AMS 2775A	Case Hardening of Titanium and Titanium Alloys
AMS 2809	Identification: Titanium and Titanium Alloy Wrought Products
AMS 4798H	Rings, Flash-Welded; Titanium and Titanium Alloys
AMS 4931	Titanium Alloy Bars, Forgings, Rings: 6Al-4V-ELI, Duplex Annealed, Fracture Toughness
AMS 4959	Titanium Alloy Wire: Ti-13.5V-11Cr-3Al
AMS 4984	Titanium Alloy Forgings; 3Al-10V-2Fe Consummable Electrode, Melted, Heat Treated, Aged; 173 ksi (1195 MPa) Tensile Strength
AMS 4985	Titanium Alloy Castings; Investment or Rammed Graphite: Ti-6Al-4V, Annealed
AMS 4986	Titanium Alloy Forgings: 3Al-10V-2Fe Consummable Electrode, Melted, Heat Treated, Overaged; 160 ksi 1105 MPa) Tensile Strength
AMS 4987	Titanium Alloy Forgings: 3Al-10V-2Fe Consummable Electrode, Melted, Single Solution, Heat Treated, Overaged; 140 ksi (965 MPa) Tensile Strength
AMS 4991	Titanium Alloy Castings, Investment: Ti-6Al-4V, Annealed
AMS 4998	Titanium Alloy Powder: 6Al-4V, Premium Quality

(a) ARP = aerospace recommended practice; AS = aerospace standards; MAM = metric aerospace recommended practice; AMS = aerospace material specification
(b) MAM is the metric version

UNS NUMBERING SYSTEM

The Unified Numbering System (UNS) is a joint effort by the Society of Automotive Engineers (SAE) and the American Society for Testing and Materials (ASTM). Its aim is to provide an extensive listing in which each metal or alloy has a unique identification number, a notation of its nominal chemical composition, and a cross reference to other specifications systems, should such exist. It also has a goal of reducing the confusion caused by multiple trade names for the same metal or alloy.

Society of Automotive Engineers, Inc.
400 Commonwealth Drive
Warrendale, PA 15096

The UNS numbers in Table I.16 follow the fourth edition (1986) of *Metals & Alloys in the Unified Numbering System*, published by SAE.

Refer to Appendix D, "Filler Metals," for UNS numbers corresponding to welding materials.

Table I.16 UNS numbering of titanium and titanium alloys

UNS No.	Chemical Compositon (%)(a)(b)	Similar Specification (c)
	Commercially pure (unalloyed) titanium	
R50250	0.2Fe; 0.03N; 0.01C; 0.015H; 0.2O; remain. Ti; 25 ksi YS	ASTM B 265, Grade 1
R50400	0.3Fe; 0.03N; 0.1C; 0.015H; 0.25O; remain. Ti; 40 YS	ASTM B 265, Grade 2
R50550	0.3Fe; 0.05N; 0.1C; 0.3H; 0.35O; remain. Ti; 55 YS (c)	ASTM B 265, Grade 3
R50700	0.5Fe; 0.05N; 0.1C; 0.015H; 0.4O; remain. Ti; 70 YS	ASTM B 265, Grade 4
	Low-alloyed titanium	
R52250	0.2Fe; 0.03N; 0.1C; 0.015H; 0.18O; 0.12 to 0.25Pd; remain. Ti; 25 YS	ASTM B 265, Grade 11
R52400	0.3Fe; 0.03N; 0.1C; 0.015H; 0.25O; 0.12 to 0.25Pd; remain. Ti; 40 YS	ASTM B 265, Grade 7
R52550	0.3Fe; 0.05N; 0.1C; 0.015H; 0.35O; 0.12Pd; remain. Ti (b)	ASTM B 367, Grade Ti-Pd 8A (casting)
	Titanium alloys	
R52700	0.5Fe; 0.05N; 0.1C; 0.01H; 0.4O; 0.12Pd; remain. Ti (b)	-
R53400	0.3Fe; 0.03N; 0.08C; 0.015H; 0.25O; 0.2 to 0.4Mo; 0.06 to 0.9Ni; remain. Ti	ASTM B 265, Grade 12
R54520	0.5Fe; 0.05N; 0.1C; 0.02H; 0.2O; 4.0 to 6.0Al; 2.0 to 3.0 Sn; remain. Ti	ASTM B 265, Grade 6
R54521	5Al; 2.5Sn; remain. Ti	AMS 4909 (Ti-5Al-2.5Sn)
R54520	5Al; 5Sn; 5Zr; remain. Ti	MIL-F-83142A, Comp. 4 (forging)
R54560	5Al; 6Sn; 2Zr; 1Mo; 0.2Si, remain. Ti	-
R54620	6Al; 2Sn; 4Zr; 2Mo; remain. Ti	AMS 4919
R54790	2Al; 11Sn; 5Zr; 1Mo; 0.2Si; remain. Ti (b)	AMS 4974
R54810	8Al; 1Mo; 1V; remain. Ti	AMS 4915
R56080	8Mn; remain. Ti	AMS 4908
R56210	6Al; 2Cb; 1Ta; 0.8Mo; remain. Ti	MIL-T-9046J
R56260	6Al; 6Sn; 4Zr; 6Mo; remain. Ti	AMS 4981
R56320	2.3 to 3.5Al; 0.25Fe; 0.02N; 0.05C; 0.013H; 0.12O, 2.0 to 3.0V; remain. Ti (b)	ASTM B 337, Grade 9 (seamless & welded pipe)
R56400	5.6 to 6.75Al; 3.5 to 4.5V; 0.4Fe; 0.05N; 0.1C; 0.015H; 0.2O; remain Ti	MIL-T-9046J, AB-1
R56401	6Al; 4V; remain. Ti (ELI)	MIL-T-9046J, AB-2
R56410 (e)	2.6 to 3.4Al; 9.0 to 11.0V; 1.6 to 2.2Fe; 0.05C; 0.015H; 0.05Ni; 0.13O; remain. Ti	AMS 4984 4986, 4987 (10V-2Fe-3Al)
R56430	4Al; 3Mo; 1V; remain. Ti	-
R56620	5.5Al; 2Sn; 5.5V; remain. Ti (b)	AMS 4918

(Continued)

Table I.16 (Continued)

UNS No.	Chemical Compositon (%)(a)(b)	Similar Specification (c)
R56740	7Al; 4Mo; remain. Ti	MIL-F-83142A, Comp. 9
R58010	3Al; 13V; 11Cr; remain. Ti	MIL-F-83142A Comp. 12
R58030	10.0 to 13.0 Mo; 4.5 to 7.5 Zr; 3.75 to 5.25 Sn; 0.35Fe; 0.05N; 0.1C; 0.02H; 0.18O; remain. Ti (b)	MIL-T-4046J, B-2
R58450	45Cb; remain Ti (b)	AMS 4982
R58640	3Al; 8V; 6Cr; 4Mo; 4Zr; remain. Ti	MIL-T-9046J, Comp. B-3
R58650	4.5 to 5.5Al; 1.5 to 2.5Sn; 1.5 to 2.5Zr; 3.5 to 4.5Cr; 3.5 to 4.5Mo; 0.3Fe; 0.4N; 0.05C; 0.0125H; 0.08 to 0.13O; 0.1Cu; 0.1Mn; remain. Y, Zr, Ti and other	AMS 4995 (5Al-2Sn-2Zr-4Cr-4Mo)
R58820	8Mo; 8V; 2Fe; 3Al; remain. Ti	MIL-T-9046J, Comp. B-4

(a) All maximums, except Ti
(b) Chemical compositions here used to aid with identification; noted specifications may vary slightly from "Similar Specifications: column.
(c) Not inclusive. See SAE's UNS volume for other related specifications
(d) "Other" elements not included in this listing which is typical and representative. Thus "remain. Ti" does not imply a lack of impurities.
(e) Not in fourth edition, but proposed for next edition.

ANSI SAFETY-RELATED STANDARDS

The basic American National Standards Institute (ANSI) publication for safety is actually issued by the sponsoring organization National Fire Protection Association (NFPA). The title is *Standard for the Production, Processing, Handling, and Storage of Titanium* (Publication NFPA 481-1982). It may be found in the latest annual edition of NFPA's Compilation series, Volume 6.

National Fire Protection Association
Batterymarch Park
Quincy, MA 02269

Topics covered include:

- Titanium sponge production

- Sponge melting

- Mill operations

- Machining and fabrications

- Scrap generation, processing, and storage

- Powder production and use

ASME CASES

The American Society of Mechanical Engineers (ASME) issues standards that are used by personnel in research, testing, and design of power-producing machines such as internal combustion engines, steam, and gas turbines, etc. They are also used to design and develop power-using machines such as refrigeration, air-conditioning equipment, etc.

American Society of Mechanical Engineers
United Engineering Center
345 East 47th Street
New York, NY 10017

ASME publishes the *Boiler and Pressure Code*, which contains the determinations of "code cases" presented to the Boiler and Pressure Vessel Committee. Refer to the volume's Section VIII, Division 1, "Pressure Vessels." The information contained in Table I.17 lists "cases" available as of Supplement No. 13, 1986.

Table I.17 ASME titanium-related cases

Case No.	Date	Title
1977	5/85	Titanium Castings, Grade C-2. Concerns the conditions under which ASTM B 367, Grade C-2 can be used in the construction of pressure vessels
1961	12/84	Steel Plates, Forgings. Concerns the conditions and requirements under which steel plates and forgings can be used in the construction of welded pressure vessels
1930	4/83	Titanium Stabilized Cr-Fe Stainless. Concerns the conditions and requirements under which titanium-stabilized Ni-Cr seamless and welded pipes can be used in Division 1 welded construction.

NATIONAL BUREAU OF STANDARDS

The United States National Bureau of Standards (NBS) makes available Standard Reference Materials (SRMs) which are samples of a material that are certified with respect to chemical composition, chemical properties, or physical properties. Each material sample bears a distinguishing name and number by which it is permanently identified. Thus each material bearing a given description is identical (within the specified limits) to every other sample bearing the same designation — with the exception of individually certified items, which are further identified by serial number.

An SRM is defined as a material or substance one or more properties of which are sufficiently well established to be used for the calibration of an apparatus, the assessment of a measurement method, or for assigning values to materials. These SRMs are certified for specific chemical or physical properties, and issued by NBS which certificates the results of the characterization.

In sum the NBS has the function to develop, produce, and distribute Standard Reference Materials that provide a basis for comparison of measurement on materials, and that aid in the control of production process.

For further information request the current NBS *Standard Reference Materials Catalog*. (This is NBS special publication 260.) Ordering information, prices, etc. are cited in it.

National Bureau of Standards
Office of Standard Reference Materials
Bldg. 222, Room B-311
Gaithersburg, MD 20899

Table I.18 is a listing of currently existing SRMs related to titanium and its alloys.

Table I.18 NBS Ti and Ti alloy SRMs (a)

SRM No. (b)	Type	Chemical composition (nominal wt. %)	Wt/unit in grams
		Titanium-base alloys, chip form (b)	
173b	6Al-4V	0.025C; 0.008Cu; 0.013Mo; (0.03)Sn; 6.36Al; 0.23Fe; 0.015N; 4.31V; 0.46Si	100
176	5Al-2.5Sn	0.015C; 0.0008Mn; 0.0003Cu; 0.003Mo; 2.47Sn; 5.16Al; 0.070Fe; 0.010N	100
650	Unalloyed A	1.55W; 0.016Mn; 0.033Cu; 0.002Mo; 0.03Sn; <0.01Al; 0.024Fe; 0.002Cr; 0.009V, 0.004Si	30
651	Unalloyed B	0.39W; 0.005Mn; 0.032Cu; 0.031Mo; 0.026Sn; <0.006Al; 0.058Fe; 0.037Cr; 0.021V; 0.011Si	30
652	Unalloyed C	0.5W; 0.046Mn; 0.081Cu; 0.039Mo; 0.053Sn; 0.039Al; 0.67Fe; 0.082Cr; 0.024V; 0.16Si	30
		Titanium-base alloys, solid form (c)	
641	8Mn (A)	6.68Mn	-
642	8Mn (B)	9.08Mn	-
643	8Mn (C)	11.68Mn	-
644	Ti-2Cr-2Fe-2Mo (A)	1.03Cr; 1.36Fe; 3.61Mo	-
646	Ti-2Cr-2Fe-2Mo (C)	3.43Cr; 2.14Fe; 1.11Mo	-
654 (d)	Ti-6Al-4V (B)	(<0.1)Mn; (0.20)Cr; (0.20)Fe; (<0.05)Mo; 6.3Al; 3.9V	-

(a) Values in parentheses are not certified, and are given for information only.
(b) Chip form = SRMs primarily for checking chemical analysis. These are furnished in units noted in rightmost column. Chips are usually in a range from 0.4 to 1.2 mm.
(c) Solid form = The 600 series of SRMs are primarily for microchemical methods of analysis such as electron probe microanalysis, spark source mass spectrometric analysis, and laser probe analysis. The thicknesses noted here are 31 mm Dx19 mm thick.
(d) This SRM applies to 31 mm Dx6.54 mm thick.

INTERNATIONAL CROSS-REFERENCE LISTING

This segment, which includes Tables I.19 and I.20, acts as a bridge between the United States titanium or titanium designations and many European designations which appear later.

It is obvious that the individual items cross-referenced are not extensive in number. In fact, most are limited to the more common commercially pure grades, or more frequently used titanium alloys. As is mentioned several times in this appendix, the technological demands placed on a country's industry greatly influence the variety of titanium and titanium alloys used, and, therefore, those that are "standardized." Many countries, lacking an advanced aerospace market, simply have no need for the more complex and/ or demanding forms of titanium.

Nevertheless, Table I.19 has value in spite of these limitations. If nothing else, it is, for those already familiar with designations used in the United States, an introduction to foreign designations. Also, it definitely helps a user relate references in product literature to a known factor

In much the same way Table I.20 offers support in cross-referencing a number of commercially pure international designations for mill products. The comparisons may be made on the basis of chemical composition.

Also refer to the Table included under SPAIN, below, for more cross-referencing information.

Table I.19 Equivalent cross-reference listing of common Ti and Ti alloys; corresponding grades or specifications

USA (a)	UK (b)	France (c)	Federal Republic of Germany (d)	AECMA (e)	Spain (f)
ASTM, Grade 1	BS TA.1	T-35	3.7024	Ti-P01	L-7001
AMS 4902, 4941, 4962, 4951; ASTM, Grade 2	BS TA.2, TA.3, TA.4, TA.5	T-40	3.7034	Ti-P02	L-7002
AMS 4900; ASTM, Grade 3	DTD 5023, 5273, 5283, 5293	T-50	3.7055	-	-
AMS 4901, 4921; ASTM, Grade 4	BS TA.6, TA.7, TA.8, TA.9	T-60	3.7064	Ti-P04	L-7004
-	BS TA.21, TA.22, TA.23, TA.24, TA.52, TA.53, TA.54, TA.55, TA.58	T-U2	-	Ti-P11	-
AMS 4909, 4910, 4924, 4926, 4953, 4966	BS TA.14, TA.15, TA.16, TA.17	T-A5E	-	Ti-P65	-
AMS 4911, 4928, 4934, 4935, 4954, 4965, 4967; ASTM, Grade 5	BS TA.10, TA.11, TA.12, TA.13, TA.28, TA.56 DTD 5363	T-A6V	3.7164	Ti-P63	L-7301
-	BS TA.45 thru TA.51, TA.57	T-A4DE	3.7184	Ti-P68	-
AMS 4974	BS TA.18, TA.19, TA.20, TA.25, TA.26, TA.27	-	-	-	-
-	DTD 5213	T-311DA	-	-	-
-	BS TA.43, TA.44	T-A6ZD	3.7154	Ti-P67	-

(a) MIL-T-9011, -9046, -9047, -14577, -46038, -46077 also apply. ASTM B 265, B 338, B 348, B 367, B 381, B 382 also apply.
(b) UK = British Standards, Aerospace Series. (See table at United Kingdom, below.) Also UK = Ministry of Defense, DTD series.
(c) Applicable French standards are AIR 9182, 9183, 9184. (See table at France, below.)
(d) These are aerospace references. DIN 17850, 17860, 17862, 17863, 17864, (WF 3.7025, 3.7035, 3.7055, 3.7065) also apply. (See table at Federal Republic of Germany, below.)
(e) AECMA recommendations.
(f) References listed here are UNE designations. (See UNE table at Spain, below, for numbers.)

Table I.20 Comparison of various specifications of titanium for mill products

Designa-tion	Chemical Composition (%) max.							Mechanical Properties		
	C	H	O	N	Fe	Pd	Total others	Tensile strength (b)	Yield strength (b)	Elong. %
JIS Class 1	-	0.015	0.15	0.05	0.20	-	-	28 to 42	≥17(a)	≥27
ASTM Grade 1	0.10	0.015	0.18	0.03	0.20	-	-	≥24.5	17.5 to 31.5	≥24
ASME Grade 1										
DIN 3·7025	0.08	0.013	0.10	0.05	0.20	-	-	30 to 42	≥18	≥30
GOST BT1-00	0.05	0.008	0.10	0.04	0.20	-	0.10 max	≥30	-	≥20
BS 19-27t/in²	-	0.0125	-	-	0.20	-	-	29 to 42	≥20	≥25
JIS Class 2	-	0.015	0.20	0.05	0.25	-	-	35 to 52	≥22 (a)	≥23
ASTM Grade 2	0.10	0.015	0.25	0.03	0.30	-	-	≥35	28 to 42	≥20
ASME Grade 2										
DIN 3·7035	0.08	0.013	0.20	0.06	0.25	-	-	≥38	≥25	≥22
GOST BT1-0	0.07	0.010	0.20	0.04	0.30	-	0.30 max	40 to 55	-	≥20
BS 25-35t/in²	-	0.0125	-	-	0.20	-	-	39 to 54	≥29	≥22
JIS Class 3	-	0.015	0.30	0.07	0.30	-	-	49 to 63	≥35(a)	≥18
ASTM Grade 3	0.10	0.015	0.35	0.05	0.30	-	-	≥45.5	38.5 to 53	≥18
ASME Grade 3										
DIN 3·7055	0.10	0.013	0.25	0.06	0.30	-	-	47 to 60	≥33	≥18
ASTM Grade 7	0.10	0.015	0.25	0.03	0.30	0.12-0.25	-	≥35	28 to 42	≥20
ASME Grade 7										
ASTM Grade 11	0.10	0.015	0.18	0.03	0.20	0.12-0.25	-	≥24.5	17.5 to 31.5	≥24

(a) Only for plate, sheet and coil
(b) (kgf/mm²)

AUSTRALIA

The Standards Association of Australia (SAA) issues standards used primarily by firms doing business in Australia and the southwest Pacific area. As yet, this organization has not issued any titanium standards.

Standards Association of Australia
Standards House
80 Arthur Street
North Sidney, NSW 2060
Australia

A representative of SAA stated that while Australia is one of the world's major titanium producers, the bulk of the ore is exported either as rutile or zircon. (Refer to Chapter 1, "Executive Summary," for details.) No titanium refining is currently carried out within the country. Because of this, there is no call for SAA to produce reference methods or standard methods of analysis for titanium.

AUSTRIA

The Austrian Standards Institute (ON) creates and publishes standards for Austria. A representative of that organization has stated that it currently does not issue standards or designations for titanium. Rather, it relies on DIN standards. (See Federal Republic of Germany.)

Osterreichisches Normungsinstitut
Heinestrasse 38
Postlach 130
A-1021, Wein, Austria

CZECHOSLOVAKIA

The Czechoslovakian Office for Standards and Measurements (CSN) is a government agency concerned with standardization, metrology, and quality assurance. The CSN was founded in 1952 and is a member of ISO. All standards are preceded by CSN or ON (sectional standards) in upper case letters.

Titanium standard publication numbers and corresponding titles are indicated in Table I.21. Note that this list dates from 1976 and is placed here only for historical purposes. A representative of CSN informed the editor in early 1988 that neither titanium nor its alloys were currently produced within the country. As the result all titanium-related standards have been cancelled.

It was further stated that in instances of importation, ASTM standards are used.

Urad pro Normalizaci a Mereni
Vaclavske Namesti 19
113 47 Praha 1,
Czechoslovakia

Table I.21 Czechoslovakian standards for Ti and Ti alloys

Standard Number	Title of Standard
CSN 42-1490	Sheet, band, and strip; technical delivery code
CSN 42-1491	Wires; technical delivery code
CSN 42-1492	Bars; technical delivery code
CSN 42-1493	Seamless tubes; technical delivery code
CSN 42-4655	Titanium, unalloyed (99.5%)
CSN 42-7390	Hot rolled sheets and strips; dimensions
CSN 42-7391	Cold-rolled sheets, bands, and strip; dimensions
CSN 42-7490	Round wires
CSN 42-7590	Hot-formed bars; dimensions
CSN 42-7591	Turned bars; dimensions
CSN 42-7790	Seamless tubes; dimensions
CSN 42-1496	Ingots; delivery code and dimensional standard
CSN 42-4656	Ingots

EUROPE/AECMA

The Association Européenne des Constructeurs de Matériel Aérospatial (AECMA) issues various standards for materials and procedures as they relate to the European aerospace industry. (Frequently these are called European Aerospace Standards.) A listing of these standards may be found in *Document AECMA/BNAE No. 5710H*. (At the time this technical guide was published, the most recent issue was January, 1987.)

The AECMA is edited and maintained by BNAE, or Bureau de Normalisation de l'Aeronautique et de l'Espace.

The standards, as a consolidated listing of individual documents, are listed in Table I.22. The standards are contained in Table I.23.
To order documents, write:

**Bureau of Normalisation de l'Aeronautique
   et de l'Espace**
8, rue Moreau-Vauthier
92100 Boulongne-Billancourt
France

Table I.22 AECMA standards: consolidated listing

Document (standard)	Ti alloy	Description
prEN2517	Ti-P63	Annealed sheets, strips, plates; $\leq$100 mm
prEN2518	Ti-P02	Bars; $\leq$200 mm, 390 to 540 MPa
prEN2519	Ti-P04	Bars; $\leq$200 mm, 540 to 740 MPa
prEN2520	Ti-P04	Forgings; $\leq$200 mm, 540 to 740 MPa
prEN2521	Ti-P11	Bars: $\leq$200 mm, 540 to 700 MPa
prEN2522	Ti-P11	Forgings; $\leq$200 mm, 540 to 700 MPa
prEN2523	Ti-P11	Bars: $\leq$75 mm, 650 to 880 MPa
prEN2524	Ti-P11	Forgings; $\leq$75 mm, 650 to 880 MPa
prEN2525	Ti-P01	Sheets, strips; $\leq$5 mm, 290 to 420 MPa
prEN2526	Ti-P02	Sheets, strips; $\leq$5 mm, 390 to 540 MPa
prEN2527	Ti-P04	Sheets, strips; $\leq$5 mm, 570 to 730 MPa
prEN2528	Ti-P11	Sheets, strips; $\leq$5 mm, 540 to 700 MPa
prEN2530	Ti-P63	Annealed bars: $\leq$150 mm, 900 to 1160 MPa
prEN2531	Ti-P63	Annealed forgings: $\leq$150 mm, 900 to 1160 MPa
prEN2532	Ti-P68	Bars; $\leq$25 mm, 1100 to 1280 MPa
prEN2533	Ti-P68	Bars; 25 to 100 mm, 1050 to 1220 MPa
prEN2534	Ti-P68	Bars; 100 to 150 mm, 1000 to 1200 MPa

Table I.23   AECMA standards; organized by grade/alloy common designations

AECMA standard No.	Product form	AECMA design.	Common design.	Chemical composition (%) max. unless noted	Mechanical properties, hardness values (a)
prEN 2518	Bars 390-540 MPa <200 mm	Ti-PO2	Commercially pure	0.2Fe; 0.08C; 0.25O; 0.06N; 0.6 total others; remain. Ti. H for forging 0.01; H for machining 0.12.	Condition: Bars for forging not heat treated; bars for machining annealed. Thickness: <200 mm (<8 in.) Tensile strength: 390-540 MPa (56-78 ksi) Yield strength (b): 290 MPa (42 ksi) Elongation: 20% min Reduction in area: 30% min
prEN 2519	Bars 540-740 MPa <200 mm	Ti-PO4	Commercially pure	0.35Fe; 0.08C; 0.4O; 0.07N; 0.6 others total; remain. Ti. H for forgings 0.01; H for machining 0.012	Condition: Bars for forging not heat treated; bars for machining annealed. Thick.: <200 mm (<8 in.) TS: 540-740 MPa (78-107 ksi) YS (b): 440 MPa (63.8 ksi) El: 15% min Red.: 25% min
prEN 2526	Sheet, strip, 390-540 MPa <5 mm	Ti-PO2	Commercially pure	Same as prEN 2518, above, but 0.12H	Condition: Annealed Thick.: <5 mm (<0.197 in.) TS: 390-540 MPa (56-78 ksi) YS (b): 290 MPa (42 ksi) El: 22% min
prEN 2527	Sheet, strip 570-730 MPa <5 mm	Ti-PO4	Commercially pure	0.35Fe; 0.08C; 0.4O; 0.07N; 0.012H; 0.6 others total; remain. Ti.	Condition: Annealed Thick.: <5 mm (<0.197 in.) TS: 570-730 MPa (82.6-105.8 ksi) YS (b): 460 MPa (66.7 ksi) El: 15% min
prEN 2520	Forgings 540-740 MPa <200 mm	Ti-PO4	Commercially pure	0.35Fe; 0.08C; 0.4O; 0.07N; 0.012H; 0.6 others total; remain. Ti.	Condition: Annealed Thick.: 200 mm (8 in.) TS: 540-740 MPa (78-107 ksi) YS (b): 440 MPa (63.8 ksi) El: 15% min Red.: 25% min

(Continued)

Table I.23 (Continued)

AECMA standard No.	Product form	AECMA design.	Common design.	Chemical composition (%) max. unless noted	Mechanical properties, hardness values (a)
prEN 2530	Bars 900–1160 MPa <80 mm <80–150 mm	Ti-P63	Ti-6Al-4V	5.5–6.75Al; 3.5–4.5V; 0.3Fe; 0.08C; 0.2O; 0.05N; 0.4 others total; remain. Ti. H for forgings 0.01; H for machining 0.012	Condition: Bars for forging not heat treat. Thick.: 150 mm (6 in.) TS: 900–1160 MPa (130.5–168 ksi) YS (b): 830 MPa (120.4 ksi) El: 10% min Red: 25% min Condition: Bars for machining annealed Thick.: 80 mm (3.2 in.) TS: 900–1160 MPa (130.5–168 ksi) YS (b): 830 MPa (120.4 ksi) El: 10% min Red.: 25% min Condition: Bars for machining annealed Thick.: 80–150 mm (3.2–6.0 in.) TS: 900–1160 MPa (130.5–168 ksi) YS (b): 830 MPa (120.4 ksi) El: 8% min Red.: 20% min
prEN 2517	Sheet, strip, plate 920–1180 MPa 5 mm 900–1160 MPa 5–100 mm	Ti-P63	Ti-6Al-4V	5.5–6.75Al; 3.5–4.5V; 0.3Fe; 0.08C; 0.2O; 0.05N; 0.012H; 0.4 total other; remain. Ti.	Condition: Annealed Thick.: of sheet; strip: 5 mm (197 in.) TS: 920–1180 MPa (113.4–171 ksi) YS (b): 870 MPa (126.2 ksi) El: 8% min Red.: N/A Condition: Annealed Thick. of plate: 5–12 mm (0.197–0.47 in.) TS: 900–1160 MPa (130.5–168.3 ksi) YS (b): 830 MPa (120.4 ksi) El: 10% min Red.: N/A Condition: Annealed Thick. of plate: 12–100 mm (0.47–4.0 in.) TS: 900–1160 MPa (130.5–168.3 ksi) YS (b): 830 MPa (120 ksi) El: 8% min Red.: 20 min

(Continued)

Table I.23 (Continued)

AECMA standard No.	Product form	AECMA design.	Common design.	Chemical composition (%) max. unless noted	Mechanical properties, hardness values (a)
prEN 2523	Bars 650-880 MPa <75 mm	Ti-P11	Ti-2.5Cu	2.0-3.0Cu; 0.2Fe; 0.08C; 0.2O; 0.05N; 0.01H; 0.4 total other; remain. Ti. H for forging 0.1 H for machining 0.012	Condition: Bars for forgings not heat treated; bars for machining solution treated and aged Thick.: <75 mm (<3 in.) TS: 650-880 MPa (94.3-127.6 ksi) YS (b): 530 MPa (76.9 ksi) El: 10% min Red.: 25% min
prEN 2521	Bars 540-700 MPa <200 mm	Ti-P11	Ti-2.5Cu	Same as prEN 2523, above, but H for forging 0.01; H for maching 0.012	Condition: Bars for forging not heat treated; bars for machining annealed Thick.: <200 mm (<8.0 in.) TS: 540-700 MPa (82.6-1011.5 ksi) YS (b): 400 MPa (58 ksi) El: 16% min Red.: 35% min
prEN 2128	Sheet, strip 540-700 MPa <5 mm	Ti-P11	Ti-2.5Cu	Same as prEN 2523, above, but H is 0.012	Condition: Annealed Thick.: <5 mm (<0.197 in.) TS: 540-700 MPa (82.6-101.5 ksi) YS (b): 460-570 MPa (66.7-82.7 ksi) El: 18% min
prEN 2522	Forgings 540-700 MPa <200 mm	Ti-P11	Ti-2.5Cu	Same as prEN 2523, above, but H is 0.012	Condition: Annealed Thick.: <200 mm (<8.0 in.) TS: 540-700 MPa (82.6-101.5 ksi) YS (b): 400 MPa (58 ksi) El: 16% min Red.: 35% min
prEN 2525	Forgings 650-880 MPa <75 mm	Ti-P11	Ti-2.5Cu	Same as prEN 2523, above, but H is 0.012	Condition: Solution treated and aged Thick.: <75 mm (<3.0 in.) TS: 650-880 MPa (94.3-127.6 ksi) YS (b): 530 MPa (76.9 ksi) El: 10% min Red.: 25% min

(Continued)

Table I.23 (Continued)

AECMA standard No.	Product form	AECMA design.	Common design.	Chemical composition (%) max. unless noted	Mechanical properties, hardness values (a)
prEN 2532	Bars 1100–1280 MPa <25 mm	Ti-P68	Ti-4Al-4Mo-2Sn-0.5Si	3.0–5.0Al; 3.0–5.0Mo; 1.5–2.5Sn; 0.3–0.7Si; 0.2Fe; 0.08C; 0.25O; 0.05N; 0.4 total other; remain. Ti. H for forging 0.01; H for machining 0.012	Condition: Bars for forging not heat treated; bars for machining solution treated and aged. Thick.: <25 mm (<1.0 in.) TS: 1100–1280 MPa (159.5–185.5 ksi) YS (b): 960 MPa (139.3 ksi) El: 9% min Red.: 20% min for forging; 25% for machining
prEN 2533	Bars 1050–1220 MPa 25–100 mm	Ti-P68	Ti-4Al-4Mo-2Sn-0.5Si	Same as prEN 2532, above	Condition: Bars for forging not heat treated; bars for machining solution treated and aged. Thick.: <25–100 mm (<1.0–4.0 in.) TS: 1050–1220 MPa (152.3–177 ksi) YS (b): 920 MPa (933.4 ksi) El: 9% min Red.: 20% min
prEN 2534	Bars 1000–1200 MPa 100–150 mm	Ti-P68	Ti-4Al-4Mo-2Sn-0.5Si	Same as prEN 2532, above	Condition: Bars for forging no heat treated; bars for machining solution treated and aged. Thick.: <100–150 mm (<4.0–6.0 in.) TS: 1000–1200 MPa (145–174 ksi) YS (b): 870 MPa (126.2 ksi) El: 9% min Red.: 20% min

(a) AECMA standards are issued in metric. English units here are conversions that have been conservatively rounded.
(b) Yield strengh = minimum and at 0.2% offset.

FEDERAL REPUBLIC OF GERMANY

The Deutsches Institut fur Normung e.V. (DIN) is an organization of approximately 130 standards committees which is sited in the Federal Republic of Germany. For titanium and titanium alloys, DIN concentrates on semifinished products. There are some English translations available for the standards noted in Table I.24. Most commonly DIN standards use these three letters as a prefix before a number grouping. However, at times, other letters are used to indicate the source of the standard. For example:

- LN - *Deutsche Luft- und Raumfahrt-Norm*

- V LN - *Vornorm*

- Vd TÜD - *Vereiningung der Technischen Überwachungs-Vereine e.V*

Copies of the DIN standards may be obtained in the United States from ANSI, foreign standards section.

Table I.25 is a listing of material performance sheets (in German: *Werkstoff-Leistungsblätter*, or WL) which may be used for more detailed information on individual product forms. These WLs are compiled in *Werkstoff-Handbuches der Deutschen Lufthart, Teil 1*. The volume may be ordered from:

Beuth Verlag GmbH
Kamekestrasse 2-8
5000 Köln 1
Federal Republic of Germany

Some notes of explanation on DIN standards may aid a reader to understand the tables included here. Thus:

- DIN 17850 and 17851 specify chemical compositions (*Zusammensetzung*) of commercially pure and alloyed titanium, respectively.

- Four distinct CP compositions are defined by DIN 17850.
  Each of these has a unique WL number: e.g. WL 3.7025.

- Two distinct titanium alloy compositions are defined by DIN 17851. Each of these has a unique WL number: e.g. WL 3.7165.

- DIN 17860, 17862, 17863, and 17864 each relates directly to an individual product form: sheet, rod, etc. Identical compositions (WL) will appear within many DINs noted here, since the product forms may be made from various chemistries.

A listing of the included tables is as follows:

- Table I.26 - Commercially pure titanium, per DIN 17850

- Table I.27 - Titanium alloys, per DIN 17851

The remaining DIN titanium standards may be found under the product forms column within these two tables.

In addition to DIN standards, there exist a few DAN, or *Deutsche Airbus Norm*. (At times these are called MBBN, or MBB, *Transport-und Verkehrsflugzeuge Norm*.) There are also ABS, or Airbus standards. These additional standards were developed by private companies in cases where the national DIN standards were found lacking. No information is available for these additional standards since they are considered "internal" documents.

for DIN:
Deutscher Normenausschus
Burggrafenstrasse 4-10
Postfach 1107
D-1000 Berlin 30
Federal Republic of Germany

for Werkstoff-Leistungsblatter
   der Deutschen Luftfahrt (W.L.):
Order document from:
Federal Defense Engineering
   and Procurement Department
Postfach 7360
5400 Koblenz 1
Federal Republic of Germany

Table I.24 DIN standards

DIN Standards	Title
DIN 1737	Filler material for pure titanium and titanium-palladium alloys; chemical composition; technical supply conditions
LN 9047	Notched-bar tensile strength testing of titanium alloys
LN 9293	Unalloyed, alloyed titanium sheets
E LN 9297	Rolled sheets and plates of titanium and titanium alloys: dimensions and masses (aerospace)
DIN 17850	Chemical compositions of unalloyed and alloyed titanium
DIN 17851	Composition of wrought titanium alloys
DIN 17860	Plate, sheet, strip titanium and titanium alloys
DIN 17862	Titanium bar and wrought titanium alloys
DIN 17863	Titanium wire
DIN 17864	Wrought titanium and titanium alloy forgings
DIN 29783	Aerospace: Precision Castings from Titanium and Titanium Alloys: Technical Conditions of Delivery
V LN 65039	Specification for aircraft sheets, plates, strips of titanium and titanium alloys
V LN 65040	Technical specification for titanium bars, rings, forging stock, forgings
LN 65047 No. 1-3	Close-tolerance aerospace bolts in titanium alloys with self-locking collars in aluminum alloy; procurement specification
LN 65072 No. 1-2	Joining elements made from titanium alloys; a technical specification
LN 65084 No. 1	General: heat treatment of titanium and titanium alloys
LN 65084 No. 2	Annealing; heat treatment of titanium and titanium alloys
LN 65084 No. 3	Stress relieving; heat treatment of titanium and titanium alloys
LN 65084 No. 4	Precipitation hardening; heat treatment of titanium and titanium alloys
VdTUV WB 230	Unalloyed titanium: · Ti I (DIN "Werkstoffnummer" or designation 3.7025) · Ti II (DIN "Werkstoffnummer" or designation 3.7035) · Ti III (DIN "Werkstoffnummer" or designation 3.7055) · Ti IV (DIN "Werkstoffnummer" or designation 3.7065)

Table I.25 Material performance sheets (WL)

WL 3.7024, Supplement 1	Unalloyed titanium, approx. 0.1% O; sheet, filler metal
WL 3.7024, Part 1	Unalloyed titanium, approx. 0.1% O; sheet
WL 3.7024, Part 2	Unalloyed titanium, approx. 0.1% O; welding materials
WL 3.7034, Supplement 1	Unalloyed titanium, approx. 0.1% O; sheet, rod, forging, stock, filler metal
WL 3.7034, Part 1	Unalloyed titanium, approx. 0.2% O; sheet
WL 3.7034, Part 2	Unalloyed titanium, approx. 0.2% O; rod, forging stock
WL 3.7034, Part 3	Unalloyed titanium, approx. 0.2% O; welding material
WL 3.7064, Supplement 1	Unalloyed titanium, approx. 0.2% O; sheet, rod, forging stock
WL 3.7064, Part 1	Unalloyed titanium, approx. 0.3% O; sheet
WL 3.7064, Part 2	Unalloyed titanium, approx. 0.3% O; rod, forging stock
WL 3.7114, Part 1	Titanium alloy: Ti-5Al-2Sn; sheet, strip, plate
WL 3.7114, Part 2	Titanium alloy: Ti-5Al-2Sn; rod, forging stock
WL 3.7124, Part 1	Titanium alloy: Ti-2Cu; sheet, strip
WL 3.7124, Part 2	Titanium alloy: Ti-2Cu; rod, forging stock
WL 3.7144, Part 1	Titanium alloy: Ti-6Al-2Sn-4Zr-2Mo; bar, forging stock
WL 3.7154, Part 1	Titanium alloy: Ti-6Al-5Zr; rod, forging stock
WL 3.7164, Supplement 1	Casting alloy: approx. Ti-6Al-4V; sheet, strip, plate, rod, forging stock
WL 3.7164, Part 1	Casting alloy: Ti-6Al-4V; sheet, strip, plate
WL 3.7164, Part 2	Casting alloy: Ti-6Al-4V; rod, forging stock
WL 3.7174, Part 1	Casting alloy: Ti-6Al-6V-2Sn; sheet, strip, plate
WL 3.7114, Part 2	Titanium alloy, approx. Ti-6Al-6V-2Sn; rod, forging stock
WL 3.7184, Part 1	Titanium alloy, approx. Ti-4Al-4Mo-2Sn; plate
WL 3.7184, Part 2	Titanium alloys, approx. Ti-4Al-4Mo-2Sn; rod, forging stock
WL 3.7264	Titanium casting alloy, approx. Ti-6Al-4V; precision casting alloy

Table I.26 DIN 17850 Chemical composition specifications for commercially pure titanium

Grade designation	Chemical composition (%) max. unless noted	Product forms/ DIN spec	Mechanical properties, hardness values (a)
Ti I	0.2Fe; 0.1O; 0.05N; 0.08C; 0.013H; remain. Ti (According to WL 3.7025)	Plate, sheet, strip DIN 17860	Condition: Annealed Thickness: <20 mm (<0.79 in.) Tensile strength: 290-410 MPa (42-59 ksi) Yield strength (b): 180 MPa (26 ksi) Elongation: 30% min
		Rod DIN 17862	Condition: Annealed Thick.: 6-100 mm (0.24-4.0 in.) TS: 290-410 MPa (42-59) ksi YS: (b): 180 MPa (26 ksi)
		Wire DIN 17863	Thick.: <6 mm (<0.24 in.) TS: 290-410 MPa (42-59 ksi) El: 30% min
		Forging DIN 17864	Condition: Annealed TS: 290-410 MPa (42-59 ksi) YS (b): 180 MPa (26 ksi) El: 20% min (25% min transverse)
Ti II	0.25Fe: 0.2O; 0.06N; 0.08C; 013H; remain. Ti (According to WL 3.7035)	Plate, sheet, strip DIN 17860	Condition: Annealed Thick.: <20 mm (<0.79 in.) TS: 390-540 MPa (56-78 ksi) YS (b): 250 MPa (36 ksi) El: 22% min
		Rod DIN 17862	Condition: Annealed Thick.: 6-100 mm (0.24-4.0 in.) TS: 390-540 MPa (56-78 ksi) YS (b): 250 MPa (36 ksi) El: 30% min
		Wire DIN 17863	Thick.: <6 mm (<0.24 in.) TS: 390-540 MPa (56-78 ksi) El: 22% min
		Forging DIN 17864	Condition: Annealed TS: 390-540 MPa (56-78 ksi) YS (b): 250 MPa (36 ksi) El: 22% min (20% min transverse)
Ti III	0.3Fe; 0.25O; 0.06N; 0.1C; 0.013H; remain. Ti (According to WL 3.7055	Plate, sheet, strip DIN 17860	Condition: Annealed Thick.:<20 mm (<0.79 in.) TS: 460-590 MPa (66-85 ksi) El: 18% min
		Rod DIN 17862	Condition: Annealed Thick.: 6-100 mm (0.24-4.0 in.) TS: 460-590 MPA (66-85 ksi) YS (b): 320 MPa (46 ksi) El: 18% min
		Wire DIN 17863	Thick.: <6 mm (<0.24 in.) TS: 460-590 MPa (66-85 ksi) El: 18% min
		Forging DIN 17864	Condition: Annealed TS: 460-590 MPa (66-85 ksi) YS (b): 320 MPa (46 ksi) El: 18% min (16% min transverse)

(Continued)

Table I.26 (Continued)

Grade designation	Chemical composition (%) max. unless noted	Product forms/ DIN spec	Mechanical properties, hardness values (a)
Ti IV	0.35Fe; 0.3O; 0.07N; 0.1C; 0.013H; remain. Ti (According to WL 3.7065)	Plate, sheet, strip DIN 17860	Condition: Annealed Thick.: <20 mm (<0.79 in.) TS: 540-740 MPa (78-107 ksi) YS (b): 390 MPa (56 ksi) El: 16% min
		Rod DIN 17862	Condition: Annealed Thick.: 6-100 mm (0.24-4.0 in.) TS: 540-740 MPa (78-107 ksi) YS (b): 390 MPa (56 ksi) El: 16% min
		Wire DIN 17863	Thick.: <6 mm (0.24 in.) TS: 540-740 MPa (78-107 ksi) El: 16% min
		Forging DIN 17864	Condition: Annealed Thick.: 540-740 MPa (78-107 ksi) YS (b): 390 MPa (56 ksi) El: 16% min (15% min transverse)

(a) DIN specifications are issued in metric. English units here are conversions that have been conservatively rounded.
(b) Yield strength = min and at 0.2% offset.

Table I.27 DIN 17851 Chemical composition specifications for titanium alloys

Alloy designation	Chemical composition (%) max. unless noted	Product forms/ DIN spec	Mechanical properties, hardness values (a)
Ti-6Al-4V	5.5-6.75Al; 3.5-4.5V; 0.08C; 0.3Fe; 0.015H; 0.05N; 0.2O; remain. Ti (According to WL 3.7165)	Plate, sheet, strip DIN 17860	Condition: Annealed Thickness: 0.8-1.5, 1.5-5.0, and 5.0-50.0 mm TS: 890 MPa (129 ksi) min YS (b): 820 MPa (119 ksi) El: For 0.8-1.5 mm = 6% min; for 1.5-5.0 mm = 8% min; for 5.0-50.0 mm = 8% min
		Rod DIN 17862	Condition: Annealed Cross-section area: <5000 mm^2 TS: 890 MPa (129 ksi) min YS (b): 820 MPa (119 ksi) El: 10% min
		Wire DIN 17860	Cross-section area: <5000 mm^2 and 5000-10 000 mm^2 TS: 890 (129 ksi) min YS (b): 820 MPa (119 ksi) El: 10% min
Ti-5Al-2Sn	4.0-6.0Al; 2.0-3.0Sn; 0.08C; 0.5Fe; 0.02H; 0.05N; 0.2O; remain. Ti (According to WL 3.7115)	Plate, sheet, strip DIN 17860	Condition: Annealed Thick.: 0.8-5.0 and 5.0-50.0 mm TS: 790 MPa (114.6 ksi) min YS (b): 760 MPa (110 ksi) El: For 0.8-5.0 mm = 6% min; for 5.0-50.0 mm = 8% min
		Rod DIN 17862	Condition: Annealed Cross-section area: <5000 mm^2 TS: 790 MPa (114.6 ksi) min YS (b): 760 MPa (110 ksi) El: 8% min
		Wire DIN 17863	Cross-section area: <10 000 mm^2 YS: 790 MPa (114.6 ksi) min YS (b): 760 MPa (110 ksi) El: 8% min

(a) DIN specifications are issued in metric. English units here are conversions that have been conservatively rounded.
(b) Yield strength = minimum and at 0.2% offset.

FINLAND

The Finnish Standards Association (SFS) consists of approximately 37 contributing organizations and acts as a centralized national organization. Its address is:

Suomen Standardisolimisliitto r.y.
Finnish Standards Association
Bulevardi 5 A 7
P. O. Box 205
00121 Helsinki 12
Finland

The SFS currently has only a single document related to titanium. This is a "TTK Directive." Its title is "The Use of Titanium as Material of Pressure Vessels (2/87/P)." It may be obtained from:

Technical Inspection Center (TTK)
P. O. Box 204
00180 Helsinki
Finland

A correspondent at the Finnish Standards Association's Information Service Department has stated that in Finland American Society for Testing and Materials (ASTM) standards are commonly used for titanium and titanium alloys.

FRANCE

The Ministère des Armées issues a limited number of "Conditions of Reception" for unalloyed sheet and for alloyed bars and forged sections and castings. These are not essentially production specifications as much as standards against which received products may be judged, upon delivery, acceptable or unacceptable.

The two basic standards are:

- AIR 9182 - Unalloyed Ti sheet

- AIR 9183 - Alloyed bars, along with forged sections and castings

Refer to Table I.28 and I.29. (AIR indicates des Reglements AIR, or AIR regulations.)

Ministere de la Defense
Delegation Generale pour l'Armement
Centre de Documentation de l'Armement (CEDOCAR)
26, boulevard Victor
75996 Paris Armees
France

In addition, there is AIR 9184, "Provisional Conditions of Receipt for Screws of Titanium Alloy." It covers both screws and bolts used in aerospace applications. The specification centers on items made from TA4M and TA6V, as covered in AIR 9183, noted above.

Some other miscellaneous AIR designations are included in Table I.30 These are either all-but obsolete or not frequently used.

The French Standards Institution (AFNOR) is a nonprofitmaking association under the supervision of the French government, as represented by the Commission for Standards. AFNOR centralizes and coordinates all activities contributing to the growth of standardization.

Association Francaise de Normalisation
Tour Europe - CEDEX 7
92080 Paris - la Defense
France

At the time of the publication of this guide, information on AFNOR standards is somewhat incomplete. What is definitely known is contained in Table I.31.

Table I.28 AIR 9182 (8/85) Terms of acceptance for unalloyed titanium sheets

Designation	Chemical composition (%) max. unless noted	Mechanical properties, hardness values (a)
T-35 ("extra mild")	0.12Fe; 0.05N; 0.08C; 0.015H; 0.04Si; 99.69Ti min	Condition: Annealed Thickness: All Tensile strength: ≤440 MPa (≤64 ksi) Yield strength: ≤882 MPa (≤127 ksi) Elongation: 30%
T-40 ("mild")	0.12Fe; 0.05N; 0.08C; 0.015H; 0.04Si; 99.69Ti min	Condition: Annealed Thick.: All TS: 392 MPa (57 ksi) or 539 MPa (78 ksi) YS: 294 MPa (42 ksi) El: 28% ≤2 mm (2 in.) 25% > 2 mm (2 in.)
T-50 ("medium hard")	0.25Fe; 0.07N; 0.08C; 0.015H; 0.04Si; 99.54Ti min	Condition: Annealed Thick.: All TS: 490 MPa (71 ksi) or 637 MPa (92 ksi) YS: 392 MPa (56 ksi) El: 24% ≤2 mm (2 in.) 20% > 2 mm (2 in.)
T-60 ("hard")	0.30Fe; 0.08N; 0.08C; 0.015H; 0.04Si; 99.56Ti min	Condition: Annealed Thick.: All TS: 588 MPa (852 ksi) or 735 MPa (107 ksi) YS: 470 MPa (68 ksi) El: 20% ≤2 mm (2 in.) 16% > 2 mm (2 in.)

(a) AIR standards are issued in metric. Stress measurements are cited in kg/mm^2 which have been converted here to MPa. English units are conversions that have been conservatively rounded.

Table I.29 AIR 9183 (11/66) Terms of acceptance for bars, shapes, and forgings of Ti alloys

Desigation	Chemical composition (%) max. unless noted	Mechanical properties, hardness values (a)
T-A6V	5.5-7.0Al; 3.5-4.5V; 0.25Fe; 0.08C; 0.012H; 0.07N; 0.20O; remain. Ti	Condition: Not heat treated Cross section: ≤5000 mm^2 and 5000-10 000 mm^2) Tensile strength: 880-1130 MPa (127-163 ksi) Yield strength: 820 MPa (119 ksi) Elongation: 10% (cross section ≤5000 mm^2) 8% (cross section 5000-10 000 mm^2)
T-A4M	3.5-5.0Al; 3.5-5.0Mn; 0.15Fe; 0.08C; 0.05N; 0.012H; 0.2O; 0.04Si; remain. Ti	Condition: Not heat treated Cross section: ≤5000 mm^2 and 5000-10 000 mm^2) TS: 930-1139 MPa (134-163 ksi) Elongation: 10% (cross section ≤5000 mm^2) 8% (cross section 5000-10 000 mm^2)

(a) AIR standards are issued in metric. English units here are conversions that have been conservatively rounded.

Table I.30   Miscellaneous AIR designations

Desig- nation	Comment
T-U2	Commonly known as Ti-2.5Cu; currently used, but not on a widespread basis
T-A5E	Commonly known as Ti-5Al-2.5Sn; widely used
T-A5DE	Commonly known as Ti-4Al-4Mo-2Sn-0.5Si; currently being used. (Also known by the British trade name IMI 550.)
T-E11DA	Commonly known by American name, Beta III. No longer used.
T-A6ZD	Commonly known as Ti-6Al-5Zr-0.5Mo-0.25Si; currently known under British trade name IMI 685. Used in European aircraft engines.

Table I.31 AFNOR standards, desigations

Standard/ publication	Desig nation	Title/comment
L 14-601	TA-6V	Forging stock, solution-treated and annealed for aerospace use
L 14-602	TA6-V	Forgings, solution-treated and annealed for aerospace use

INDIA

The Bureau of Indian Standards is responsible for issuing the country's national standards. Although the agency currently has seven titanium-related standards, only three are related to the metals industry.

Bureau of Indian Standards
Manak Bhavan
9 Bahadur Shah Zafar Marg
New Delhi 110002
India

Table I.32 lists the three titanium-related standards.

Table I.32 Indian Ti and Ti alloy standards

Standard	Title
IS: 5347, Part 3	Unalloyed titanium
IS: 5347, Part 4	Wrought titanium, 6Al-4V alloy
IS: 8369	Titanium carbide powder for hard metals

INTERNATIONAL/ISO

The International Organization for Standardization (ISO) is a specialized international agency with a goal of promoting worldwide standardization to facilitate international exchange of goods and services. As of 1987, ISO standards centered on only two aspects of titanium:

• Determination of Ti in ores or substances

• Implants for surgery

The implant standards may be found in ISO reference TC 150, ISO 5832/2-1978. Reference the *ISO Catalog* for the most recent year. (English-French editions are available.)

International Organization for Standardization
Case Postale 56
1211 Geneve 20,
Switzerland

ISRAEL

The Standards Institute of Israel (SII) writes and develops standards for all areas of industry, ranging from metallurgy to pesticides. Some of these standards are declared official government standards. These become obligatory, and no commodity may be manufactured, sold, used, exported, or imported unless it complies with the standard.

The Institute reports that, as of the date of the publication of this volume, there are no SII titanium standards.

Standards Institute of Israel
42 University Street
Tel Aviv 69977
Israel

JAPAN

The basic standards reference for titanium in Japan is the *JIS Handbook; Non-ferrous Metals and Metallurgy*, published by the Japanese Standards Association. English versions are available. When updated and republished, the title includes a notation of the year. (At the time this technical guide was published, the most recent edition was 1986.) Refer to the *JIS Handbook* for specification details.

Japanese Industrial Standards Committee
Ministry of International Trade & Industry
33rd Mori Bldg. 3-8-21
Torakomon 2-chome
Minato-ku, Tokyo 107
Japan

Books may be ordered from:

Standards Information Service
2-5 Toranomon 2-chome
Minato-ku, Tokyo 107
Japan

The more general standards are listed in Table I.33.

Within Tables I.34 thru I.40, the relevant standard number for individual product form is noted in the title. The "designation" (symbol) for each type of product, along with chemical compositions, are also included. A listing of the included tables is as follows:

- Table I.34 - H 2151, Ti Sponge

- Table I.35 - H 2152, Compressed Ti Sponge

- Table I.36 - H 4600, Ti Sheet (Plate) and Strip

- Table I.37 - H 4630, Ti Pipes (Tubes) for Ordinary Piping

- Table I.38 - H 4631, Ti Tubes (Pipes) for Heat Exchangers

- Table I.39 - H 4650, Ti Rods (Bars)

- Table I.40 - H 4670, Ti Wire

A second, minor standards-issuing organization exists. It is the Japan Titanium Society (JTS), founded in 1952 to promote the titanium and zirconium industry. Its members include approximately 27 companies engaged in the production of ore, sponge, or mill products.

Japan Titanium Society
Daishin Bldg., 9
2-chome, Kanda-Nishiki-cho
Chiyoda-ku
Tokyo 101, Japan

The Society's standards are listed in Table I.41.

Table I.33 JIS general standards

JIS No.	Description
H 0511	Method of Measurement for Brinell Hardness of Titanium Sponge
H 1610	Sampling Methods for Titanium
H 1611	General Rule for Chemical Analysis of Titanium
H 1612	Methods for Determination of Nitrogen in Titanium
H 1613	Methods for Determination of Manganese in Titanium
H 1614	Methods for Determination of Iron in Titanium
H 1615	Methods for Determination of Chlorine in Titaniun
H 1616	Methods for Determination of Magnesium
H 1617	Methods for Determination of Carbon in Titanium
H 1618	Methods for Determination of Silicon in Titanium
H 1619	Methods for Determination of Hydrogen in Titanium
H 1620	Methods for Determination of Oxygen in Titanium
H 1621	Methods for Determination of of Palladium in Titanium Alloys
H 1622	Methods for Determination of Aluminum in Titanium Alloys
H 1623	Methods for Determination of Sodium in Titanium
H 1630	Methods for Emission Spectrochemical Analysis of Titanium
H 1651	Titanium Sponge
H 1652	Compressed Titanium
H 3331	Titanium Welding Wires

Table I.34 JIS H 2151 Ti sponge (a)

Symbol (b)	Class	\multicolumn{12}{c}{Chemical composition %}	Hardness HB 10/1500
		Fe	Cl	Mn	Mg	Na	Si	N	C	H	O	Ti	
TS-105 M	Class 1 M (c)	0.10 max	0.10 max	0.01 max	0.06 max	–	0.03 max	0.02 max	0.03 max	0.005 max	0.08 max	99.6 min	105 max
TS-105 S	Class 1 S (d)	0.03 max	0.15 max	0.01 max	–	0.10 max	0.03 max	0.01 max	0.03 max	0.010 max	0.08 max	99.6 min	105 max
TS-120 M	Class 2 M	0.15 max	0.12 max	0.02 max	0.07 max	–	0.03 max	0.02 max	0.03 max	0.005 max	0.12 max	99.4 min	106 thru 120
TS-120 S	Class 2 S	0.05 max	0.20 max	0.02 max	–	0.15 max	0.03 max	0.01 max	0.03 max	0.010 max	0.12 max	99.4 min	106 thru 120
TS-140 M	Class 3 M	0.20 max	0.15 max	0.05 max	0.08 max	–	0.03 max	0.03 max	0.03 max	0.005 max	0.15 max	99.3 min	121 thru 140
TS-140 S	Class 3 S	0.07 max	0.20 max	0.05 max	–	0.15 max	0.03 max	0.03 max	0.03 max	0.015 max	0.15 max	99.3 min	121 thru 140
TS-160 M	Class 4 M	0.20 max	0.15 max	0.05 max	0.08 max	0.15	0.03 max	0.03 max	0.03 max	0.005 max	0.25 max	99.2 min	141 thru 160
TS-160 S	Class 4 S	0.07 max	0.20 max	0.05 max	–	max	0.03 max	0.03 max	0.03 max	0.015 max	0.25 max	99.2 min	141 thru 160

(a) All as manufactured
(b) TS = Titanium sponge. For 105, 120, etc., see right column
(c) M = Magnesium reduction process
(d) S = Sodium reduction process

Table I.35 JIS H 2152 Compressed Ti sponge

Symbol	Class	\multicolumn{9}{c}{Chemical composition %}
		Fe	Cl	Mn	Mg	Si	N	C	H	Ti
TC-1 (a)	Class 1	0.60 or less	0.15 or less	0.03 or less	0.10 or less	0.04 or less	0.03 or less	0.05 or less	0.005 or less	99.0 or over
TC-2	Class 2	2.0 or less	0.15 or less	0.05 or less	0.50 or less	0.10 or less	0.10 or less	0.10 or less	0.005 or less	97.0 or less

(a) TC = Titanium compressed (sponge)

Table I.36 JIS H 4600 Ti sheet (plate) and strip

Symbol				Chemical composition %				
Sheet (a)	Strip (b)	Class	Finishing process	H	O	N	Fe	Ti
TP 28 H	TR 28 H	Class 1	Hot-rolled	0.013 max	0.15 max	0.05 max	0.20 max	Remain.
TP 28 C	TR 28 C		Cold-rolled					
TP 35 H	TR 35 H	Class 2	Hot-rolled	0.013 max	0.20 max	0.05 max	0.25 max	Remain.
TP 35 C	TR 35 C		Cold-rolled					
TP 49 H	TR 49 H	Class 3	Hot-rolled	0.013 max	0.30 max	0.07 max	0.30 max	Remain.
TP 49 C	TR 49 H		Cold-rolled					

(a) TP = Titanium sheet (plate)
   H = Hot-rolled
(b) TR = Titanium strip
   C = Cold-rolled

Table I.37 JIS H 4630 Ti pipes (tubes) for ordinary piping

Symbol (a)	Class	Manufacturing process	Finishing process	Chemical composition %				
				H	O	N	Fe	Ti
TTP 28 E	Class 1	Seamless pipe	Hot-extruded	0.015 max	0.15 max	0.05 max	0.20 max	Remain.
TTP 28 D			Cold-drawn					
TTP 28 W		Welded pipe	As welded					
TTP 28 WD			Cold-drawn					
TTP 35 E	Class 2	Seamless pipe	Hot-extruded	0.015 max	0.20 max	0.05 max	0.25 max	Remain.
TTP 35 D			Cold-drawn					
TTP 35 W		Welded pipe	As welded					
TTP 35 WD			Cold-drawn					
TTP 49 E	Class 3	Seamless pipe	Hot-extruded	0.015 max	0.30 max	0.07 max	0.30 max	Remain.
TTP 49 D			Cold-drawn					
TTP 49 W		Welded pipe	As welded					
TTP 49 WD			Cold-drawn					

(a) TTP = Titanium pipe for piping
   E = Hot-extruded, seamless pipe
   D = Cold-drawn, seamless pipe
   W = As welded pipe
   WD = Cold-drawn, welded pipe

Table I.38   JIS H 4631 Ti tubes (tubes) for heat exchangers

Symbol (a)	Class	Manufacturing process	Finishing process	Chemical composition %				
				H	O	N	Fe	Ti
TTH 28 D	Class 1	Seamless tube	Cold-drawn	0.015 max	0.15 max	0.05 max	0.20 max	Remain.
TTH 28 W		Welded tube	As welded					
TTH 28 WD			Cold-drawn					
TTH 35 D	Class 2	Seamless tube	Cold-drawn	0.015 max	0.20 max	0.05 max	0.25 max	Remain.
TTH 35 W		Welded tube	As welded					
TTH 35 WD			Cold-drawn					
TTH 49 D	Class 3	Seamless tube	Cold-drawn	0.015 max	0.30 max	0.07 max	0.30 max	Remain.
TTH 49 W								
TTH 49 WD		Welded tube	As welded					
			Cold-drawn					

(a) TTH = Titanium tube (pipe), heat exchanger
D = Cold-drawn, seamless tube
W = As welded tube
WD = Cold-drawn, welded tube

Table I.39   JIS H 4650 Ti bars (rods)

Symbol (a)	Class	Finishing process	Chemical composition %				
			H	O	N	Fe	Ti
TB 28 H	Class 1	Hot worked	0.015 max	0.15 max	0.05 max	0.20 max	Remain.
TB 28 C		Cold-drawn					
TB 35 H	Class 2	Hot worked	0.015 max	0.20 max	0.05 max	0.25 max	Remain.
TB 35 C		Cold-drawn					
TB 49 H	Class 3	Hot worked	0.015 max	0.30 max	0.07 max	0.30 max	Remain.
TB 49 C		Cold-drawn					

(a) TB = Titanium bar, including rods
E = Hot-worked
D = Cold-drawn

Table I.40 JIS H 4670 Ti wire

Symbol (a)	Class	Chemical composition %				
		H	O	N	Fe	Ti
TW 28	Class 1	0.015 max	0.15 max	0.05 max	0.20 max	Remain.
TW 35	Class 2	0.015 max	0.20 max	0.05 max	0.25 max	Remain.
TW 49	Class 3	0.015 max	0.30 max	0.07 max	0.30 max	Remain.

(a) TW = Titanium wire

Table I.41 Japan Titanium Society's industrial standards (TIS)

TIS number	Description
7402	Methods for Determination of Palladium in Titanium Alloys
7403	Methods for Determination of Aluminium in Titanium Alloys
7504	General Rules for Standards of Japan Titanium Society
7505	Titanium Castings
7606	Methods for Determination of Vanadium in Titanium Alloys
7607	Titanium Forgings
7708	Methods for Determination of Manganese in Titanium Alloys
7809	Methods for Determination of Tin in Titanium Alloys
7810	Methods for Determination of Chromium Titanium Alloys
7811	Methods for Determination of Iron in Titaniun Alloys
7912	Titanium-Palladium Alloy Sheets, Plates and Strip
7913	Titanium-Palladium Alloy Pipes and Tubes for Ordinary Piping
7914	Titanium-Palladium Alloy Pipes and Tubes for Heat Exchanger
7915	Titanium-Palladium Alloy Rods and Bars
7916	Titanium-Palladium Alloy Wires
8117	Methods for Determination of Copper in Titanium Alloys
8218	Eddy Current Testing Method of Thin Wall Welded Titanium Tubing
8319	Radiographic Testing Methods for Welds of Thin Titanium Plates and Thin Wall Tubes
8320	Methods for Determination of Tantalum in Titanium Alloys
8321	Methods for Determination of Niobium in Titanium Alloys
8422	Methods for Determination of Zirconium in Titanium Alloys
8423	Methods for Determination of Molybdenum in Titanium Alloys

KOREA

Korean Standards Association (KSA) is a nonprofit, nongovernmental institution established under the Industrial Standardization Act for the purpose of contributing to the development of national economy through development of scientific technology and enhancement of productivity by promoting industrial standardization and quality control.

Korean Standards Association
International Standards Section
105-153 Kongduck-dong, Mapo-ku
Seoul, 121 Korea

Major functions include research for, and development of, industrial standardization and quality control; publication and dissemination of Korean Industrial Standards; education and training; consultation and guidance for industrial standardization and QC; production of industrial standardization and QC-related publications; mutual cooperation with international standards body and with quality control organizations; and distribution of information on overseas standards.

The Korean Industrial Standard is a national standard established based on Industrial Standardization Act and is abbreviated to the KS. Korean standards cover industrial and mineral products.

Korean standards related to titanium are listed in Table I.42. These are identical to Japanese standards. Thus the JIS standards, noted under the Japan paragraph, above, may be referenced for details. The following list provides correlations:

- Ti sponge, D2353: see Table I.34

- Compressed Ti sponge, D2354: see Table I.35

- Ti pipe (tube), ordinary piping, D5574: see Table I.37

- Ti pipe (tube), heat exchanger, D5575: see Table I.38

- Ti bars and rods, D5604: see Table I.39

- Ti wire, D5576: see Table I.40

Table I.42 Korean Ti and Ti alloy standards

Standard No.	Title
D 0032	Testing method for Brinell hardness of sponge titanium
D 1769	General rules for technical analysis of titanium
D 2353	Titanium sponge
D 2354	Compressed titanium sponge
D 5574	Titanium pipes and tubes for ordinary piping
D 5575	Titanium pipes and tubes for heat exchangers
D 5576	Titanium wires
D 5577	Titanium flat mill products, rod, and wire
D 5604	Titanium rods and bars
D 7030	Titanium welding wires
D 3059	General rules for chemical analysis of titanium ores
D 3063	Determination of titanium ores

NETHERLANDS

The Netherlands Normalization Institute (NNI) is composed of approximately 3000 individual firms and companies, along with about 200 other organizations. This association helps prepare Dutch standards and cooperates in the development of international standards. The address is:

Nederlands Normalisatie-Instituut
Postbus 5059
2600 GB Delft
The Netherlands

A correspondent at NNI states that currently there are no Dutch standards that deal with titanium and titanium alloys. Those of the International Organization for Standardization (ISO) are used. (See ISO under International.)

PAN AMERICA

The Pan America Standards Commission (COPANT) is a non-profit, autonomous group that promotes technical standardization in countries throughout the American continent. It also serves as a liaison and coordinating body among national standards institutions. COPANT has as members the following countries: Brazil, Panama, Venezuela, Mexico, Dominican Republic, Bolivia, Centralamerica (Guatemala), Columbia, Chile, Equador, Paraguay, Argentina, Peru, Trinidad, Tobago, and Uruguay.

Pan American Standards Commission
Avenida Julio A. Roca 651
Piso 3° Sectores 9 y 10
(1322) Buenos Aires,
Argentina

A correspondent at COPANT has written that the organization currently does not have any standards involving titanium.

PEOPLE'S REPUBLIC OF CHINA

The People's Republic of China (that is, mainland China), according to a correspondent, has no national standards-issuing organization responsible for titanium. However, titanium and titanium alloys are produced and made available from three nationally owned facilities. Located in Sh'ansi Provence, the largest is the Bao Ji Nonferrous Metals Plant.

The data cited in Table I.43 was supplied by Nonferrous Metals (U.S.A.), Inc., a subsidiary of China National Nonferrous Metals Import and Export Corp. As such, it represents a *de facto* national standard/specification if only because these are the only designations.

A correspondent has stated that the titanium industry in the People's Republic produces according to any national standard specified in an order. Thus the issue of Chinese standards and specifications is not of prime importance.

Table I.43 Mainland China's Ti and Ti alloy chemical compositions, product forms (a)

Designation	Chemical composition, %	Impurities content, max.	Product forms
		Commercially pure Ti	
CP, Grade I	-	0.15Fe; 0.10Si, 0.05C; 0.03N; 0.015H; 0.15O	Ingot, plate, sheet, slab, strip, bar, wire, tubing, forgings
CP, Grade II	-	0.30Fe; 0.15Si; 0.10C; 0.05N; 0.015H; 0.25O	Ingot, plate, sheet, slab, strip, bar, wire, tubing, forgings, castings
CP, Grade III	-	0.30Fe; 0.15Si; 0.10C; 0.05N; 0.015H; 0.30O	Ingot, plate, sheet, slab, bar, wire, tubing, forgings
		Ti alloys	
Ti-2Al-1.5Mn	1.0-2.5Al; 0.8-2.0Mn; base Ti	0.4Fe; 0.15Si; 0.10C; 0.05N; 0.015H; 0.15O	Ingot, plate, sheet, slab, bar, tubing
Ti-3Al	2.0-3.3Al; base Ti	0.30Fe; 0.15Si; 0.10C; 0.05N; 0.015H; 0.15O	Ingot, bar
Ti-3Al-1.5Mn	2.0-3.5Al; 0.8-2.0Mn; base Ti	0.4Fe; 0.15Si; 0.10C; 0.05N; 0.015H; 0.15O	Ingot, plate, sheet, slab, bar, tubing
Ti-3Al-2.5Mn	2.5-3.5Al; 2.0-3.0V; base Ti	0.30Fe; 0.15Si; 0.08C; 0.05N; 0.015H; 0.12O	Ingot
Ti-5Al-2.5Sn	4.0-5.5Al; 2.0-3.0V; base Ti	0.3Fe; 0.15Si; 0.10C; 0.05N; 0.015H; 0.20O	Ingot, plate, sheet, slab, bar, wire, tubing
Ti-6Al-4V	5.5-6.8Al; 3.5-4.5V; base Ti	0.30Fe; 0.15Si; 0.10C; 0.50N; 0.015H; 0.15O	Ingot, plate, sheet, slab, bar, wire, tubing, forgings, castings
Ti-6Al-6V-2Sn-0.5Cu-0.5Fe	5.5-6.5Al; 1.5-2.5Sn; 5.5-6.5V; 0.35-1.0Fe; 0.35-1.0Cu; base Ti	0.00Fe; 0.15Si; 0.10C; 0.04N; 0.015H; 0.20O	Ingot, plate, sheet, slab
Ti-6.5Al-3.5 Mo-2.5Sn-0.3Si	5.8-6.8Al; 2.8-3.8Mo; 1.8-2.8Sn; 0.2-0.4Si	0.40 Fe; 0.10C; 0.05N; 0.015H; 0.15O	Bar
Ti-8Al-1Mo-1V	7.5-8.5Al; 0.75-1.25Mo; 0.75-1.25V; base Ti	0.30Fe; 0.15Si; 0.10C; 0.04N; 0.015H; 0.15O	Ingot
Ti-32Mo	31-35Mo; base Ti	0.25Fe; 0.10Si; 0.10C; 0.04N; 0.012H; 0.12O	Castings

(a) No mechical properties are cited here, since they are varied and dependent on a number of physical factors. Furthermore, in many cases the physical properties can be customized for the user or the market.

PORTUGAL

The Portuguese Institute for Quality, formerly the General Directorate of Quality (DGQ), is the national organization for standardization, conformity assessment, and metrology. The five existing standards related to titanium all involve methods of determination of the amounts of Ti within other metals or materials.

Instituto Portugues da Qualidade
Rua Jose Estevao, 83A
1199 Lisboa CODEX
Portugal

A correspondent at the Institute states that Portugal also uses EURONORM and ISO standards for titanium.

REPUBLIC OF CHINA

The National Bureau of Standards (CNS), a governmental agency affiliated with the Ministry of Economic Affairs, is in charge of establishing national standards for the Republic of China (Taiwan). The Bureau concerns itself with four primary areas: standards, weights and measures, patents, and trademarks.

National Bureau of Standards
102, Kwan-Fu South Road
Taipei, (Taiwan) 10553
Republic of China

An English version of the *CNS Catalog (1987)* is available on request. It contains almost 11 600 standards' titles divided into 24 sections, or topic areas. The Bureau carries on a continuing effort to translate the existing standards into English. At the end of 1986 there were about 1500 available.

As of 1987 the *Catalog* lists only titanium dioxide, titanium tetrachloride, and titanium trichloride standards.

REPUBLIC OF SOUTH AFRICA

The South African Bureau of Standards (SABS) was officially established by the South African government in 1945, although work in the area of standardization by other organizations began in the early 1900s. SABS has written that, as of the date of the publication of this volume, it has no standards involving titanium.

South African Bureau of Standards
Private Bag X191
Pretoria 0001
Republic of South Africa

SPAIN

The national standards-issuing agency of Spain is the National Institute of Rationalization and Normalization (IRANOR). Currently the existing titanium-related standards cover only wrought products, a situation reflected in the scope of the tables appearing here.

Instituto National de Rationalizacion y
   Normalizacion
Fernandez de la Hoz, 52
28010 Madrid, Spain

Each standard is identified by a 3-letter alpha group followed by a 5-letter numerical group: UNE 38-711.

There is a general, overview standard which explains the organizing principles around which the actual titanium and titanium alloy standards are grouped. This is "Titanium and Titanium Alloy for Forging; Generalities" (UNE 38-700). Other related documents are:

- UNE 38-701 "Wrought Titanium and Titanium Alloys; Official Equivalencies"

- UNE 38-702 "Wrought Titanium and Titanium Alloys; Commercial Equivalencies"

Table I.44 is a complete listing of titanium-related standards as of the publication of this volume. Table I.45 contains a sampling of the details of the more common titanium and titanium alloy standards covered by UNE.

One Spanish correspondent has suggested that the major Spanish aerospace company generally uses Aerospace Material Specifications (AMS). This individual also offered the opinion that UNE titanium specifications are very limited and are generally used only in nonaerospace applications.

Another correspondent has written that currently UNE 38-716 (Ti-6Al-4V) and UNE 38-722 (Ti-2.5Cu) are the most frequently used standards. He also claims that DIN, ASTM, and AFNOR are frequently consulted and/or used.

With respect to Spain's Ministry of Defense (INTA), National Institute of Aerospace Technology, has stated that this organization does not have any standards related to titanium. (This organization's address is listed here for user information.)

Ministerio de Defensa
Instituto Nacional de Tecnica Aerospacial
Carretera de Ajalvir, km. 4
Torrejon de Ardoz, Spain

Table I.44 Listing of Spanish Ti and Ti alloys standards, designations for forgings

UNE standard	UNE standard	UNE symbolic designation	Equivalent cross-references
		Commercially pure titanium	
38-711	L-7001	Ti 99.6	AECMA Ti-01; DIN 17850, 17851, 17860, 17862, 17863, 17864; DIN/Werkstoffnummer 3.7024; ASTM B 265, B 337, B 338, B 367, B 381; MIL-T-9064H, MIL-T-81556; AIR 9182; BS (aicraft) 1TA.1; DTD 5033B, 5073, 5013B.
38-712	L-7002	Ti 99.4	AECMA Ti-02; DIN 17850, 17851, 17862, 17863, 17864; DIN/Werkstoffnummer 3.7034; AMS 4902B, 4941, 4942, 4951C; ASTM B 265, B 337, B 338, B 348, B 367, B 381, B 382 (Grade 2); MIL-T-7093 (Class 3), MIL-T-81556 I (Comp. B), MIL-T-9046 F, MIL-T-9047 (Rev. E., Comp. 1), MIL-T-12117 (Class 50); AIR 9182 (Comp. T-40); BS (aircraft) 2TA.2, 2TA.3, 2TA.4, 2TA.5; DTD 5183, 5003A, 5023A; JIS H 4600, H 4630, H 4631, H 4650, H 4670; GOST VT-1-0.
38-713	L-7003	Ti 99.2	-
38-714	L-7004	Ti 99.0	AECMA Ti-PO4; DIN 17860, 17862, 17863, 17864, DIN/Werkstoffnummer 3.7064; AMS 4901H, 4921H, 4921D; ASTM B 265, B 381 (Grade 4); MIL-T-7993A (Class 1), MIL-T-9010 (Class 1), MIL-T-9046C (Class C), MIL-T-9047F (Class 1); AIR 9182 (T-60); BS 3003 (Part 9, Grade 5), BS (aircraft) 2TA.6, 2TA.7, 2TA.8, 2TA.9; DTD 5063A, 5093.
38-715	L-7021	Ti 99.4 Pd	ASTM B 265, B 337, B 338, B 348, B 381.
		Alpha and superalpha alloys	
38-716	L-7101	Ti-5Al-Sn	-
38-717	L-7102	Ti-8Al-V-Mo	AECMA Ti-P66; AMS 4915, 4916, 4933, 4955, 4972, 4973; MIL-T-9046H, MIL-9047F, MIL-T-81556, MIL-T-81588, MIL-F-8312.
38-718	L-7103	Ti-6Al-4Zr-Mo-Sn	AMS 4919, 4975B, 4976A, 4981; MIL-T-9046 (Type III, Comp. G), MIL-T-9047E (Comp. II), MIL-F-83142.
38-719	L-7104	Ti-6Al-5Zr-Si	-
38-720	L-7105	Ti-11Sn-Zr-Al	-
38-721	L-7106	Ti-6Al-Nb	-
		Alpha plus a compound alloy	
38-722	L-7501	Ti-2.5Cu	-
		Alpha plus beta alloys	
38-723	L-7301	Ti-6Al-4V	AECMA Ti-P63; DIN 17851, 17860, 17862, 17864, DIN/Wekstoffnummer 3.7164; AMS 4906, 4907B, 4911D, 4928H, 4930A, 4934A, 4935D, 4954B, 4956, 4965D, 4967E; ASTM B 265, B 348, B 381; MIL-T-9047F (Comp. 6 and 7), MIL-T-46047A, MIL-T-81556 (Type III, Comp. A and B); AIR 9183 (T-A6V); BS TA.10, TA.11, TA.12, TA.13, TA.28; DTD 5163, 5173, 5303, 5313, 5323.
38-724	L-7302	Ti-4Al-Mo-V	-

(Continued)

Table I.44 (Continued)

UNE standard	UNE standard	UNE symbolic designation	Equivalent cross-references
38-725	L-7303	Ti-6Al-6V-Sn	AECMA Ti-P65; DIN/Werkstoffnummer 3.7174; ASM 4918E, 4936A, 4971B, 4978A, 4979A; MIL-T-9046F (Type III, Comp. E), MIL-T-9047E (Comp. 8), MIL-T-46035, MIL-T-81556, MIL-F-83142 (Type III, Comp. 8).
38-726	L-7304	Ti-8Mn	AMS 4908D; MIL-T-9064 (Type III, Comp. A).
38-727	L-7305	Ti-7Al-Mo	-
38-728	L-7306	Ti-6Al-6Mo-Zr-Sn	-
Beta alloys			
38-729	L-7701	Ti-13V-Cr-Al	AMS 4917B, 4959; MIL-T-9046H (Type IV, Comp. A), MIL-T-9047F (Comp. 12).
38-730	L-7702	Ti-12Mo-Zr-Sn	-

Table I.45 Selected UNE standards for various wrought products (a)

UNE designation	UNE symbolic designation	Chemical composition (%)	Mechanical properties, hardness values
L-7001	Ti 99.6	0.20Fe; 0.08C; 0.20O; 0.05N; 0.0125H; 99.6 Ti, min	Condition: Annealed sheet, plate, strip, bar, wire, wrought, and extruded products Thickness: 0.4 to 100 mm, but in ranges related to specific products Tensile strength: 290 MPa min (13 ksi) (b) Yield strength: 180 MPa max (26 ksi) Elongation: 25% minimum, except wire (30%) and extruded products (20 and 18%)
L-7002	Ti 99.4	0.25Fe; 0.08C; 0.25O; 0.05 N; 0.0125H; 99.4 Ti, min	Condition: Annealed sheet, plate, strip, bar, wire, wrought and extruded products Thick.: 2 mm, or less (sheet, plate, strip only) TS: 390 MPa min (56 ksi) YS: 275 MPa min (40 ksi) El: 22% (sheet, plate, strip); 20% (bars, wire); 18% (wrought, extruded wires)
L-7004	Ti 99.0	0.40Fe; 0.10C; 0.40O; 0.07Ni; 0.0125H; 99.0Ti, min	Condition: Annealed sheet, plate, strip, bar, wire, wrought and extruded products Thick.: 2 mm, or less (sheet, plate, strip only) TS: 540 MPa min (79 ksi) YS: 440 MPa min (64 ksi) El: 16%
L-7021	Ti 99.4, Pd	0.12-0.25 Pd; 0.25 Fe; 0.025O; 0.0125H; 0.05N; 0.08C; remain. Ti	Condition: Annealed sheet, plate, strip, bar, wire, wrought and extruded products Thick.: 2 mm, or less (sheet, plate, strip only) TS: 390 MPa min (57 ksi) YS: 275 MPa min (40 ksi) El: 22% (sheet, plate, strip); 20% (bars, wire); 18% (wrought, extruded products)
L-7102	Ti-8Al-V-Mo	7.35-8.35Al;0.75-1.25V; 0.75-1.25Mo; 0.30Fe; 0.08C; 0.12O; 0.05N; 0.015OH; others each 0.10; others total 0.40; remain Ti	Condition: Annealed, duplex, and triplex annealed sheet, plate, strip, bars and wrought products; annealed extruded products Thick.: Vary widely (c) TS: Vary widely (c) YS: Vary widely (c) El: Vary widely (10-7% minimum)

(Continued)

Table I.45 (Continued)

UNE designation	UNE symbolic designation	Chemical composition (%)	Mechanical properties, hardness values
7103	Ti-6Al-4Zr-Mo-Sn	5.50-6.50Al;1.80-2.20Mo; 1.80-2.20Sn; 3.60-4.40Zr; 0.25Fe; 0.12O; 0.015OH; 0.05N; 0.05C; others each 0.10; others total 0.40; remain. Ti	Condition: Annealed sheet, plate, strip Thick.: Vary in ranges between 0.6 and less than 75mm TS: 930 MPa min YS: 860 MPa min El: 6% min (sheet); 8% min (plate, strip) Condition: Tempered and artifically aged bars, wire, wrought products Thick.: Varies widely in ranges between 0.6 to 100 mm TS: Vary widely (c) YS: Vary widely (c) El: Vary widely (8-4% minimum)
L-7301	Ti-6Al-4V	5.50-6.75Al;3.50-4.50V; 0.3Fe; 0.20; 0.125H; 0.05N; 0.10C; 0.10 others each; 0.40 others total; remain Ti	Condition: Annealed or tempered and artificially aged sheet, plate, strip, bars, wrought and extruded products Thick.: Varies widely in specific ranges from less than 0.20 thru 100 mm TS: Vary widely (c) TS: Vary widely (c) El: Vary widely (10-4% min)
L-7303	Ti-6Al-6V-Sn	5.0-6.0Al; 5.0-6.0V; 1.5-2.5Sn; 0.35-1.00Cu; 0.35-1.00Fe; 0.20O; 0.0125H; 0.04N; 0.05C; 0.10 others each; others total 0.40; remain. Ti	Condition: Annealed sheet, plate, strip, annealed or tempered and artificially aged bars, wrought and extruded products Thick.: Vary widely in ranges between less than 50 and 100 mm TS: Vary widely (c) TS: Vary widely (c) El: Vary widely (10.8-6.4%min)
L-7304	Ti-8Mn	6.5-9.0Mn; 0.50Fe; 0.20O; 0.015H; 0.05N; 0.08C; 0.10 others each; 0.40 others total; remain. Ti	Condition: Annealed sheet, plate Thick.: Less than 1.8 mm (sheet); 1.8-4.8 mm (plate) TS: 860 MPa min (124 ksi) YS: 760 MPa min (110 ksi) El: 9% min
L-7701	Ti-13V-Cr-Al	2.5-3.5Al;12.5-14.5V; 10.0-12.0Cr; 0.35Fe; 0.180; 0.020H; 0.05N; 0.05C; 0.10 others each; 0.40 others total; remain. Ti	Condition: Tempered or tempered and artificially aged sheet, plate, strip; tempered and artificially aged wire, bars and wrought products Thick.: Vary widely in ranges from less than 0.5 to less than 175 mm TS: Vary widely (c) YS: Vary widely (c) El: Vary widely (9-3% min)

(a) This table contains a sampling of the more common titanium and titanium alloy standards covered by UNE.
(b) Wire has an additions TS factor of 490 MPa (71 ksi).
(c) The standards contain such a wide variety of factors in direct relation to types of products and diameters that the standard should be consulted for details.

TURKEY

The Turkish Standards Institute (TSE) is a government agency dedicated to the preparation and publication of the country's standards. (It is also a member of ISO.)

At the time of the publication of this guide a correspondent stated that titanium-related standards were "in preparation." The individual also stated that, in cases when Turkey does not have an applicable standard, DIN, BSI (United Kingdom), or ISO are used.

Turk Standardlari Enstitusu
Necatibey Caddesi 112
Bakanliklar, Ankara
Turkey

UNITED KINGDOM

The basic specifications reference for titanium in the United Kingdom is the British Standards Institution's (BSI) *Standards Catalogue*. When updated and republished, the title includes the year as a suffix. (At the time this guide was published, the most recent edition was 1987.)

Within Table I.46, included here, the specification standards are listed by reference number, and a quick description of the content is also noted. For details, including chemical compositions and designations, refer to the individual specification publication.

Copies of individual BSI specifications may be obtained from:

British Standards Institution
Sales Department
Linford WoodMilton Keynes, MK14 6LE
England

In addition, there is a second standards-issuing agency for military purposes. This is the Ministry of Defense's Directorate of Standardization. These standards are prefaced with the letters DTD.

Ministry of Defense
Directorate of Standardization
First Avenue House
High Holborn
London WC1V 6HE

Table I.46 British Standards (a)

Specification	Date	Description
2TA.1 Amnd. 4628, Jul. 1984 Amnd. 4919, Sep. 1985	1974 (1980)	Sheet and strip of commercially pure titanium. (Tensile strength 290-420 MPa.)
2TA.2 Amnd. 4618, Jul. 1984 (b) Amnd. 4918, Sept. 1985	1973 (1980)	Sheet and strip of commercially pure titanium. (Tensile strength 390-540 N/mm^2.)
2TA.3 Amnd. 3322, Nov. 1980	1973 (obsolete)	Bar and section for machining of commercially pure titanium. (Tensile strength 390-540 N/mm^2.)
2TA.4 Amnd. 3323, Nov. 1980	1973 (obsolete)	Forging stock of commercially pure titanium. (Tensile strength 390-540 N/mm^2.)
2TA.5 Amnd. 3324, Nov. 1980	1973 (obsolete)	Forgings of commercially pure titanium. (Tensile strength 390-540 N/mm^2.)
2TA.6 Amnd. 4619, Jul. 1984 Amnd. 4917, Sept. 1985	1973 (1980)	Sheet and strip of commercially pure titanium. (Tensile strength 570-730 N/mm^2.)
2TA.7	1973 (1980)	Bar and section for machining of commercially pure titanium. (Tensile strength 540-740 N/mm^2.)
2TA.8	1973 (1980)	Forging stock of commercially pure titanium. (Tensile strength 540-740 N/mm^2.)
2TA.9	1973 (1980)	Forgings of commercially pure titanium. (Tensile strength 540-740 N/mm^2.)
2TA.10 Amnd. 3771, Oct. 1981	1974 (1982)	Sheet of titanium-aluminum-vanadium alloy. (Tensile strength 960-1270 MPa.)
2TA.11	1974 (1980)	Bar and section for machining of titanium-aluminum-vanadium alloy. (Tensile strength 900-1160 MPa; limiting ruling strength 150 mm.)
2TA.12	1974 (1980)	Forging stock of titanium-aluminum-vanadium alloy (Tensile strength 900-1160 MPa; limiting ruling section 150 mm.)
TA.13	1974 (1980)	Forgings of titanium-aluminum-vanadium alloy. (Tensile strength 900-1160 MPa; limiting ruling section 150 mm.)
TA.14	1968	(Withdrawn)
TA.15	1968	(Withdrawn)
TA.16	1968	(Withdrawn)
TA.17	1968	(Withdrawn)
TA.18	1968	(Withdrawn)
TA.19	1968	(Withdrawn)
TA.20	1968	(Withdrawn)
2TA.21 Amnd. 1392, Apr. 1974	1973 (1981)	Specification for sheet and strip of titanium-copper alloy. (Tensile strength 540-700 N/mm^2.)
2TA.22	1973 (1980)	Specification for bar and section for machining of titanium-copper alloy. (Tensile strength 540-770 N/mm^2.)
2TA.23	1973 (1980)	Specification for forging stock of titanium-copper alloy. (Tensile strength 540-770 N/mm^2.)
TA.24	1973 (1980)	Specification for forgings of titanium-copper alloy. (Tensile strength 540-770 N/mm^2.)
TA.25	1968	(Withdrawn)
TA.26	1968	(Withdrawn)
TA.27	1968	(Withdrawn)
TA.28 Amnd. 2134, Oct. 1976 Amnd. 3755, Apr. 1984	1974	Specification for forging stock and wire of titanium-aluminum-vanadium alloy. (Tensile strength 1100-1300 MPa; limiting ruling section 20 mm.) (Primarily intended for fastener manufacturing per 'A' series of British Standards.)
TA.29	1969	(Withdrawn; replaced by TA 45, 1973)

(Continued)

Table I.46 (Continued)

Specification	Date	Description
TA.30	1969	(Withdrawn; replaced by TA 47, 1973)
TA.31	1969	(Withdrawn; replaced by TA 48, 1973)
TA.32	1969	(Withdrawn; replaced by TA 46, 1973)
TA.33	1969	(Withdrawn; replaced by TA 47, 1973)
TA.34	1969	(Withdrawn; replaced by TA 48, 1973)
TA.35	1969	(Withdrawn; replaced by TA 46, 1973 and TA 49, 1973)
TA.36	1969	(Withdrawn; replaced by TA 47, 1973 and TA 50, 1973)
TA.37	1969	(Withdrawn; replaced by TA 48, 1973 and TA 51, 1973)
TA.38	1971	Bar for machining of titanium-aluminum-molybdenum-tin-silicon-carbon alloy. (Tensile strength 1250-1420 N/mm^2; limiting ruling section 25 mm.)
TA.39	1971	Forging stock of titanium-aluminum molybdenum-tin-silicon-carbon alloy. (Tensile strength 1250-1420 N/mm^2; limiting ruling section 25 mm.)
TA.40	1971	Bar for machining of titanium-aluminum-molybdenum-tin-silicon-carbon alloy. (Tensile strength 1250-1375 N/mm^2; limiting ruling section over 25 mm thru 75 mm.)
TA.41	1971	Forging stock of titanium-aluminum-molybdenum-tin-silicon-carbon alloy. (Tensile strength 1205-1375 N/mm^2; limiting ruling section over 25 mm thru 75 mm.)
TA.42	1971	Forgings of titanium-aluminum-molybdenum-tin-silicon-carbon alloy. (Tensile strength 1205-1375 N/mm^2; limiting ruling section over 25 mm thru 75 mm.)
TA.43 Amnd. 1623, Nov. 1974 Amnd. 3424, Dec. 1980	1972 (1981)	Forging stock of titanium-aluminum-zirconium-molybdenum-silicon alloy. (Tensile strength 990-1140 N/mm^2; limiting ruling section 65 mm.)
TA.44 Amnd. 1624, Nov. 1974 Amnd. 3425, Dec. 1980	1972 (1981)	Forgings of titanium-aluminum-zirconium-molybdenum-silicon alloy. (Tensile strength 990-1140 N/mm^2; limiting ruling section 65 mm.)
TA.45 Amnd. 3426. Dec. 1980	1973 (1981)	Bar and section for machining of titanium-aluminum-molybdenum-tin-silicon alloy. (Tensile strength 1100-1280 N/mm^2; limiting ruling section 25 mm.) Replaces TA 29, 1969.
TA.46 Amnd. 3427, Dec. 1980	1973 (1981)	Bar and section for machining of titanium-aluminum-molybdenum-tin-silicon alloy. (Tensile strength 1050-1220 N/mm^2; limiting ruling section over 25 mm thru 100 mm.) Replaces TA 32, 1969 and TA 35, 1969.
TA.47 Amnd. 3428, Dec. 1980	1973 (1981)	Forging stock of titanium-aluminum-molybdenum-silicon-tin alloy. (Tensile strength 1050-1220 N/mm^2; limiting ruling section 100 mm.) Replaces TA 30, 1969; TA 33, 1969; and TA 36, 1969.
TA.48 Amnd. 3429, Dec. 1980	1973 (1981)	Forgings of titanium-aluminum-molybdenum-silicon-tin alloy. (Tensile strength 1050-1200 N/mm^2; limiting ruling section 100 mm.) Replaces TA 31, 1969; TA 34, 1969; and TA 37, 1969

(Continued)

Table I.46 (Continued)

Specification	Date	Description
TA.49 Amnd. 3430, Dec. 1980	1973 (1981)	Bar and section for machining of titanium-aluminum-molybdenum-silicon-tin alloy. (Tensile strength 1000-1200 N/mm^2; limiting ruling section over 100 mm thru 150 mm.) Replaces TA 35, 1969
TA.50 Amnd. 3431, Dec. 1980	1973 (1981)	Forging stock of titanium-aluminum-molybdenum-silicon-tin alloy. (Tensile strength 1000-1200 N/mm^2; limiting ruling section over 100 mm thru 150 mm.) Replaces TA 36, 1969.
TA.51 Amnd. 3432, Dec. 1980	1973 (1981)	Forgings of titanium-aluminum-molybdenum-tin-silicon alloy. (Tensile strength 1000-1200 N/mm^2; limiting ruling section 100 mm thru 150 mm.) Replaces TA 37, 1969.
TA.52 Amnd. 3433, Dec. 1980	1973 (1981)	Sheet and strip of titanium-copper alloy. (Tensile strength 690-920 N/mm^2.)
TA.53 Amnd. 3434, Dec. 1980	1973 (1981)	Bar and section for machining of titanium-copper alloy. (Tensile strength 650-880 N/mm^2; limiting ruling section 75 mm.)
TA.54 Amnd. 3435, Dec. 1980	1973 (1981)	Forging stock of titanium-copper alloy. (Tensile strength 650-880 N/mm^2; limiting ruling section 75 mm.)
TA.55 Amnd. 3436, Dec. 1980	1973 (1981)	Forgings of titanium-copper alloy. (Tensile strength 650-880 N/mm^2; limiting ruling section 75 mm.)
TA.56 Amnd. 3756, Sep. 1981	1974 (1981)	Plate of titanium-aluminum-vanadium alloy. (Tensile strength 895-1150 MPa; maximum thickness 100 mm.)
TA.57 Amnd. 3757, Sep. 1981	1974 (1981)	Plate of titanium-aluminum-molybdenum-tin-silicon alloy. (Tensile strength 1030-1220 MPa; maximum thickness 65 mm.
TA.58 Amnd. 3753, Sep. 1981	1974 (1981)	Plate of titanium-copper alloy. (Tensile strength 520-640 MPa; maximum thickness 10 mm.)
TA.59	1980	Sheet and strip of titanium-aluminum-vanadium alloy. (Tensile strength 920 MPa to 1180 MPa.)
Inspection Procedures		
2TA.100 Amnd. 1557, Sep. 1974 Amnd. 2495, Mar. 1978 Amnd. 3307, Nov. 1980 Amnd. 4620, Jul. 1984	1973	Inspection and testing of wrought titanium and titanium alloys
Medical Applications		
BS 3531, Part 1, 2	1980	Covers surgical implants

(a) As of 1987.
(b) Amnd.xxxx designates the basic specification was ammended and the issue date.

USSR

Soviet titanium and titanium alloys are frequently designated by an industry or organization for or by which the material was developed. Thus, by contrast with the formats of other national standards, Soviet designations may begin with a wide variety of letter or number prefixes. By way of example the following listing is included, but note that the Cyrillic letters have been transliterated into Roman:

- VT for the aviation industry

- OT for defense

- 48 for shipbuilding

- IRM for the State Institute of Rare Metals

- AT for the Institute of Metallurgy imeni Baykov

There are GOST (State Standards) standards for titanium sponge, but, as far as is known, there are no GOST standards for pure titanium and titanium alloys. However, OST standards apparently exist for these materials. (OST, or All-Union Standards, were superseded by GOST in 1940.)

Soviet titanium can be grouped into unalloyed Ti, powder, sponge, corrosion-resistant alloys, casting alloys, and welding rod types. The items included in Table I.47 are a sampling of the approximately 190 known titanium and titanium alloy designations.

It is very difficult, if not impossible, to identify equivalent United States designations for the Soviet items. The two countries have different raw materials available, separate industry needs, and individual alloying traditions.

Table I.47 Soviet titanium and titanium alloy designations; a selection

Grade/ alloy	Designation	Common name	Chemical composition (%) max. unless noted	Mechanical properties hardness values, product forms (a)
Com- mercially pure	VT1-1	–	0.07C; 0.11-0.15Fe; 0.12Si; remain. Ti	Ultimate tensile strength: 510 MPa (74 ksi) Tensile yield strength: 441 MPa (64 ksi) Elongation: 30% Reduction in area: 50% Forms: Sheet, bar, forgings, extrusions, plate
	VT1D-1	–	Not available	UTS: 510 MPa (74 ksi) TYS: Not available El: 27% Red.: 50% Forms: Sheet, bar, forgings, extrusions, plate
	VT1-2	–	Not available	UTS: 634 MPa (92 ksi) TYS: Not available El: 25% Red.: 45% Forms: Sheet, bar, forgings, extrusions, plate
Alpha alloys	VT-5	Ti-4.5Al-5Al	4.3-6.2Al;0.1C;0.3Fe;0.015 Si;0.15O;0.04N;0.015H; remain. Ti	Condition: Not heat treated UTS: 855 MPa (124 ksi) TYS: Not available El: 18% Red.: 37% Forms: Sheet, forg- ings, extrusion, tubing
	VT-5-1	Ti-5Al-2.5Sn	4.3-6.0Al;0.10C;0.30Fe; 0.15Si;0.02O;0.05N;0.015H; 2.0-3.0Sn;remain. Ti	Condition: Annealed YTS: 848 MPa (123 ksi) TYS: 730 MPa (106 ksi) El: 10% Red.: Not available Forms: Sheet, forg- ings, extrusions
	48-OT3	Ti-4Al-0.1Si- 0.1Fe-0.005B	3.3-4.3Al;0.04C; 0.1Fe; 0.1Si;0.002H;0.0005B; remain. Ti	Condition: Annealed YTS: 703 MPa (102 ksi) TYS: Not available El: 9% Red.: 23% Forms: Sheet, plate, bar

(Continued)

Table I.47 (Continued)

Grade/ alloy	Designation	Common name	Chemical composition (%) max. unless noted	Mechanical properties hardness values, product forms (a)
	IRM-1	Ti-4Al-4Nb	3.0-5.0Al;4.0-5.0Nb; remain. Ti	Condition: Annealed UTS: 889 MPa (129 ksi) TYS: 751 MPa (109 ksi) El: 17% Red.: 50% Forms: Sheet, welding wire
	AT-2-1	Ti-Zr-(Mo/Nb/ V)	1.5Mo or 1.5Nb or 1.5V;0.055Fe;0.01Si;0.12O; 0.038N;0.005H;2.0-3.0Zr; remain. Ti	Condition: Annealed UTS: 786 MPa (114 ksi) TYS: Not available El: 21% Form: Sheet
	AT2-2	Ti-Zr-(Mo-Nb)	0.8-1.8Mo or 2.0-3.0V;2.5Zr;remain. Ti	Condition: Annealed UTS: 744 MPa (108 ksi) TYS: Not available El: 24% Form: Sheet
Alpha + beta	VT-2	Ti-1.6Al-2.5Cr	1.0-2.0Al;2.0-3.0Cr;0.5Fe; remain. Ti	Condition: Not heat treated UTS: 979 MPa (142 ksi) TYS: Not available El: 8% Red.: Not available Forms: Forgings, extrusions
	VT-3	Ti-4.6Al-2.5Cr	4.0-5.0Al;2.0-3.0Cr;0.1C; 0.3Fe;0.2Si;0.2O;0.05N; 0.015H;remain. Ti	Condition: Annealed UTS: 1034 MPa (150 ksi) TYS: 930 MPa (135 ksi) El: 13% Red.: 32% Forms: Forgings, extrusions
	VT-3-1	Ti-4.6Al-2Cr-1.7Mo-0.5Fe	5.5-7.0Al;0.8-2.3Cr;2.0-3.0 Mo;0.1C;0.2-0.7Fe;0.15-0.4 Si;0.2O;0.05N;0.015H; remain. Ti	Condition: Annealed UTS: 1048 MPa (152 ksi) TYS: 951 MPa (138 ksi) El: 13% Red.: 32% Forms: Forgings, extrusions
	VT-4	Ti-4.6Al-1.5Mn	4.0-5.2Al;1.0-2.0Mn;0.05C; 0.3Fe;0.15Si;0.15O;0.05N; 0.015H;remain. Ti	Condition: Not heat treated UTS: 834 MPa (121 ksi) TYS: Not available El: 19% Red.: 25% Forms: Sheet, forgings, extrusions

(Continued)

Table I.47 (Continued)

Grade/ alloy	Designation	Common name	Chemical composition (%) max. unless noted	Mechanical properties hardness values, product forms (a)
VT-6	Ti-6Al-4V		5.5-7.0Al;4.2-6.0V;0.05C; 0.15Fe;0.15Si;0.15O;0.04N; 0.015H;remain. Ti	Condition: Annealed UTS: 930 MPa (135 ksi) TYS: 834 MPa (121 ksi) El: 11% Red.: 38% Forms: Sheet, bars, forgings
VT-8	Ti-6Al-3Mo		6.0-7.3Al;2.8-3.0Mo;0.01C; 0.4Fe;0.2-0.4Si;0.2O;0.05N; 0.01H;remain. Ti	Condition: Annealed UTS: 1103 MPa (160 ksi) TYS: 1007 MPa (146 ksi) El: 12% Red.: 40% Forms: Bars, forgings
VT-14	Ti-4Al-3Mo-1V		3.5-4.5Al;2.5-3.5Mo;0.7-1.5V;0.1C;0.4Fe;0.15Si; 0.15O;0.5N;0.015H; remain. Ti	Condition: Aged UTS: 565 MPa (82 ksi) TYS: Not available El: 8% Form: Sheet
VT-16	Ti-2Al-7Mo		1.8-3.8Al;4.5-5.5Mo;4.0-5.5V;remain. Ti	Condition: Aged UTS: 1296 MPa (188 ksi) TYS: Not available El: 7% Form: Sheet
OT-4	Ti-3Al-1.5Mn		3.5-5.0Al;0.8-2.0Mn;0.1C; 0.4Fe;0.15Si;0.15O;0.05N; 0.015H;remain. Ti	Condition: Annealed UTS: 758 MPa (110 ksi) TYS: 581 MPa (85 ksi) El: 28% Form: Sheet
OT-4-1	Ti-1.7Al-1.4Mn		1.0-2.5Al;0.8-2.0Mn;0.1C; 0.4Fe;0.15Si;0.15O;0.05N; 0.015H;remain. Ti	Condition: Annealed UTS: 655 MPa (95 ksi) TYS: 558 MPa (81 ksi) El: 20% Form: Sheet
IRM-2	Ti-4Al-4Nb-0.1 Re		5.0-6.0Al;4.0-5.0Nb;0.1-2.0Re; remain. Ti	Condition: Annealed UTS: 923 MPa (134 ksi) TYS: 868 MPa (126 ksi) El: 12% Red.: 43% Forms: Sheet, welding wire

(Continued)

Table I.47 (Continued)

Grade/ alloy	Designation	Common name	Chemical composition (%) max. unless noted	Mechanical properties hardness values, product forms (a)
	IRM-3	Ti-4Al-3.5Mo	3.0-5.0Al;3.0-5.0Mo; remain. Ti	Condition: Annealed UTS: 855 MPa (124 ksi) TYS: 717 MPa (104 ksi) El: 15% Red.: 54% Forms: Sheet, welding wire
	IRM-4	Ti-3.5Al-3.5Mo -0.1Re	3.0-4.0Al;3.4Mo;0.1Re; remain. Ti	Condition: Annealed UTS: 896 MPa (130 ksi) TYS: 820 MPa (119 ksi) El: 11% Red.: 53% Forms: Sheet, welding wire
	AT-3	Ti-3Al-0.7Cr- 0.4Fe-0.2Si- 0.01B	2.5-3.5Al;1.5Cr;1.5Fe; 1.5Si;0.3N;0.01H;remain. Ti	Condition: Annealed UTS: 792 MPa (115 ksi) TYS: 772 MPa (112 ksi) El: 18% Form: Sheet
	AT-4	Ti-4Al-0.6Cr- 0.23Fe;0.4Si- 0.01B	3.5-5.0Al;0.4-0.9Cr;0.25- 0.6Fe;0.25-0.6Si;0.3N; 0.01H; remain. Ti	Condition: Annealed UTS: 944 MPa (137 ksi) TYS: 903 MPa (131 ksi) El: 18% Form: Sheet
	AT-6	Ti-6Al-0.6Cr- 0.4Fe;0.3Si- 0.01B	5.0-6.0Al;1.5Cr;1.5Fe; 1.5Si;0.3N;0.01H;0.01B; remain. Ti	Condition: Annealed UTS: 1110 MPa (161 ksi) YS: 1082 MPa (157 ksi) El: 13% Form: Sheet
	AT-8	Ti-7Al-0.6Cr- 0.2Fe-0.3Si- 0.01B	6.5-8.0Al;0.4-0.9Cr;0.25- 0.6Fe;0.25-0.6Si;0.3N; 0.01H; remain. Ti	Condition: Annealed UTS: 1151 MPa (167 ksi) TYS: 1089 MPa (158 ksi) El: 15% Form: Sheet
Beta	VT-15	Ti-3Al-6.5Mo- 11Cr	2.3-3.6Al;9.5-11.0Cr;6.8- 8.0Mo;0.1C;0.01Si;0.12O; 0.03N;0.15H;1.5Zr remain. Ti	Condition: Aged UTS: 786 MPa (214 ksi) TYS: Not available El: 4% Form: Sheet

(a) Tensile properties at room temperature.

Appendix J

Designations, Applications, Properties

INTRODUCTION

This appendix, along with Table J.1, supports two functions:

- Provides a listing of the majority of titanium designations commonly and currently used in the United States

- Provides an initial guideline for the selection of an individual titanium grade or alloy

Neither the listing nor the typical applications are definitive. Designers and applications engineers should work closely with vendors to identify those grades and types which best support the individual application.

SUPPLIERS' DATA SHEETS

The selection of a specific type of titanium or titanium alloy should not be made merely on the basis of the suggested and typical applications noted in Table J.1, since a number of other important factors directly affect final choice. Material suppliers must be consulted for advice and for product data sheets.

Many companies supply data sheets and associated technical publications which provide the kind of details on composition, performance, and test results that are beyond the scope of this technical guide. The following topics are among the most important ones that should be discussed on product literature:

- General characteristics

- Composition

- Physical constants

Table J.1 Common designations, typical applications (a)

Designation (b)	Class	Variation	Typical Applications
Commercially pure (CP) titanium			
ASTM Grade 1	Unalloyed	May be written as Ti Grade 1: or 99.5 Ti (UNS R50250)	Airframes; chemical processing; distillation; condenser tubing; marine and seawater parts; plate-type heat exchangers; cold spun or pressed parts; platinized anodes; pickling baskets. Cryogenic vessels. Also electrolytic production of copper, manganese, nickel and other metals. Pulp and paper production. Has high formability. May be available with low iron content for extra corrosion resistance.
ASTM Grade 2	Unalloyed	May be written as Ti Grade 2: or 99.2 Ti (UNS R50400)	Airframes; aircraft engines; machine, chemical parts; heat exchangers, condenser, evaporator, distillation tubing. Also electrolytic productions of copper, manganese, nickel and other metals. Pulp and paper production. Has higher strength and formability than Grade 1. May be available with low iron content for extra corrosion resistance.
ASTM Grade 3	Unalloyed	May be written as Ti Grade 3: or 99.1 Ti (UNS R50550)	Chemical, marine airframe, aircraft parts which require formability, strength, weldability, corrosion resistance. Heat exchangers; condenser, desalinization, tubing; pickling baskets. Also electrolytic production of copper, manganese, nickel and other metals. Pulp and paper production. Contrasted with Grades 1 and 2, only good formability and moderate strength.
ASTM Grade 4	Unalloyed	May be written as Ti Grade 4; or 99.0 Ti (UNS R50700)	Chemical, marine, airframe, aircraft engine parts; heat exchangers; surgical implants; high-speed fans; gas compressors. Corrosive waste-disposal wells. Also electrolytic production of copper, manganese, nickel and other metals. Pulp and paper production. Has corrosion resistance, and the highest strength of all unalloyed grades. Only moderate formability.
Low-alloyed titanium			
ASTM Grade 7	Light alloy	May be written as Ti Grade 7; 99.2 Ti; 0.15Pd (UNS R52400)	Good corrosion resistance for chemical processing industry applications in which the liquid medium is mildly reducing or varies between oxidizing and reducing. Palladium improves resistance to crevice corrosion. Has good formability.
ASTM Grade 11	Light alloy	May be written as Ti Grade 11; 99.5 Ti; 0.15Pd or Ti-Pd (UNS R52250)	Similar to Grade 7.
ASTM Grade 12	Light alloy	May be written as Ti Grade 12; 98.9Ti; 0.8Ni, 0.3Mo (No UNS No.)	Similar to Grade 7. (See Code 12, below.)
Titanium alloys			
1Al-8V-5Fe	-	-	-

(Continued)

Table J.1 (Continued)

Designation (b)	Class	Variation	Typical Applications
2.25Al-11Sn-5Zr-1Mo	Alpha ("near alpha")	Usually written 11Sn-1Mo-2.5Al-5Zr-0.2Si. Also contains 0.2 Si which may be noted in designation.	High-strength fasteners. Airframes, blades, discs, wheels, spacers and fasteners for turbine engines.
3Al-1.5Mn	-	-	-
3Al-2.5V	Alpha-beta	Also called ASTM Grade 9.	Aircraft hydraulic tubing and fittings; foil. Combines strength, weldability, formability. Excellent cold formability plus higher tensile strength than strongest unalloyed grade.
3Al-2Fe-10V	-	Frequently written 10V-2Fe-3Al.	Aircraft structural components requiring toughness at high strengths.
3Al-8Mo-8V-2Fe	Beta	May be written 8Mo-8V-2Fe-3Al. May be abbrev. 8-8-2-3.	For fasteners.
3Al-8V-6Cr-4Mo-4Zr	Beta	May be abbrev. 3-8-6-4-4.	Annealed: parts requiring formability and corrosion resistance. Solution plus age. High-strength fasteners; high-strength aircraft sheet parts;springs.
3Al-13V-11Cr	Beta	May be written 13V-11Cr-3Al. May be abbrev. 13-11-3	High-strength fasteners; aerospace components; honey-comb panel. Good formability; heat treatable. Very high strength at room and moderate temperatures. Limited suitability above 316ºC (600ºF).
3Al-15V-3Cr-3Sn	Beta	Usually written 15V-3Cr-3Sn-3Al. May be abbrev. 15-3-3, or Ti-15-3.	Annealed and/or solution plus age: cold formable; ageable sheet and strip alloy for airframes. High-strength fasteners; torsion bars; aerospace components. Downhole applications. Some manufacturers claim it offers an economical cold forming alternative to hot forming. Cold forming capabilities similar to unalloyed grades.
4Al-3Mo-1V	Alpha-beta	-	Aircraft uses requiring high-strength and elevated-temperature stability.
4Al-4Mo-2Sn	Alpha-beta	May be called Ti550.	Forging alloy for engines and airframes.
5Al-2.5Sn	Alpha	Also called ASTM Grade 6.	Weldable alloy for forgings and sheet metal parts such as aircraft engine compressor blades and ducting; steam turbine blades; rocket engine, ordnance components. Good oxidation resistance and strength at 316º to 538ºC (600º to 1000ºF). Good stability at elevated temperatures. Can be cold-worked only and used only in annealed condition.
5Al-2.5Sn-ELI	Alpha	-	Special alloy for cryogenic vessels operating down to -254ºC (-425ºF) and for pressure vessels for temperatures below -196ºC (-320ºF)
5Al-2Sn-2Zr-4Mo-4Cr	Alpha-beta	-	Jet engine parts.
5Al-5Sn-5Zr	Alpha	-	-

(Continued)

Table J.1 (Continued)

Designation (b)	Class	Variation	Typical Applications
5Al-2Sn-2Zr-4Mo-4Cr	Alpha-beta ("near beta")	May be written T-17	High-strength fasteners.
5Al-5Sn-2Zr-2Mo	Alpha	-	-
5Al-6Sn-2Zr-1Mo	Alpha ("near alpha")	May also contain 0.25Si	High strength and creep resistance to 482°C (900°F). For aircraft, jet engine components.
6Al-2Cb-1Ta-1Mo	Alpha ("near alpha")	-	High toughness; moderate strength; cold resistance to ambient temperature, sea water stress corrosion and hot salt. Good weldability. Used for deep diving undersea.
6Al-2Nb-1Ta-0.8Mo	Alpha	-	Plate for naval ship building; submersible hulls; pressure vessels; other high-toughness applications.
6Al-2Sn-2Zr-2Mo-2Cr-0.25Si	Alpha-beta	-	Good high-temperature strength, stability. For aircraft sheet parts such as skins, stiffeners, ribs, webs.
6Al-2Sn-4Zr-2Mo	Alpha ("near alpha")	May be abbrev. 6-2-4-2	Parts and cases for high-temperature jet engines and compressor blades. Airframe skin components. Used where high creep strength is required. High-performance automotive valves.
6Al-2Sn-4Zr-6Mo	Alpha-beta	-	Intermediate temperature components for advanced jet engines; discs, spacers, seals.
6Al-4V	Alpha-beta	May be abbrev. 6-4, or ASTM Grade 5. Known as the "workhorse" Ti alloy	May be used over a broad range of temperatures from cryogenic to about 427°C (800°F) for long-term applications. May be used in both annealed and solution-treated and aged conditions. Rocket motor cases; blades, discs and rings for aircraft turbines, compressers. Structural forgings; fasteners; pressure vessels; gas, chemical pumps; cryogenic parts; ordnance equipment; marine components; steam turbine blades; space capsule components; helicopter rotor hubs; critical forgings requiring high strength-to-weight ratios. Downhole explorations; logging equipment; springs and hubs; implants in humans.
6Al-4V-ELI	Alpha-beta	May be abbrev. 6-4-ELI	High pressure and cryogenic vessels operating down to -251°C (-320°F). Applications where fracture toughness and fatigue strength are crucial: e.g., aircraft and structural components, and biomedical.
6Al-6V-2Sn	Alpha-beta	May be abbrev. 6-6-2	Has higher strength potential than Ti-6Al-4V, but at the sacrifice of weldability and toughness. In most cases welding not recommended. Rocket motor cases; ordnance cases; structural aircraft parts; landing gear; nuclear reactor components. Wireline logging; logging equipment. Responds well to heat treatments. Good hardenability. In annealed conditoin, excellent combination of strength and workability. In solution-treated or solution-treated plus aged condition, some manufacturers offer very

(Continued)

Table J.1 (Continued)

Designation (b)	Class	Variation	Typical Applications
			high ksi tensile strength, but with accompanying degradation in flatness, forming and fracture toughness. May be used in solution treated or aged conditions.
7Al-4Mo	Alpha-beta	-	Aircraft, jet engines.
7Al-12Zr	-	-	-
8Al-1Mo-1V	Alpha ("near alpha")	May be abbrev. 8-1-1	Used primarily at elevated temperatures to 454°C (850°F). Airframe and jet engine parts requiring high strength. Good creep and toughness properties. Fan blades, compressor blades, seals, rings. Good weldability. Some manufacturers claim this has lowest density of any commercial titanium-base alloy. Generally used in annealed condition, but some manufacturers have developed special duplex annealing procedures.
8Mn	Alpha-beta	-	Aircraft sheet parts such as skins, stiffeners, ribs, webs.
10V-2Fe-3Al	Alpha ("near alpha")	-	A forging alloy for airframe structural components requiring toughness at high strengths. Some manufacturers claim that it maybe used for near-net-shape forging and, also, that it develops high strength and toughness in heavy sections. This, it is claimed, implies lower-temperature isothermal forging using economical die materials.
11.5Mo-6Zr-4.5Sn	Beta	Usually written this way, but also known as ASTM Grade 10.	Aircraft sheet components; structural parts; skins, etc.; fasteners. Very high strength at room temperature. Good stability to 371°C (700°F).
Code 12	Alpha	Proprietary product	Developed by TIMET specifically for the process industries and is similar to the Ti-Pd light alloys. Has improved resistance to crevice corrosion in hot brines. Extends titanium's usefulness in harsh environments. Included in ASTM specifications as Grade 12: Ti with 3% Mo, 0.8% Ni.

(a) Applications are typical and general. Consult vendor for final suitability.
(b) Alloys are ordered with Al first for ease of use. Many Ti alloys may use other element first; e.g. Ti-13.5V-11Cr-3Al.

- Mechanical and physical properties
- Product forms available
- Suggested applications
- Corrosion resistance
- Type and results of heat treatment
- Machining, grinding information
- Workability ratings

- Weldability, if a consideration

- Cleaning techniques

- Equivalent specifications, if necessary

Although all of these topics have been discussed in general ways throughout this guide, because of subtle and/or major differences introduced by each manufacturer, data sheets must be consulted before a selection is made.

By way of example, detailed typical mechanical and physical properties for selected grades and alloys are included in Tables J.2 and J.3, respectively.

Note: Table J.1 was contributed by **Ronald W. Schutz**, Supervisor, Industrial Product Development and Corrosion Research, TIMET Corp., Henderson, NV.

Table J.2 Mechanical properties of wrought titanium alloys (average)

Nominal composition, %	Condition	Room temperature Tensile strength MPa	ksi	Yield strength MPa	ksi	Elongation, %	Reduction in area, %	Average mechanical properties / Extreme temperatures Test temperature °C	°F	Tensile strength MPa	ksi	Yield strength MPa	ksi	Elongation, %	Reduction in area, %	Charpy impact strength J	ft·lb	Hardness
Commercially pure																		
99.5 Ti	Annealed	331	48	241	35	30	55	315	600	152	22	97	14	32	80	...	...	120 HB
99.2 Ti	Annealed	434	63	345	50	28	50	315	600	193	28	117	17	35	75	43	32	200 HB
99.1 Ti	Annealed	517	75	448	65	25	45	315	600	234	34	138	20	34	75	38	28	225 HB
99.0 Ti	Annealed	662	96	586	85	20	40	315	600	310	45	172	25	25	70	20	15	265 HB
99.2 Ti(a)	Annealed	434	63	345	50	28	50	315	600	186	27	110	16	37	75	43	32	200 HB
98.9(b)	Annealed	517	75	448	65	25	42	205	400	345	50	248	36	37	...	...	...	...
								315	600	324	47	207	30	32	...	...	...	...
Alpha alloys																		
5 Al, 2.5 Sn	Annealed	862	125	807	117	16	40	315	600	565	82	448	65	18	45	26	19	36 HRC
5 Al, 2.5 Sn (low O$_2$)	Annealed	807	117	745	108	16	...	-195	-320	1241	180	1158	168	16	...	27	20	35 HRC
								-255	-423	1579	229	1420	206	15	...	...	...	...
Near alpha																		
8 Al, 1 Mo, 1 V	Duplex annealed	1000	145	951	138	15	28	315	600	793	115	621	90	20	38	32	24	35 HRC
								425	800	738	107	565	82	20	44	...	...	...
								540	1000	621	90	517	75	25	55	...	...	...
11 Sn, 1 Mo, 2.25 Al, 5.0 Zr, 0.2 Si	Duplex annealed	1103	160	993	144	15	35	315	600	896	130	758	110	20	44	...	...	36 HRC
								425	800	827	120	676	98	22	48	...	...	...
								540	1000	758	110	586	85	24	50	...	...	...
6 Al, 2 Sn, 4 Zr, 2 Mo	Duplex annealed	979	142	896	130	15	35	315	600	772	112	586	85	16	42	...	...	32 HRC
								425	800	703	102	586	85	21	55	...	...	...
								540	1000	648	94	489	71	26	60	...	...	...
5 Al, 5 Sn, 2 Zr, 2 Mo, 0.25 Si	975 °C (1785 °F) (½ h), AC + 595 °C (1100 °F) (2 h), AC	1048	152	965	140	13	...	315	600	793	115	565	82	15	...	...	...	...
								425	800	779	113	531	77	17	...	...	...	...
								540	1000	689	100	503	73	19	...	...	...	...
6 Al, 2 Cb, 1 Ta, 1 Mo	As rolled 2.5 cm (1 in.) plate	855	124	758	110	13	34	315	600	586	85	462	67	20	...	31	23	30 HRC
								425	800	517	75	414	60	20	...	...	...	...
								540	1000	483	70	379	55	20	...	...	...	...
6 Al, 2 Sn, 1.5 Zr, 1 Mo, 0.35 Bi, 0.1 Si	Beta forge + duplex anneal	1014	147	945	137	11	...	480	900	724	105	586	85	15	...	...	...	...

(a) Also contains 0.2 Pd. (b) Also contains 0.8 Ni and 0.3 Mo.
Source: Titanium Metals Corp. of America and RMI Co.

(Continued)

Table J.2 (Continued)

Nominal composition, %	Condition	Room temp. Tensile strength MPa	ksi	Yield strength MPa	ksi	Elongation, %	Reduction in area, %	Extreme temp. °C	°F	Tensile strength MPa	ksi	Yield strength MPa	ksi	Elongation, %	Reduction in area, %	Charpy impact strength J	ft·lb	Hardness
Alpha-beta alloys																		
8 Mn	Annealed	945	137	862	125	15	32	315	600	717	104	565	82	18	...	...	...	...
3 Al, 2.5 V	Annealed	689	100	586	85	20	...	315	600	483	70	345	50	25	...	...	...	...
6 Al, 4 V	Annealed	993	144	924	134	14	30	315	600	724	105	655	95	14	35	19	14	36 HRC
								425	800	669	97	572	83	18	40			
								540	1000	531	77	427	62	35	50			
	Solution + age	1172	170	1103	160	10	25	315	600	862	125	703	102	10	28			41 HRC
								425	800	800	116	621	90	12	35			
								540	1000	655	95	483	70	22	45			
6 Al, 4 V (low O$_2$)	Annealed	896	130	827	120	15	35	160	320	1517	220	1413	205	14	...	24	18	35 HRC
6 Al, 6 V, 2 Sn	Annealed	1069	155	1000	145	14	30	315	600	931	135	807	117	18	42	18	13	38 HRC
	Solution + age	1276	185	1172	170	10	20	315	600	979	142	896	130	12	28			42 HRC
7 Al, 4 Mo	Solution + age	1103	160	1034	150	16	22	315	600	976	127	745	108	18	50	18	13	38 HRC
								425	800	848	123	717	104	20	55			
6 Al, 2 Sn, 4 Zr, 6 Mo	Solution + age	1269	184	1172	170	10	23	315	600	1020	148	841	122	18	55			42 HRC
								425	800	951	138	758	110	19	67			
								540	1000	848	123	655	95	19	70			
6 Al, 2 Sn, 2 Zr, 2 Mo, 2 Cr, 0.25 Si	Solution + age	1276	185	1138	165	11	33	315	600	979	142	807	117	14	27			
10 V, 2 Fe, 3 Al	Solution + age	1276	185	1200	174	10	19	205	400	1117	162	1048	152	13	33			
								315	600	1103	160	979	142	13	42			
Beta alloys																		
13 V, 11 Cr, 3 Al	Solution + age	1220	177	1172	170	8	...	315	600	883	128	793	115	19	...			
	Solution + age	1276	185	1207	175	8	...	425	800	1103	160	827	120	12	...			
8 Mo, 8 V, 2 Fe, 3 Al	Solution + age	1310	190	1241	180	8	...	315	600	1131	164	979	142	15	...	11	8	40 HRC
3 Al, 8 V, 6 Cr, 4 Mo, 4 Zr	Solution + age	1448	210	1379	200	7	...	315	600	1034	150	896	130	20	...	10	7.5	40 HRC
								425	800	938	136	758	110	17	...			
11.5 Mo, 6 Zr, 4.5 Sn	Annealed	883	128	834	121	15	...	315	600	724	105	655	95	22	...			
	Solution + age	1386	201	1317	191	11	...	315	600	903	131	848	123	16	...			42 HRC

(a) Also contains 0.2 Pd. (b) Also contains 0.8 Ni and 0.3 Mo.
Source: Titanium Metals Corp. of America and RMI Co.

Table J.3 Physical properties of wrought titanium alloys (average)

Nominal composition, %	Average physical properties													
	Coefficient of linear thermal expansion, μm/m · K (μin./in./°F)							Modulus of elasticity(a)		Modulus of rigidity(a)		Poisson's ratio (a)	Density(a)	
	20-100 °C (70-212 °F)	20-205 °C (70-400 °F)	20-315 °C (70-600 °F)	20-425 °C (70-800 °F)	20-540 °C (70-1000 °F)	20-650 °C (70-1200 °F)	20-815 °C (70-1500 °F)	GPa	10^6 psi	GPa	10^6 psi		Mg/m³	lb/in.³
Commercially pure														
99.5 Ti	8.6 (4.8)	...	9.2 (5.1)	...	9.7 (5.4)	10.1 (5.6)	10.1 (5.6)	102.7	14.9	38.6	5.6	0.34	4.51	0.163
99.2 Ti	8.6 (4.8)	...	9.2 (5.1)	...	9.7 (5.4)	10.1 (5.6)	10.1 (5.6)	102.7	14.9	38.6	5.6	0.34	4.51	0.163
99.1 Ti	8.6 (4.8)	...	9.2 (5.1)	...	9.7 (5.4)	10.1 (5.6)	10.1 (5.6)	103.4	15.0	38.6	5.6	0.34	4.51	0.163
99.0 Ti	8.6 (4.8)	...	9.2 (5.1)	...	9.7 (5.4)	10.1 (5.6)	10.1 (5.6)	104.1	15.1	38.6	5.6	0.34	4.51	0.163
99.2 Ti(b)	8.6 (4.8)	...	9.2 (5.1)	...	9.7 (5.4)	10.1 (5.6)	10.1 (5.6)	102.7	14.9	38.6	5.6	0.34	4.51	0.163
98.9(c)	...	...	...	...	...	...	...	...	...	102.7	14.9	...	4.54	0.164
Alpha alloys														
5 Al, 2.5 Sn	9.4 (5.2)	...	9.5 (5.3)	...	9.5 (5.3)	9.7 (5.4)	10.1 (5.6)	110.3	16.0	...	...	...	4.48	0.162
5 Al, 2.5 Sn (low O_2)	9.4 (5.2)	...	9.5 (5.3)	...	9.7 (5.4)	9.9 (5.5)	10.1 (5.6)	110.3	16.0	...	...	...	4.48	0.162
Near alpha														
8 Al, 1 Mo, 1 V	8.5 (4.7)	...	9.0 (5.0)	...	10.1 (5.6)	10.3 (5.7)	...	124.1	18.0	46.9	6.8	0.32	4.37	0.158
11 Sn, 1 Mo, 2.25 Al, 5.0 Zr, 1 Mo, 0.2 Si	8.5 (4.7)	...	9.2 (5.1)	...	9.4 (5.2)	...	...	113.8	16.5	...	...	...	4.82	0.174
6 Al, 2 Sn, 4 Zr, 2 Mo	7.7 (4.3)	...	8.1 (4.5)	...	8.1 (4.5)	...	...	113.8	16.5	...	...	...	4.54	0.164
5 Al, 5 Sn, 2 Zr, 2 Mo, 0.25 Si	...	...	...	...	...	...	10.3 (5.7)	113.8	16.5	...	...	0.326	4.51	0.163
6 Al, 2 Cb, 1 Ta, 1 Mo	...	...	...	...	...	9.0 (5.0)	...	113.8	17.5	...	...	...	4.48	0.162

(a) Room temperature. (b) Also contains 0.2 Pd. (c) Also contains 0.8 Ni and 0.3 Mo.
Source: Titanium Metals Corp. of America and RMI Co.

(Continued)

Table J.3 (Continued)

Nominal composition, %	Coefficient of linear thermal expansion, μm/m · K (μin./in./°F)							Modulus of elasticity(a)		Modulus of rigidity(a)		Poisson's ratio (a)	Density(a)	
	20-100 °C (70-212 °F)	20-205 °C (70-400 °F)	20-315 °C (70-600 °F)	20-425 °C (70-800 °F)	20-540 °C (70-1000 °F)	20-650 °C (70-1200 °F)	20-815 °C (70-1500 °F)	GPa	10⁶ psi	GPa	10⁶ psi		Mg/m³	lb/in.³
6 Al, 2 Sn, 1.5 Zr, 1 Mo, 0.35 Bi, 0.1 Si	...	...	...	...	...	...	...	...	...	...	...	...	...	...
Alpha-beta alloys														
8 Mn	8.6 (4.8)	9.2 (5.1)	9.7 (5.4)	10.3 (5.7)	10.8 (6.0)	11.7 (6.5)	12.6 (7.0)	113.1	16.4	48.3	7.0	...	4.73	0.171
3 Al, 2.5 V	9.5 (5.3)	...	9.9 (5.5)	...	9.9 (5.5)	...	...	106.9	15.5	...	...	...	4.48	0.162
6 Al, 4 V	8.6 (4.8)	9.0 (5.0)	9.2 (5.1)	9.4 (5.2)	9.5 (5.3)	9.7 (5.4)	...	113.8	16.5	42.1	6.1	0.342	4.43	0.160
6 Al, 4 V (low O₂)	8.6 (4.8)	9.0 (5.0)	9.2 (5.1)	9.4 (5.2)	9.5 (5.3)	9.7 (5.4)	...	113.8	16.5	42.1	6.1	0.342	4.43	0.160
6 Al, 6 V, 2 Sn	9.0 (5.0)	...	9.4 (5.2)	...	9.5 (5.3)	...	...	110.3	16.0	...	...	...	4.54	0.164
7 Al, 4 Mo	9.0 (5.0)	9.2 (5.1)	9.4 (5.2)	9.7 (5.4)	10.1 (5.6)	10.4 (5.8)	11.2 (6.2)	113.8	16.5	44.8	6.5	...	4.48	0.162
6 Al, 2 Sn, 4 Zr, 6 Mo	9.0 (5.0)	9.2 (5.1)	9.4 (5.2)	9.5 (5.3)	(5.3)	...	...	113.8	16.5	...	...	...	4.65	0.168
6 Al, 2 Sn, 2 Zr, 2 Mo, 2 Cr, 0.25 Si	...	...	9.2 (5.1)	...	...	...	...	122.0	17.7	46.2	6.7	0.327	4.57	0.165
10 V, 2 Fe, 3 Al	...	...	...	...	...	...	...	111.7	16.2	...	...	...	4.65	0.168
Beta alloys														
13 V, 11 Cr, 3 Al	9.4 (5.2)	...	10.1 (5.6)	...	10.6 (5.9)	...	...	101.4	14.7	42.7	6.2	0.304	4.84	0.175
8 Mo, 8 V, 2 Fe, 3 Al	...	...	...	...	...	...	...	106.9	15.5	...	...	...	4.84	0.175
3 Al, 8 V, 6 Cr, 4 Mo, 4 Zr	...	...	...	9.68 (5.38) (to 900 °F)	...	...	...	105.5	15.3	...	...	...	4.82	0.174
11.5 Mo, 6 Zr, 4.5 Sn	...	...	...	...	...	...	...	103.4	15.0	...	...	...	...	...

(a) Room temperature. (b) Also contains 0.2 Pd. (c) Also contains 0.8 Ni and 0.3 Mo.
Source: Titanium Metals Corp. of America and RMI Co.

Index